T0328632

Self-assessment Q&A in Clinical Laboratory Science, III

Self-assessment Q&A in Clinical Laboratory Science, III

Edited by

Alan H.B. Wu
Professor of Laboratory Medicine, School of Medicine,
San Francisco, CA, United States

Published in cooperation with AACC

ELSEVIER

AACC | PRESS

Elsevier
Radarweg 29, PO Box 211, 1000 AE Amsterdam, Netherlands
The Boulevard, Langford Lane, Kidlington, Oxford OX5 1GB, United Kingdom
50 Hampshire Street, 5th Floor, Cambridge, MA 02139, United States

About AACC
Dedicated to achieving better health through laboratory medicine, AACC brings together more
than 50,000 clinical laboratory professionals, physicians, research scientists, and business
leaders from around the world focused on clinical chemistry, molecular diagnostics, mass
spectrometry, translational medicine, lab management, and other areas of progressing laboratory
science. Since 1948, AACC has worked to advance the common interests of the field, providing
programs that advance scientific collaboration, knowledge, expertise, and innovation. For more
information, visit www.aacc.org.

Library of Congress Cataloging-in-Publication Data
A catalog record for this book is available from the Library of Congress

British Library Cataloguing-in-Publication Data
A catalogue record for this book is available from the British Library

ISBN: 978-0-12-822093-1

For information on all Elsevier publications
visit our website at https://www.elsevier.com/books-and-journals

Publisher: Stacy Masucci
Senior Editorial Project Manager: Susan Ikeda
Production Project Manager: Stalin Viswanathan
Cover Designer: Miles Hitchen

Typeset by SPi Global, India

Working together
to grow libraries in
developing countries

www.elsevier.com • www.bookaid.org

Contents

Section B
Therapeutic drug monitoring and toxicology

Section C
Molecular diagnostics and infectious diseases

Section D
Other clinical laboratory sections

Section E
General laboratory topics

Contributors

Numbers in parentheses indicate the pages on which the authors' contributions begin.

Zane D. Amenhotep (295), University of California, San Francisco, CA, United States

Wayne B. Anderson (195), University of Rochester, Rochester, NY, United States

Jason M. Baron (427), Massachusetts General Hospital, Boston, MA, United States

Carey-Ann D. Burnham (277), Washington University School of Medicine, St. Louis, MO, United States

Jing Cao (27), Department of Pathology, University of Texas Southwestern Medical Center, Dallas, TX, United States

Janine D. Cook (139), Substance Abuse Mental Health Services Administration, Rockville, MD, United States

Gyorgy Csako (363), National Institute of Health, Washington, DC, United States

Sarah Delaney (161), Department of Laboratory Medicine and Pathology, Mayo Clinic, Rochester, MN, United States

Anand S. Dighe (427), Massachusetts General Hospital, Boston, MA, United States

Shu-Ling Fan (89), UMass Memorial Medical Center, Worcester, MA, United States

Wieslaw Furmaga (75,99,207), Department of Pathology, University of Texas Health Science Center at San Antonio, San Antonio, TX; University of California, San Francisco, CA, United States

Putuma P. Gqamana (195), University of Rochester, Rochester, NY, United States

Dina N. Greene (397), Kaiser Permanente, Seattle, WA, United States

Neil Harris (113), University of Florida, Gainesville, FL, United States

Ibrahim Hashim (131), University of Texas Southwestern, Dallas, TX, United States

Erika M. Hissong (325), Weill Cornell Medicine, New York, NY, United States

Paul J. Jannetto (161), Department of Laboratory Medicine and Pathology, Mayo Clinic, Rochester, MN, United States

Michael Karasick (89), University of Rochester, Rochester, NY, United States

Adil I. Khan (439), Temple University, Philadelphia, PA, United States

Rasoul A. Koupaei (377), California Department of Public Health, Richmond, CA, United States

Kent Lewandrowski (125), Massachusettes General Hospital, Boston, MA, United States

Chuanyi Mark Lu (351), University of California, San Francisco, CA, United States

Maximo J. Marin (37), Department of Pathology and Laboratory Medicine, Keck School of Medicine, University of Southern California, Los Angeles, CA, United States

Yvette McCarter (267), University of Florida, Jacksonsville, FL, United States

Qing H. Meng (49), MD Anderson Cancer Center, Houston, TX, United States

James H. Nichols (131), Vanderbilt University, Nashville, TN, United States

Anthony Okorodudu (113), University of Texas Medical Branch, Galveston, TX, United States

Octavia M. Peck Palmer (63), University of Pittsburgh School of Medicine, Pittsburgh, PA, United States

Hanna Rennert (325), Weill Cornell Medicine, New York, NY, United States

Luke Rodda (177), Office of the Chief Medical Examiners, San Francisco, CA, United States

Lusia Sepiashvili (315), Hospital for Sick Children, University of Toronto, Toronto, ON, Canada

Christine L.H. Snozek (223), Mayo Clinic Arizona, Scottsdale, AZ, United States

Carole A. Spencer (131,139), University of Southern California, Los Angeles, CA, United States

John Toffaletti (113), Duke University, Durham, NC, United States

Nam Tran (235,253,267), University of California, Davis, CA, United States

Greg Tsongalis (235,245), Dartmouth Hitchock Hospital, Hanover, NH, United States

Priya D. Velu (325), Weill Cornell Medicine, New York, NY, United States

Jeffrey Whitman (253), University of California, San Francisco, CA, United States

Xander M.R. van Wijk (37), Department of Pathology, Pritzker School of Medicine, The University of Chicago; The University of Chicago Medicine and Biological Sciences, Chicago, IL, United States

Alison Woodworth (139), University of Kentucky, Lexington, KY, United States

Alan H.B. Wu (3,13,75,89,113,125,131,139,177,195,215,235,245,315,337,351,363, 389,397,407,417,439), University of California, San Francisco, CA, United States

Melanie L. Yarbrough (277), Washington University School of Medicine, St. Louis, MO, United States

He Sarina Yang (207), Weill Cornell Medicine, New York, NY, United States

Brandy Young (195), University of Rochester, Rochester, NY, United States

Y. Victoria Zhang (89,195), University of Rochester, Rochester, NY, United States

Zhen Zhao (325), Weill Cornell Medicine, New York, NY, United States

Reviewers

Justin Volk, Office of the Chief Medical Examiners, San Francisco, CA, United States

Kelsa West, Office of the Chief Medical Examiners, San Francisco, CA, United States

A.S. Danielle Dhari, Office of the Chief Medical Examiners, San Francisco, CA, United States

Jirair Gevorkyan, Office of the Chief Medical Examiners, San Francisco, CA, United States

Sue Pearring, Office of the Chief Medical Examiners, San Francisco, CA, United States

Ruben Luo, University of California, San Francisco, CA, United States

Yu Jada Zhang, University of California, San Francisco CA, United States

Preface

Since the First Edition of *Clinical Chemistry Self-Assessment* was published by the American Association for Clinical Chemistry (AACC) in 1982 by Marge A. Brewster, a clinical chemist at the University of Arkansas. I joined Dr. Brewster and Charles Turley for authorship of the Second Edition in 1989. When I joined the staff at Hartford Hospital, I convinced (coerced) my colleagues there (Robert Moore, Greg Tsongalis, and Robert Burnett) to coauthor the Third Edition in year 2000. The title of this edition was changed from *Clinical Chemistry* to *Clinical Laboratory Science*, to reflect questions that cover more of the field of laboratory medicine and to include emerging topics. In 2008 rather than a publish a Fourth Edition, we decided to add new questions to complement those found in the Third Edition, and labeled the book *Self-Assessment in Clinical Laboratory Science II*.

Now, more than a decade later, I found it appropriate to produce what is essentially the Fifth Edition, entitled, *Self-Assessment Q&A in Clinical Laboratory Science III*. As the AACC is dropped out of the book publishing business, this edition is produced by Elsevier Science. Like the Fourth Edition, this book is written by an army of content experts in our field. Some chapters on general clinical chemistry have been reproduced from prior editions with updated references. These questions reflect the ever increasing scope and range of topics that we as clinical laboratory professionals have to deal with on a regular basis.

Alan H.B. Wu, San Francisco, CA

Section A

General clinical chemistry

Chapter 1

Antiquated and novel clinical laboratory tests

Alan H.B. Wu
University of California, San Francisco, CA, United States

1. Which of the following tests may be unnecessary in light of grain supplementation that occurs in North America?
   a. mean corpuscular volume
   b. vitamin B12
   c. serum folate
   d. homocysteine
   e. methylmalonic acid
2. Which of the following test is largely obsolete with regards to pancreatic disease?
   a. amylase
   b. lipase
   c. serum trypsin
   d. CA19-9
   e. secretin
3. Which test is most appropriate for testing fetal lung maturity, replacing all of the others listed?
   a. lecithin:sphingomyelin ratio
   b. foam stability index
   c. phosphatidyl glycerol
   d. lamellar body count
   e. FLM test (surfactant-to-albumin ratio)
4. Which cardiac marker is largely antiquated?
   1. myoglobin
   2. creatine kinase-MB
   3. lactate dehydrogenase isoenzymes
   4. total creatine kinase
      a. 1,2,3
      b. 1,3
      c. 2,4

Self-assessment Q&A in Clinical Laboratory Science, III. https://doi.org/10.1016/B978-0-12-822093-1.00001-6

    d. 4 only

    e. all of the above

5. Which of the following is recommended for vitamin D testing?

    a. general population screening

    b. routine measurement of both hydroxy and dihydroxyvitamin D

    c. use of mass spectrometry for D2 and D3 speciation to monitor the effect of supplementation

    d. analysis of the C-3 25(OH)D3 epimer in neonates

    e. analysis of the C-3 25(OH)D3 epimer in adults

6. Which of the following analytes is not yet standardized by a recognized body?

    a. creatinine

    b. hemoglobin A1c

    c. cholesterol

    d. glucose

    e. parathyroid hormone

7. Which of the following is correct regarding the differences between the MDRD and CKD-EPI equations for estimated glomerular filtration rate?

    a. The CKD-EPI equation is more accurate at 60 mL/min.

    b. The CKD-EPI equation accounts for Asians and Hispanics.

    c. The CKD-EPI equation uses cystatin C.

    d. The MDRD equation significantly overestimates GFR among individuals with normal kidney function.

    e. Neither equation has biases for patients with renal transplant.

8. The MDRD and CKD-EPI use a correction factor for subjects who are African American. On average, this is because African Americans

    a. have larger kidneys

    b. have larger muscle mass

    c. have large fat volume

    d. eat more red meat

    e. have higher likelihood of reduced kidney function

9. What of the following is true regarding prostatic acid phosphatase today (PAP)?

    a. It is synonymous with prostate-specific antigen (PSA).

    b. Ratio of PAP/PSA better separates benign prostatic hypertrophy from prostate cancer.

    c. Recent data suggests that PAP may be better than PSA for screening.

    d. Recent data suggests that PAP may be better than PSA for predicting metastatic disease.

    e. No recurring role for PAP.

10. Which of the following is true regarding the use of the erythrocyte sedimentation rate (ESR) vs C-reactive protein (CRP)?

    a. Results are concordant 95% of the time, obviating the need for both testing.

b. ESR is more sensitive as an early marker of inflammation.

c. ESR is less sensitive than CRP in patients with bacterial infections.

d. CRP results from an increase in fibrinogen.

e. CRP has a longer half-life.

11. Which of the following tests is rarely used regarding a workup of pernicious anemia?

a. Schilling test

b. methylmalonic acid

c. mean corpuscular volume

d. serum B12 level

e. red cell folate level

12. Which of the following statements do not provide rationale for measuring methylmalonic acid (MMA) and homocysteine for megaloblastic anemia?

a. MMA is an early indicator of B12 deficiency.

b. Serum vitamin B12 levels do not represent functional stores.

c. Homocysteine is elevated in B12 deficiency.

d. MMA concentrations decline with vitamin B12 treatment.

e. MMA is not increased in folate deficiency.

13. Which of the following has not been studied as a marker for traumatic brain injury?

a. S100b

b. myelin basic protein

c. ubiquitin C-terminal hydrolase L1

d. CK-MM Isoenzyme

e. brain-derived neurotrophic factor

14. Which of the following traumatic brain injury markers is released as the result of microvascular injury?

a. glial fibrillary acidic protein

b. neuron specific enolase

c. matrix metalloproteinase

d. tau protein

e. αII-spectrin

15. What may the role of detecting autoantibodies to brain proteins after traumatic brain injury (TBI)?

a. early detection

b. differentiation between health and mild TBI

c. the presence of autoantibodies themselves produce adverse events

d. predicting onset of TBI-induced Parkinson syndrome

e. their absence excludes the need for a head CT scan

16. Which of the following is not a clinical application of measuring anti-Müllerian hormone levels?

a. prediction the age of menopause

b. aid in the diagnosis of polycystic ovary syndrome

c. assessment of ovarian function before and after gynecologic surgeries

    d. prediction of ectopic pregnancy

    e. predicting the risk for developing ovarian hyperstimulation syndrome

17. Which of the following is true regarding anticyclic citrullinated peptide antibodies with regards to rheumatoid arthritis?

    a. is highly specificity

    b. arrive late after rheumatoid arthritis onset

    c. do not contribute to the pathophysiology

    d. are positive in all cases

    e. increases are more prevalent among the younger patients

18. What is the medical utility of fetal fibronectin testing?

    a. fetal lung maturity

    b. prediction of spontaneous preterm birth

    c. ectopic pregnancy

    d. diagnosis of endometriosis

    e. diagnosis of polycystic ovary syndrome

19. What is the value of measuring total bile salts during pregnancy?

    a. detection of congenital heart malformations

    b. identification of Down syndrome

    c. intrauterine malignancy

    d. identification of preeclampsia

    e. detection of obstetric cholestasis

20. What is hepcidin?

    a. transport protein for iron

    b. iron regulatory hormone for gut absorption and release from storage sites

    c. major iron storage protein

    d. hormone that stimulates reabsorption from the renal tubule

    e. drug used to treat iron deficiency

21. What is the medical value of hepcidin measurements?

    1. differentiate between iron deficiency and anemia of chronic disease

    2. iron refractory iron deficiency anemia

    3. aid in the diagnosis of hemochromatosis

    4. sickle cell disease

        a. 1,3

        b. 1,2,3

        c. 2,4

        d. 4 only

        e. all of the above

22. Which of the following is not used as a serologic test for celiac disease?

    a. antigliadin antibodies

    b. antiendomysial antibodies

    c. antitransglutaminase antibodies

    d. antimitochondrial antibodies

    e. deamidated gliadin peptides

23. Which of the following conditions is not a cause of false positive results for use of celia antibodies?
   a. chronic liver disease
   b. Patient is not on a gluten-free diet when tested.
   c. enteric infections
   d. hypergammaglobulinemia
   e. congestive heart failure

24. Which of the following is not a test for acute kidney injury (AKI)?
   a. neutrophil gelatinase-associated lipocalin
   b. miRNA-122
   c. proenkephalin
   d. kidney injury molecule-1
   e. metalloproteinases-2 and insulin-like growth factor-binding protein

25. Which of the following is an essential attribute for a biomarker of acute kidney injury?
   a. increased 24–48 h before creatinine
   b. equations available for estimating glomerular filtration rate
   c. predicts the need for renal transplant
   d. no difference in results between ancestry
   e. predicts renal transplant rejection

26. Creatinine testing is performed prior to contrast imaging with gadolinium to reduce the risk of
   a. systemic nephrogenic fibrosis
   b. acute kidney injury
   c. lupus nephritis
   d. scleroderma
   e. anaphylactic reactions

27. What is the best rationale for intraoperative parathyroid hormone analysis?
   a. Prediction of postsurgical parathyroid hormone function.
   b. Determine that additional tumor is present.
   c. PTH assays are more analytically accurate when conducted at the point-of-care.
   d. Eliminates the need for repeat surgical resection.
   e. Supersedes the need for a frozen biopsy of the excised gland.

28. Which parathyroid hormone testing criteria is used to determine that the procedure was successful?
   a. a reduce ratio of intact to N-terminal PTH
   b. a 90% reduction of PTH within 10 min of resection relative to baseline
   c. a 90% reduction of PTH within 30 min of resection relative to baseline
   d. a 50% reduction of PTH within 10 min of resection of PTH relative to baseline
   e. depends on the number of glans removed

Answers

1. c. Folate deficiency is rare after the supplementation of grain, even among the homeless. *McMullin MF. Homocysteine and methylmalonic acid as indicators of folate and vitamin B12 deficiency in pregnancy. Clin Lab Haematol 2001;23:161–5. Joelson DW, et al. Diminished need for folate measuremens among indigent populations in the post folic acid supplementation era. Arch Pathol Lab Med 2007;131:477–80.*

2. a. Amylase is less specific than lipase and is largely redundant. CA19-9 is a pancreatic tumor marker. Secretin and trypsin are useful for some diseases that affect the pancreas including steatorrhea and as cystic fibrosis. *Barbieri JS, et al. Amylase testing for acute pancreatitis. J Hosp Med 2016;5:366–8. https://doi.org/10.1002/jhm.2544.*

3. d. The lamellar body count can be measured using a hematology analyzer. All of the others are largely no available or should not be used. *Lu J, et al. Lamellar body counts performed on automated hematology analyzers to assess fetal lung maturity. Lab Med 2008;38:419–23.*

4. a. Total creatine kinase can be used to determine reinfarction and detection of skeletal muscle disease. *Wu AHB, et al. Antiquated tests within the clinical pathology laboratory. Am J Manag Care 2010; https://www. ajmc.com/journals/issue/2010/2010-09-vol16-n09/ajmc_10sep_wu_xcl_ e220to227.*

5. c. General population screening is not recommended. The 1,25(OH)2 testing is recommended only for those who have renal disease. The D3 epimer is largely inactive and present in low concentrations in adults but higher in neonates. Either vitamin D2 or D3 can be supplemented. *Karras SN, et al. The road no so travelled: should measurement of vitamin D epimers during pregnancy affect our clinical decisions? Nutrients 2017;9:90. https://doi. org/10.3390/nu9020090.*

6. e. Standardization enables interpretation of test results generated from different assays and laboratories. PTH is currently not standardized. *American Association for Clinical Chemistry. The need to harmonize clinical laboratory test results July 2015. White paper.*

7. a. The CKD-EPI is more accurate between 60 and 100 mL/min whereas with the MDRD, no value should be given if the eGFR is >60 mL/min. *Murata K, et al. Relative performance of the MDRD and CKD-EPI equations for estimating glomerular filtration rate among patients with varied clinical presentations. Clin J Am Soc Nephrol 2011;6:1963–72.*

8. b. Muscle mass affects the creatinine concentration, which on the average, is higher for African Americans. *Delanaye P, et al. Are the creatinine-based equations accurate to estimate glomerular filtration rate in African American populations? Clin J Am Soc Nephrol 2011;6:906–12.*

9. e. Recent data have shown prognostic value of PAP, especially after surgery. But there are no clinical practice groups that have recommended

revival of PAP testing. *Xu H, et al. Prostatic acid phosphastase (PAP) predicts prostate cancer progress in a population-based study: the renewal of PAP? Dis Mark 2019;10. https://doi.org/10.1155/2019/7090545.*

10. c. ESR and CRP are both markers of inflammation but CRP is considered more sensitive and therefore overall, more useful especially for infections. CRP declines faster than ESR in blood. *Bray C, et al. Erythrocyte sedimentation rate and C-reactive protein measurements and their relevance in clinical medicine. Wisc Med J 2016;115:317–21.*

11. a. Pernicious anemia is caused by a deficiency of vitamin B12 due to poor dietary absorption due to the absence of intrinsic factor. The Schilling test involves injection of radiolabeled B12 and measuring urinary excretion, but is not widely available. *Ramphul K, et al. Schilling Test. [Updated 2019 Jun 9]. In: StatPearls [Internet]. Treasure Island (FL): StatPearls Publishing; 2019 Jan-. Available from: https://www.ncbi.nlm.nih.gov/books/NBK507784/.*

12. e. Vitamin B12 is a cofactor required in the conversion of methylmalonyl CoA to succinyl CoA. An increase in MMA is an early indicator of vitamin B12 deficiency and folate deficiency. MMA is falsely increased in renal disease. *Hannibal L et al. Biomarkers and algorithms for the diagnosis of vitamin B12 deficiency. Front Mol Biosci 2016;2. https://doi.org/10.3389/fmolb.2016.00027.*

13. d. The CK-BB isoenzyme is being investigated as a TBI marker but not the MM isoenzyme that originates from skeletal muscles. *Gan ZS, et al. Blood biomarkers for traumatic brain injury: a quantitative assessment of diagnostic and prognostic accuracy. Front Neurol 2019;10:446. https://doi.org/10.3389/fneur.2019.00446.*

14. a. Traumatic brain injury triggers a complex array of events beginning from disruption of metabolism, microvessels and synaptic membrane dysfunction, cell leakage, cellular necrosis, and cerebral ischemia. GFAB is released from microvesicles and is one of the earlier biomarkers of TBI. *Dambinova,et al. Gradual return to play: potential role of neurotoxicity biomarkers in assessment of concussions severity. J Mol Biom Diag 2013, s3. https://doi.org/10.4172/2155-9929.S3-003.*

15. c. One theory is that exposure of proteins from the central nervous system may illicit an autoimmune response and may produce injury in a similar manner as in autoimmune diseases. *Raad M, et al. Autoantibodies in traumatic brain injury and central nervous system trauma. Neurosci 2014;281:16–23.*

16. d. Anti-Mullerian is a glycoprotein hormone related to inhibin and actin from the transforming growth factor beta family. *Oh SR, et al. Anti-Müllerian hormone is a peptide growth factor and a marker of ovarian reserve. Clinical application of serum anti-Müllerian hormone in women. Clin Exp Reprod Med 2019;46:50–59.*

17. a. Anti-CCP antibodies are highly specific, early indicator, and cause bone and joint damage, but there are rheumatoid arthritis patients who have negative results. *Cader MZ, et al. The relationship between the presence of anti-cyclic citrullinated peptide antibodies and clinical phenotype in very early rheumatoid arthritis. BMC Musculoskelet Disord 2010;11:187. https://doi.org/10.1186/1471-2474-11-187.*

18. b. During pregnancy, the presence of vaginal fetal fibronectin is an indicator of disruption of the maternal-fetal interface, and the initial onset of labor. *Kiefer DG, et al. The utility of fetal fibronectin in the prediction and prevention of spontaneous preterm birth. Ref Ostet Gynecol 2008;1: 106–12.*

19. d. Bile salts are end products of cholesterol catabolism. When present, it can cause postpartum hemorrhage, preterm labor, intrapartum fetal distress, and stillbirth. *Egan N, et al. Reference standard for serum bile acids in pregnancy. Brit J Obstet Gynecol 2012;119:493–8.*

20. b. Hepcidin regulates the rate of iron efflux into plasma. During iron overload, hepcidin is upregulated preventing release of iron from enterocytes. *Ganz T, et al. Hepcidin, a key regulator of iron metabolism and mediator of anemia of inflammation. Blood 2003;102:783–8.*

21. c. Hepcidin is useful for both iron overload and deficiency. Iron refractory iron deficiency anemia is due to a mutation and patients do not respond to iron treatment. *Girelli D, et al. Hepcidin in the diagnosis of iron disorders. Blood 2016;127:2809–13.*

22. d. Antimitochondrial antibodies are useful for diagnosis of biliary cirrhosis. *Singh A, et al. Non-invasive biomarkers for celiac disease. J Clin Med 2019;8:885. https://doi.org/10.3390/jcm8060885.*

23. b. Patients on a gluten-free or reduced-gluten diet will produce a false negative result for many of these antibodies. *Singh A, et al. J Clin Med 2019;8:885. https://doi.org/10.3390/jcm8060885.*

24. b. miRNA-122 is a specific liver injury marker. Each of the others has been studied as AKI markers. *Konukoglu D, et al. Biomarkers for acute kidney injury. Int J Med Biochem 2018;1:80–7.*

25. a. Creatinine is a renal function marker that rises within 1-2 days after acute kidney injury. AKI biomarkers are useful if they are increased within the first few hours after AKI. *Konukoglu D, et al. Biomarkers for acute kidney injury. Int J Med Biochem 2018;1:80–7.*

26. a. Patients with reduced clearance of gadolinium can produce systemic nephrogenic fibrosis that produces thickening and hardening of the skin and is similar in appearance to scleroderma. Acute kidney injury occurs with iodide contrast imaging and less so with gadolinium. Anaphylactic reactions can occur with contrast injection but is not predicted by creatinine measurement. *Schieda N, et al. Gadolinium-based contrast agents in kidney disease: a comprehensive review and clinical practice guideline* *issued by the Canadian Association of Radiologists. Can J Kidney Health Dis 2018;5. https://doi.org/10.1177/2054358118778573.*

27. b. Due to the short half-life of PTH in blood, estimated to be 3–5 min, residual activity indicates the presence of residual tumor that should be resected. *Carter AB, et al. Intraoperative testing for parathyroid hormone: a comprehensive review of the use of the assay and the relevant literature. Arch Path Lab Med 2003;127:1424–42.*

28. d. If there is substantial residual PTH, a full bilateral exploration of the neck is warranted to search for additional tumors. *Carter AB, et al. Intraoperative testing for parathyroid hormone: a comprehensive review of the use of the assay and the relevant literature. Arch Path Lab Med 2003;127:1424–42.*

Chapter 2

Cardiac markers

Alan H.B. Wu
University of California, San Francisco, CA, United States

1. NT-pro BNP is
   a. inactive metabolite of BNP
   b. active metabolite of BNP
   c. inactive metabolite of proBNP
   d. active metabolite of proBNP
   e. inactive metabolite of atrial natriuretic peptide (ANP)
2. Which of the following is true regarding BNP and NT-proBNP?
   a. Compensated heart failure have higher values than decompensated
   b. Increased to the same extent in chronic obstructive pulmonary disease
   c. Diastolic failure lower than systolic failure
   d. No age- or sex-related differences in healthy subjects
   e. No correlation to the 6-minute walk test
3. Which of the following is the most accurate regarding the origin of the natri-uretic peptides (BNP and ANP) in blood?
   a. BNP originates from the brain and cardiac ventricles
   b. BNP originates from the cardiac ventricles, ANP from the cardiac atria
   c. BNP originates from the left ventricles, ANP from the right ventricles
   d. BNP originates from the cerebral ventricles
   e. BNP originates from the heart, ANP from the kidneys
4. What is true regarding the use of neprilysin inhibitors in an HF patient?
   a. BNP increased, NT-proBNP decreased
   b. BNP decreased, NT-proBNP increased
   c. Both are increased
   d. Both are decreased
   e. Both stay the same

5. Which of the following is true regarding the use of BNP/NT-proBNP to guide heart failure treatment for heart failure patients?
   a. This has become part of standard practice
   b. BNP is effective but NT-proBNP is not
   c. Guided therapy is limited to angiotensin-converting enzyme inhibitors
   d. Various randomized trials have produced mixed result
   e. Must be used in conjunction with high-sensitivity cardiac troponin

6. Which of the following is not part of the diagnostic criteria for acute myocardial infarction according to the 4th Universal Definition?
   a. rising or falling pattern of cardiac troponin
   b. electrocardiographic changes
   c. intimal thickness of the carotid artery
   d. pathologic findings
   e. evidence of occlusion upon angiography

7. Which of the following is not correct regarding types of AMIs?
   a. Type I is associated with plaque rupture
   b. Cocaine use can produce a type II AMI
   c. Cardiac troponin testing can be used to differentiate between types
   d. Type 1 and type 2 are treated differently
   e. Demand ischemia characterizes type 2 AMI

8. Which of the following are the criteria for determining if a troponin assay qualifies as high sensitivity?
   a. Values at the 99th percentile produce a 10% imprecision
   b. Values at the 99th percentile produce a 20% imprecision
   c. At least 50% of healthy subjects detected above the assay's limit of detection
   d. At least 90% of healthy subjects detected above the assay's limit of detection
   e. 99% of patients with AMI detected

9. Which of the following is the biggest advantage for implementing high-sensitivity troponin?
   a. more accurate risk stratification for adverse cardiac events
   b. more accurate diagnosis of ST-segment elevation myocardial infarction
   c. obviates the need for testing the natriuretic peptides
   d. earlier rule out of myocardial infarction
   e. aid in the selection of the most appropriate therapy

10. Biological variation studies have shown that cardiac troponin has a low index of individuality. Which of the following cannot be concluded?
    a. Reference ranges for troponin will be of limited value

b. Serial testing will be essential for interpretation of results for AMI diagnosis

c. An individual's own baseline value during health could be useful if there were standardization among troponin assays

d. CK-MB is more valuable for AMI rule out

e. Either a delta change or absolute change in troponin over time can be used

11. Which of the following is least likely to cause an increase in troponin in the absence of cardiac injury?

    a. renal failure

    b. cardiac failure

    c. pulmonary embolus

    d. sepsis

    e. untreated cancer

12. What is the value of measuring troponin in a patient who presents to the emergency department with chest and has an ST-segment elevation on an electrocardiogram?

    a. No value for acute testing

    b. Troponin determines whether or not a cardiac catheterization is required

    c. Troponin differentiates between types I and II AMI

    d. Troponin is useful for 30-day risk stratification for future adverse events

    e. Troponin determines the likelihood of the patient developing chronic heart failure

13. Which of the following is correct regarding recommendations for the reporting units for cardiac troponin?

    a. report results of all assays in ng/mL

    b. report results of all assays in ng/L

    c. report high sensitivity in ng/L, conventional assays in ng/mL

    d. report point-of-care troponin in ng/mL, high sensitivity in ng/L

    e. dependent on the manufacturer's recommendations

14. According to the International Federation of Clinical Chemistry, which epitopes within the troponin protein should the antibodies be directed toward?

    a. Two antibodies directed toward the central part

    b. One antibody directed toward the C-terminus and one toward the N-terminus

    c. One antibody directed toward the C-terminus and one toward the central part

    d. One antibody directed toward the N-terminus and one toward the central part

    e. Three antibodies, one at the C-terminus, N-terminus, and one toward the central part

15. Which of the following describes the clearance kinetics of cardiac markers in successful versus unsuccessful revascularization?

    a. No change in the marker vs time profile

    b. Earlier release in successful than unsuccessful reperfusion

    c. Earlier release in unsuccessful reperfusion is due to periprocedural MI

    d. Release kinetics dependent on the type of AMI (type I vs II)

    e. Early release in unsuccessful reperfusion is due to reperfusion injury

16. What is the current state of standardization for cardiac troponin assays?

    a. Troponin T and troponin I are standardized to each other

    b. There is no standardization of cTnI, but results are harmonized to each other

    c. Standardization will not be possible as there are no industry-wide agreement on antibodies used (epitopes or number of antibodies)

    d. There is standardization between point-of-care and the central lab cTnI within a company

    e. Although there is no standardization, troponin I results between manufacturers are currently within 20% of each other

17. The following electrocardiogram was obtained from a patient who presents to an emergency department with chest pain. This patient went straight to the cardiac catheterization laboratory for revascularization. Which of the following is false?

    a. ST-segment elevation in leads V2 through V5

    b. ST-depression in leads III and afF

    c. indicative of an anterolateral AMI

    d. prolongation of the QT interval

    e. absence of a left bundle branch block

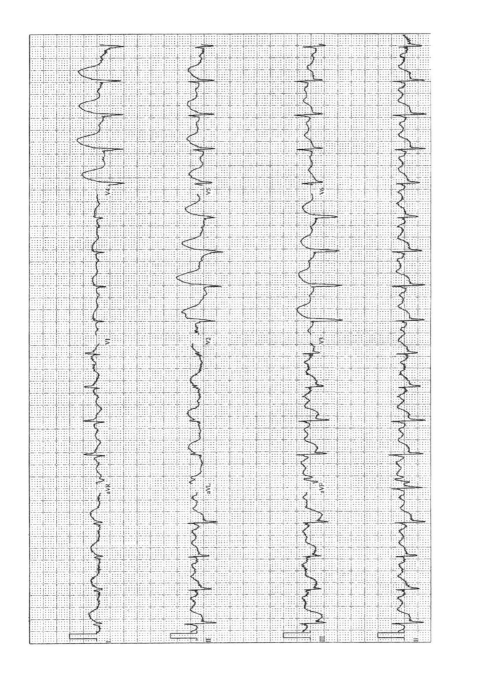

18. Which of the following is a marker of early myocardial infarction?
    a. soluble ST2
    b. CD40 ligand
    c. growth differentiation factor-15
    d. pregnancy-associated plasma protein-A
    e. copeptin
19. Some have opined that troponin should replace the need for CK-MB test-ing. Which of the following might counter that argument?
    a. Since troponin remains increased for many days after CK-MB, the latter could be used to detect the presence of a reinfarction
    b. It is more accurate to measure the size of infarcted tissue with CK-MB
    c. CK-MB has more sensitivity for AMI than troponin
    d. CK-MB is more stable than troponin in vitro
    e. Reagents for troponin assays are substantially more expensive than CK-MB
20. The following figure is a plot of cardiac biomarker release vs time after myocardial infarction. Label the curves appropriately.
    a. A = myoglobin, B = CK-MB, C = troponin I, D = lactate dehydroge-nase, E = troponin T
    b. A = CK-MB, B = myoglobin, C = lactate dehydrogenase, D = troponin I, E = troponin T
    c. A = myoglobin, B = CK-MB, C = troponin I, D = troponin T, E = lactate dehydrogenase
    d. A = lactate dehydrogenase, B = aspartate aminotransferase, C = CK-MB, D = troponin I, E = troponin T
    e. A = CK-MB, B = myoglobin, C = troponin C, D = troponin I, E = lactate dehydrogenase

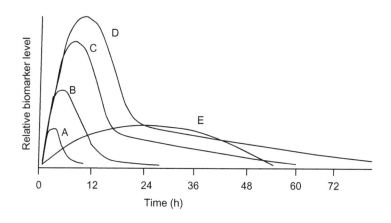

21. The following figure represents the thin filament of muscle fiber. What is the correct labeling?

a. A=actin, B=troponin complex, C=tropomyosin
b. A=myosin, B=troponin complex, C=tropomyosin
c. A=actin, B=myosin, C=troponin complex
d. A=troponin complex, B=myosin, C=actin
e. A=troponin complex, B=tropomyosin, C=actin

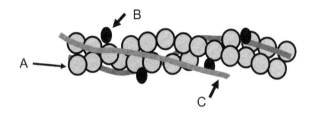

22. Galectin-3 and soluble ST-2 are novel heart failure biomarkers. Unlike the natriuretic peptides, they are not influenced by volume overload. Which of the following medical applications would this attribute give these markers an advantage?

a. aid in the diagnosis of acute myocardial infarction
b. outpatient monitoring for success of antiheart failure medications
c. diagnosis of heart failure from the emergency department
d. risk stratification for future adverse cardiac events
e. as a replacement for the echocardiogram

23. After intensive discussion with ED resident about all possible interpretation of cardiac markers results, he asks you to summarize the most important information about laboratory diagnosis of cardiac diseases. What is considered to be true about cardiac markers?

a. In myocardial infarct LD is "flipped" and LD2 > LD1
b. In the normal heart, CK-MB consists of 15%–20% of the total CK and this percentage is greater in the right than in the left ventricle
c. In patient with myocardial infarct even small secondary increase in cTnI after initial diagnostic peak is interpreted as a reinfarction
d. The concentrations of cTnT and cTnI must be at least two times higher than 99th percentile to indicate damage to myocytes and myocardial infarct

24. Following are angiograms of patient undergoing an elective cardiac catheterization. The "Before" image shows a major blockage of the left anterior descending (LAD) coronary artery. Two minor collaterals are present near the blockage that forms as a compensatory mechanism to supply blood to that portion of the heart. The interventional cardiologist decides to open the artery with a balloon catheter. The "After" image shows a patent LAD artery. One of the collateral arteries is now blocked.

Before

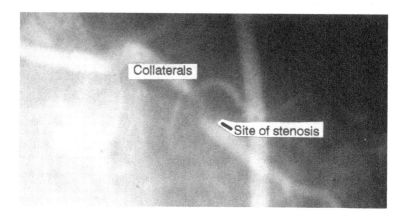

After

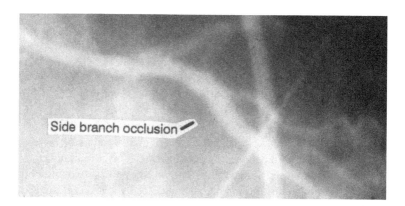

Which of the following is true?
a. This patient has suffered postprocedural myocardial infarction
b. This is sufficient to release a small amount of cardiac troponin that can be detected with a high-sensitivity assay
c. This will stimulate release of B-type natriuretic peptide
d. CK-MB but not troponin will be increased
e. A postprocedural electrocardiogram will demonstrate significant changes

Answers

1. c. Both BNP (active) and NT-proBNP (inactive) are cleaved from proBNP and released into the circulation in equimolar concentrations. *Mair, J. Biochemistry of B-type natriuretic peptide: where are we now? Clin Chem Lab Med 2008;46(11):1507–14.*

2. c. In diastolic failure, now labeled as heart failure with preserved ejection fraction (HFpEF), BNP levels are lower than in systolic failure, now labeled as heart failure with reduced ejection fraction (HFrEF). Are of the others are false. *Harada E, et al. B-type natriuretic Peptide in heart failure with preserved ejection fraction: relevance to age-related left ventricular modeling in Japanese. Circ J 2017;81:1006–13.*

3. b. BNP originates from both ventricles and ANP from both atria. While the original name for BNP was "brain natriuretic peptide," blood concentrations originate from the heart. *Weber M, et al. Role of B-type natriuretic peptide (BNP) and NT-proBNP in clinical routine. Heart 2006;92 (6):843–9.*

4. a. Neprilysin inhibitors affect the metabolism of BNP, therefore values are slightly higher immediately after treatment. This leads to improvement of heart function, as reflected by a decrease in NT-proBNP. *Mair J. Clinical utility of BNP for monitoring patients with chronic HF treated with -sacubitril-valsartan. Am Coll Cardiol, Latest in cardiology. https://www.acc.org/latest-in-cardiology/articles/2016/12/19/08/06/clinical-utility-of-bnp-for-monitoring-patients-with-chronic-hf-treated-with-sacubitril-valsartan.*

5. d. A meta-analysis has shown that BNP/NT-proBNP can be used to guide therapy but this has not been clinically adopted. *Porapakkham P, et al. B-type natriuretic peptide-guided heart failure therapy. A meta-analysis. Arch Intern Med 2010;170:507–14.*

6. c. Imaging is part of the Universal Definition of AMI but not of the carotid artery. *Thygesen K, et al. Fourth universal definition of myocardial infarction. Eur Heart J 2018;40:237–69.*

7. c. Troponin is increased in both types of AMIs although the concentrations in type 1 are usually higher than type 2. *Thygesen K, et al. Fourth universal definition of myocardial infarction. Eur Heart J 2018;40:237–69.*

8. c. The designation of high sensitivity is an analytical determination from healthy subjects. Prior generation of troponin assays was inadequately sensitive enough to detect troponin in blood of healthy subjects. *Apple FS. A new season for cardiac troponin assays: it's time to keep a score card. Clin Chem 2009;55:1303–6.*

9. d. High-sensitivity assays enable rule out of myocardial infarction earlier, even with the admission specimen only. *Body R, et al. Diagnostic accuracy of a high-sensitivity cardiac troponin assay with a single serum test in the emergency department. Clin Chem 2019;65:1006–14.*

10. d. The index of individuality is the ratio of the intra-individual divided by the inter-individual variation. A low index of <0.6 indicates that reference values will not be useful and serial testing becomes important. Very low values for troponin are useful for AMI rule out on the admission sample. *Wu AHB, et al. Short- and long-term biological variation in cardiac troponin I with a high-sensitivity assay: implications for clinical practice. Clin Chem 2009;55:52–8.*

11. e. Certain chemotherapeutic such as doxorubicin can cause cardiac injury. However, cancer by itself, unless it has metastasized to the heart, should not increase troponin. *Wu AHB, et al. National Academy of Clinical Biochemistry Laboratory Medicine Practice Guidelines: use of cardiac troponin and the natriuretic peptides for etiologies other than acute coronary syndromes and heart failure. Clin Chem 2007;53:2086–96.*

12. e. There is no value in patients with an STEMI for diagnosis and minimal value for risk stratification. These individuals should be sent for emergent cardiac catheterization. *Cediel G, et al. Prognostic value of new-generation troponins in ST-segment-elevation myocardial infarction in the modern era: the RUTI-STEMI Study. J Am Heart Assoc 2017;6 (12):e007252. Published 2017 Dec 23. doi:10.1161/JAHA.117.007252.*

13. c. According to the Academy of the American Association for Clinical Chemistry, the reporting units should be dependent on the generation of the assay. Since high-sensitivity troponin produces the lowest results, the ng/L should be used with values reported in whole numbers. *Wu AHB, et al. Clinical laboratory practice recommendations for use of cardiac troponin in acute coronary syndrome: expert opinion from the Academy of the American Association for Clinical Chemistry and the Task Force on Clinical Applications of Cardio Bio-markers of the International Federation of Clinical Chemistry and Laboratory Medicine. Clin Chem 2018;64:645–55.*

14. a. Troponin degrades when released into the circulation at the C-terminal and N-terminal ends of the protein. Therefore most assays use antibodies that are directed toward the central portion. Some commercial troponin assays do use three antibodies, but not one directed at the N-terminus. *Panteghini M, et al. Quality specifications for cardiac troponin assays. International Federation of Clinical Chemistry and Laboratory Medicine (IFCC). Clin Chem Lab Med 2001;39:174–8.*

15. b. In patients with successful reperfusion, such as angioplasty, revascularization facilitates egress directly into the circulation and a washout of cardiac biomarkers. *Tanasijevic MJ, et al. Myoglobin, creatine-kinase-MB and cardiac troponin-I 60-minute ratios predict infarct-related artery patency after thrombolysis for acute myocardial infarction: results from the Thrombolysis in Myocardial Infarction study (TIMI) 10B. J Am Coll Cardiol. 1999;34(3):739–47.*

16. c. Standardization of cTnI assays will not occur unless all manufacturers agree on the antibodies used in their assay, which is highly unlikely. *Apple FS. Counterpoint: Standardization of cardiac troponin I assays will not occur in my lifetime. Clin Chem. 2012 Jan;58(1):169–71. doi:10.1373/clinchem.2011.166165.*

17. d. The yellow highlighted areas indicate ST elevation myocardial infarction (STEMI) and blue areas indicate ST-segment depression indicative of myocardial ischemia and MI when it exceeds 5 mm (as in this case). STEMIs do not require troponin testing. ST segment elevations in leads V3 and V4 are indicative of an anterolateral AMI. A left bundle branch block can indicate underlying heart disease. Prolongation of the QT interval can cause arrhythmias. *Akbar H. Acute myocardial infarction ST elevation (STEMI). StatPearls [Internet]. https://www.ncbi.nlm.nih.gov/books/NBK532281/Image courtesy of Wikimedia commons.*

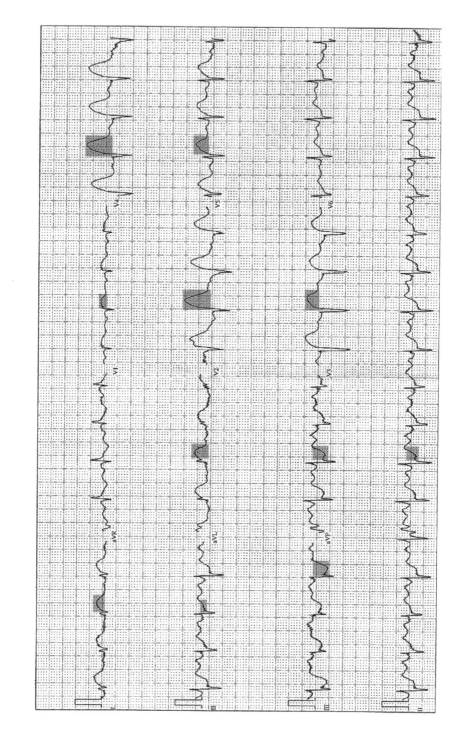

18. e. Copeptin is the C-terminal portion of arginine vasopressin and is released before troponin after myocardial infarction. owever, the peptide is not as specific. *Maisel AS, et al. Copeptin helps in the early detection of patients with acute myocardial infarction: primary results of the CHO-PIN Trial (Copeptin Helps in the early detection Of Patients with acute myocardial INfarction). J Am Coll Cardiol 2013;62:150–60.*

19. a. In theory, reinfarctions can be more readily detected with CK-MB than troponin because CK-MB returns to baseline, although published cases have suggested that a secondary rise is observed with troponin as well. *Apple FS, et al. Cardiac troponin and creatine kinase MB monitoring during in-hospital myocardial reinfarction. Clin Chem 2005;51:460–3.*

20. c. Myoglobin is the smallest of the cardiac biomarkers and is released and cleared the fastest. LDH is larger and is released slower than myoglobin and CK-MB. The troponins have a prolonged biphasic release (cytosolic and structural component) with troponin T producing higher results. *Jaffe AS, et al. Biomarkers in acute cardiac disease–the present and the future. J Am Coll Cardiol 2006;48:1–11.*

21. a. The troponin complex of C, T, and I subunits resides on the tropomyosin strand of the actin thin filament. It regulates contraction by binding to the myosin light chain, attached to the myosin thick filament, enabling the two filaments to slide past each other during contraction and back again during relaxation. *Danek J. Troponin levels in patients with stable CAD. Cor et Vasa 2017;59:e229–34.*

22. b. The natriuretic peptides can be released with volume overload therefore are not as useful as galectin-3 and soluble ST-2 that are released only with cardiac remodeling. While they can also be used for risk stratification, the natriuretic peptides are also useful for this purpose. *Meijers WC, et al. ST2 and galectin-3: Ready for prime time? EJIFCC 2016 Aug 1;27(3):238–52. PMID: 27683537; PMCID: PMC5009948.*

23. b. In the normal heart, CK-MB consists of 15%–20% of the total CK and this percentage is greater in the right than in the left ventricle. LD is composed of H (characteristic for heart) and M (characteristic for skeletal muscle) units, which can be combined into five isoenzymes. These isoenzymes are marked from LD_1 (HHHH) to LD_5 (MMMM). In healthy subject the concentration of $LD2 > LD1$ which is characteristically "flipped" ($LD1 > LD2$) in patients with myocardial infarct. Troponins (Tn) are part of the regulatory complex located in the thin filament of striated muscle. The elements of this complex are tropomyosin-binding subunit (TnT, 37 kDa), inhibitory cardio specific subunit (TnI, 24 kDa), and calcium-binding subunit (TnC, 18 kDa). The concentration of cTnI after myocardial infarct shows peak, which drops quickly and follows by a plateau or even a small secondary increase. This small peak should not be interpreted as evidence of reinfarction. The 99th percentile of the cTnT and cTnI 99th in healthy population is around 0.04 ng/mL. The concentration above these thresholds is consistent with damage to the myocytes. *Pincus MR, et al. Henry's Clinical Diagnosis and Management by Laboratory Methods; 22nd edition 2011, pp. 246–7.*

24. b. Side branch occlusions occasionally occur after coronary intervention and are largely an insignificant finding. Measuring serum with high-sensitivity cardiac troponin assay may reveal a minor increase relative to the precatheterization value. *Buturak A, et al. Rise of serum troponin levels following uncomplicated elective percutaneous coronary interventions in patients without clinical and procedural signs suggestive of myocardial necrosis. Postepy Kardiol Interwencyjnej 2016;12(1):41–8. doi:10.5114/pwki.2016.56948.*

Chapter 3

Cardiovascular disease markers

Jing Cao
Department of Pathology, University of Texas Southwestern Medical Center, Dallas, TX, United States

1. Which of the following biomarker contains apolipoprotein B100?
   a. oxidized phospholipids
   b. lipoprotein(a)
   c. triglyceride
   d. secretory phospholipase A2
   e. high-density lipoprotein
2. Select the correct description of proprotein convertase subtilisin/kexin type 9.
   a. PCSK9 is expressed only in liver and small intestine.
   b. PCSK9 inhibitors are prescribed for patients with type IIA familial hypercholesterolemia.
   c. PCSK9 promotes the degradation of hepatocyte LDL receptors.
   d. Blocking PCSK9 results in decreased recycling of LDL receptors.
   e. Inhibition of PCSK9 leads to decreased removal of LDL from circulation.
3. Which statement is true regarding systemic inflammation?
   a. The thrombus that occludes the coronary artery may result in myocardial ischemia and trigger endothelial cells to release proinflammatory cytokines.
   b. Proinflammatory cytokines include tumor necrosis factor (TNF)-α, interleukin (IL)-1β, IL-6, and cardiac troponin.
   c. High-sensitivity C-Reactive Protein (hs-CRP) and C-Reactive Protein are secreted by different cell types at different stages of the inflammatory process.
   d. Elevated hs-CRP levels predict long-term cardiovascular disease risk.
   e. Only proinflammatory markers but not antiinflammatory markers have been found to be associated with cardiovascular disease risk.

4. Select the upstream event of platelet activation in the thrombosis process in acute coronary syndrome.
   a. atherosclerotic plaque rupture
   b. release of platelet microparticles
   c. activation of coagulation cascade
   d. formation of stable fibrin clot
   e. platelet aggregation
5. Which of the following biomarkers promotes myocardial fibrosis?
   a. cardiac troponin I
   b. transforming growth factor-β
   c. cardiac troponin T
   d. interleukin-6
   e. brain natriuretic peptide
6. Which of the following suppresses inflammation in cardiac injury?
   a. growth differentiation factor-15
   b. soluble ST2
   c. matrix metalloproteinase-9
   d. transforming growth factor-β
   e. all of the above
7. Certain micro-RNAs (miRNA) play a role in the pathophysiology of cardiovascular disease. Which of the following statements is TRUE regarding miRNA?
   a. miRNAs are coding RNAs.
   b. Majority of miRNAs suppress target protein synthesis.
   c. miRNAs regulate gene transcription.
   d. miRNAs are most abundant in erythrocytes.
   e. Most miRNAs are circulating extracellularly.
8. Development of thrombus after atherosclerotic plaque rupture may result from failure of spontaneous endogenous thrombolysis. Which of the following is NOT a regulator of fibrinolysis?
   a. tissue-type plasminogen activator
   b. plasminogen activator inhibitor-1
   c. cystatin C
   d. α-2 antiplasmin
   e. D-dimer
9. Which of the following statements is FALSE regarding risk markers for adverse cardiovascular events in patients with stable coronary artery disease?
   a. Raised levels of hemodynamic stress markers such as NT-proBNP are associated with increased risk of adverse cardiovascular events in patients with stable coronary artery disease.

    b. Traditional CVD risk factors predict secondary events in patients with history of coronary artery disease.

    c. cTnT can be used in diagnosis and monitoring of acute myocardial infarction but not as a risk marker to predict future event.

    d. Sociodemographic characteristics are major determinants of risk of developing acute coronary syndrome (ACS) in these patients.

    e. Cardiovascular and noncardiovascular comorbidities such as renal disease are associated with risk of adverse cardiovascular events.

10. Select the pathophysiological process involved in acute coronary syndrome.

    a. Rupture of atherosclerotic plaque triggers platelet adhesion to subendothelial components.

    b. Formation of occlusive thrombus recruits monocytes and neutrophils to interact with platelets and further causes the release of proinflammatory cytokines.

    c. Impaired myocardial contractility and hemodynamics results in myocardial stretch, leading to renal disturbances.

    d. Coronary artery obstruction leading to myocardial ischemia and cause recruitment of neutrophils and monocytes.

    e. All of the above.

11. Which of the following statements is FALSE regarding copeptin?

    a. Copeptin is more predictive of myocardial infarction in patients presenting with chest pain than troponin.

    b. Copeptin is the C-terminal part of antidiuretic hormone.

    c. It is a marker of vasopressin secretion.

    d. Level of copeptin indicates status of hemodynamic stress.

    e. Copeptin presents at about equimolar concentration as vasopressin.

12. Select a lipid biomarker that shows the greatest difference in fasting vs nonfasting blood samples.

    a. LDL-C

    b. triglyceride

    c. ApoB

    d. Lp(a)

    e. HDL-C

13. Which of the following statements is FALSE regarding small dense LDL-cholesterol?

    a. Small dense LDL-cholesterol is associated with risk of cardiovascular disease independently of traditional lipid risk factors.

    b. Elevated level of small dense LDL-cholesterol is associated with raised triglyceride and decreased HDL-C levels in diabetes, obesity, and metabolic syndrome.

  c. Measurement of small dense LDL-cholesterol is available through electrophoresis, automated immunoassay, or ultracentrifugation.

  d. Smaller LDL particles are less atherogenic than their larger, more buoyant counterparts.

  e. Circulating small dense LDL undergoes atherogenic modifications including desialylation, glycation, and oxidation.

14. Select the biomarker that is NOT lowered by statin treatment.

  a. cholesterol

  b. LDL-C

  c. hsCRP

  d. lipoprotein(a)

  e. non-HDL-C

15. Which statement is FALSE regarding hsCRP?

  a. CRP is an acute-phase reactant produced predominantly in hepatocytes.

  b. Analyses from large-scale clinical trials have suggested using a hsCRP cut point of 2 mg/L for defining increased CVD risk.

  c. hsCRP is commonly used clinically together with traditional CVD risk factors to improve risk estimates.

  d. hsCRP has little within-individual variation even over acute inflammatory conditions.

  e. Due to the poor sensitivity and low negative predictive value, CRP is not used to rule out CVD.

16. What is the method to measure GlycA?

  a. nuclear magnetic resonance (NMR)

  b. immunoassay

  c. HPLC

  d. electrophoresis

  e. all of the above

17. Which of the following properties applies to lipoprotein (a)?

  a. Lipoprotein(a) has density close to high-density lipoprotein (HDL).

  b. Apolipoprotein (a) is covalently linked to apoB in lipoprotein(a).

  c. Dietary pattern is a critical determinant of lipoprotein(a) levels.

  d. Assays of lipoprotein(a) are standardized and are expressing lipoprotein(a) levels in the preferred mass concentration unit.

  e. None of the above.

18. Which of the following statements is TRUE regarding LDL-C?

  a. The Friedewald equation assumes very-low-density lipoprotein cholesterol concentration is equivalent to 25% of the triglyceride value.

  b. The reference method for LDL-C measurement, beta quantification, is based on two-dimensional gel electrophoresis.

  c. Newer LDL-C calculations such as the Sampson method estimate LDL-C more accurately, especially in patients with high triglycerides.

  d. The Friedewald equation is not valid when triglyceride is over 350 mg/dL.

e. None of the above.
19. Which of the following statements is FALSE regarding lipoprotein?
   a. Triglyceride-rich lipoproteins include very-low-density lipoprotein and chylomicrons.
   b. Lipoprotein lipase converts triglyceride-rich lipoprotein into cholesterol-rich lipoproteins.
   c. LDL population is heterogeneous in size and density.
   d. Very-low-density lipoprotein contains ApoA and ApoCIII.
   e. Triglyceride levels are a causal biomarker of cardiovascular disease risk.
20. Select the category of fat/fatty acids that is associated with lower risk of cardiovascular disease.
   a. saturated fat
   b. trans fat
   c. free fatty acids
   d. dairy fat
   e. long chain omega-3 fatty acids

Answers
1. b. Lipoprotein(a) contains two covalently linked apolipoproteins, ApoB100 and Apo(a). *Byun YS, et al. TNT Trial Investigators. Relationship of oxidized phospholipids on apolipoprotein B-100 to cardiovascular outcomes in patients treated with intensive versus moderate atorvastatin therapy: the TNT trial. J Am Coll Cardiol 2015;65:1286–95.* https://doi.org/10.1016/j.jacc.2015.01.050.
2. c. Answer
   a. PCSK9 is ubiquitously expressed in many tissues and cell types.
   b. The first two PCSK9 inhibitors, alirocumab and evolocumab, were approved by the U.S. Food and Drug Administration in 2015 for lowering LDL-particle concentrations when statins and other drugs were not sufficiently effective or poorly tolerated. c. PCSK9 promotes the degradation of hepatocyte LDL receptors. d,e. If PCSK9 is blocked, more LDLRs are recycled and are present on the surface of cells to remove LDL particles from the extracellular fluid. *Weinreich M, et al. Antihyperlipidemic therapies targeting PCSK9. Cardiol Rev 2014;22(3):140–6.*
3. d. Answer
   a. The *thrombus* that occludes the coronary artery may result in myocardial ischemia, and trigger leukocytes to release proinflammatory cytokines. b. Proinflammatory cytokines include tumor necrosis factor (TNF)-α, interleukin (IL)-1β, IL-6, but not cardiac troponin. c. High-sensitivity C-Reactive Protein (hs-CRP) and C-Reactive Protein are the same molecule measured by different assays. e. Both proinflammatory markers and antiinflammatory markers (e.g., TGF-β) have been found to be associated with cardiovascular disease risk. *Swirski FK, et al. Leukocyte behavior in atherosclerosis, myocardial infarction, and heart failure. Science. 2013;339:161–6.*

https://doi.org/10.1126/science.1230719. *Prabhu SD, et al. The biological basis for cardiac repair after myocardial infarction: from inflammation to fibrosis. Circ Res 2016;119:91–112.* https://doi.org/10.1161/CIRCRESAHA.116.303577.

4. a. After atherosclerotic plaque rupture, platelets adhere to exposed subendothelial *components*, such as collagen and von Willebrand factor, which promotes platelet activation and aggregation. Platelet microparticles are released upon platelet activation. Platelet activation and aggregation lead to the activation of the coagulation cascade and the formation of a stable crosslinked fibrin clot. *Storey RF, et al. The central role of the P(2T) receptor in amplification of human platelet activation, aggregation, secretion and procoagulant activity. Br J Haematol 2000;110:925–34. Vajen T, et al. Microvesicles from platelets: novel drivers of vascular inflammation. Thromb Haemost 2015;114:228–36.* https://doi.org/10.1160/TH14-11-0962.

5. b. Acute cardiac injury induces myocardial infiltration of leukocytes and proliferation of resident fibroblasts, which differentiate into myofibroblasts. The initial inflammatory response is then suppressed, in part, by antiinflammatory cytokines, and transforming growth factor (TGF)-β activates *myofibroblasts*, which promotes collagen synthesis. TGF-β has a major role in the formation of a collagenous scar and myocardial fibrosis. *Prabhu SD, et al. The biological basis for cardiac repair after myocardial infarction: from inflammation to fibrosis. Circ Res 2016;119:91–112. Siwik DA, et al. Interleukin-1beta and tumor necrosis factor-alpha decrease collagen synthesis and increase matrix metalloproteinase activity in cardiac fibroblasts in vitro. Circ Res 2000;86:1259–65.*

6. e. Answer

a. Growth differentiation factor-15 belongs to the TGF-β cytokine superfamily and acts *on* TGF-β receptor I and TGF-β receptor II, which inhibits neutrophil integrin activation and thus reduces neutrophil recruitment. b. Soluble ST2 receptor is a member of the IL-1 receptor family. Soluble ST2 blocks activation of ST2 receptor by acting as a decoy to prevent ligation by IL-33, and subsequently induction of TH 2 cytokine response. c. Matrix metalloproteinase-9 degrades extracellular matrix components, and it also degrades cytokines and other damage-associated molecular pattern (DAMP), which may limit inflammation. d. TGF-β predominantly acts to suppress inflammation and promote tissue repair. *Kempf T, et al. GDF-15 is an inhibitor of leukocyte integrin activation required for survival after myocardial infarction in mice. Nat Med 2011;17:581–8. Januzzi JL Jr. ST2 as a cardiovascular risk biomarker: from the bench to the bedside.*

J Cardiovasc Transl Res 2013;6:493–500. Iyer RP, et al. MMP-9 signaling in the left ventricle following myocardial infarction. Am J Physiol Heart Circ Physiol 2016;311:H190–8.

7. b. miRNAs are small noncoding RNAs that regulate gene expression. miR-NAs interact with *specific* mRNA to regulate their translation, largely by suppression of protein synthesis. Most miRNAs are located intracellularly while platelets and platelet microparticles are an abundant circulating source. *Condorelli G, et al. MicroRNAs in cardiovascular diseases: current knowledge and the road ahead. J Am Coll Cardiol 2014;63:2177–87. Willeit P, et al. Circulating microRNAs as novel biomarkers for platelet activation. Circ Res 2013;112:595–600.*

8. c. Key regulators of fibrinolysis include tissue-type plasminogen activator (tPA), plasminogen activator inhibitor-1, α-2 antiplasmin, α-2-antiplasmin–plasmin complex, D-dimer, soluble fibrin, thrombin activatable fibrinolysis inhibitor, and Lp(a). *Okafor ON, et al. Endogenous fibrinolysis: an important mediator of thrombus formation and cardiovascular risk. J Am Coll Cardiol 2015;65:1683–99.*

9. c. In patients with stable coronary artery disease, the main determinants of risk of developing an ACS are related to sociodemographic characteristics, pattern of coronary artery disease progression, traditional risk factors, cardiovascular and noncardiovascular comorbidities, heart rate, and levels of creatinine, hemoglobin, and white cell count. Markers of cardiac injury and hemodynamic stress, such as cTnT, NT-proBNP, MR-proANP, are associated with an increased risk of cardiovascular events and death. *Rapsomaniki E, et al. Prognostic models for stable coronary artery disease based on electronic health record cohort of 102,023 patients. Eur Heart J 2014;35:844–52. Sabatine MS, et al. Evaluation of multiple biomarkers of cardiovascular stress for risk prediction and guiding medical therapy in patients with stable coronary disease. Circulation 2012;125:233–40.*

10. e. The pathogenesis of acute coronary syndromes is driven by atherosclerosis, thrombosis, *inflammation*, and cardiac injury. Foam cells derived from macrophages and lymphocytes play a central role in the development of a lipid-rich athcrosclerotic plaque with a necrotic core. Rupture or erosion of an atherosclerotic plaque triggers platelet adhesion to subendothelial components, resulting in the formation of an occlusive thrombus, which also recruits monocytes and neutrophils. Platelet-leukocyte interactions cause the release of proinflammatory cytokines and recruited neutrophils also release neutrophil extracellular traps. Myocardial ischemia, caused by coronary artery obstruction, leads to the recruitment of neutrophils and monocytes toward chemokines. Leukocyte adhesion molecules then mediate transmigration of leukocytes. Impaired myocardial contractility and hemodynamics results in myocardial stretch, leading to

consequent renal disturbances. *O'Donoghue ML, et al. Multimarker risk stratification in patients with acute myocardial infarction. J Am Heart Assoc 2016;5:e002586.*

11. a. Copeptin is the C-terminal part of arginine vasopressin (also known as antidiuretic hormone). It is equimolar to vasopressin and therefore acts as a 1:1 marker of vasopressin secretion. Arginine vasopressin has a central role in hemodynamic fluid balance. Copeptin is a marker of hemodynamic stress and has been proven useful in the diagnosis of acute chest pain. Although copeptin is predictive of MI in patients presenting with chest pain, it is less predictive than troponin. *Morgenthaler NG, et al. Assay for the measurement of copeptin, a stable peptide derived from the precursor of vasopressin. Clin Chem 2006;52:112–9. O'Malley RG, et al. Prognostic performance of multiple biomarkers in patients with non-ST-segment elevation acute coronary syndrome: analysis from the MERLIN-TIMI 36 trial. Metabolic efficiency with ranolazine for less ischemia in non-ST-elevation acute coronary syndromes-Thrombolysis In Myocardial Infarction-36. J Am Coll Cardiol 2014;63:1644–53.*

12. b. When comparing random nonfasting lipid profiles with fasting profiles, the maximal mean changes at 1–6 h after habitual meals are +0.3 mmol/L (26 mg/dL) for triglycerides, 20.2 mmol/L (8 mg/dL) for total cholesterol, 20.2 mmol/L (8 mg/dL) for LDL-cholesterol, +0.2 mmol/L (8 mg/dL) for calculated remnant cholesterol, 20.2 mmol/L (8 mg/dL) for calculated non-HDL cholesterol; concentrations of HDL cholesterol, apolipoprotein A1, apolipoprotein B, and lipoprotein(a) are not affected by fasting/nonfasting status. *Nordestgaard BG, et al. Fasting is not routinely required for determination of a lipid profile: clinical and laboratory implications including flagging at desirable concentration cut-points—a joint consensus statement from the European Atherosclerosis Society and European Federation of Clinical Chemistry and Laboratory Medicine. Eur Heart J. 2016;37(25):1944–58.*

13. d. LDL consists of several subclasses of particles with different sizes and densities, including large buoyant and intermediate and small dense LDLs. sdLDL has a greater atherogenic *potential* than other larger LDL subfractions and that sdLDL cholesterol proportion is a better marker for prediction of cardiovascular disease than that of total LDL-C. Circulating sdLDL readily undergoes multiple atherogenic modifications in blood plasma, such as desialylation, glycation, and oxidation, that further increase its atherogenicity. Small dense LDL-cholesterol is associated with raised TG and decreased HDL-c levels in adiposity and diabetes. The development of homogeneous immunoassays makes possible clinical testing of small dense LDL-cholesterol. *Philipp AG, et al. Small, dense LDL, an update. Curr Opin Cardiol 2017;32(4):454–59.*

14. d. Meta-analysis reveals that statins significantly increase plasma Lp(a) levels despite that it lowers total *cholesterol*, LDL-C, and inflammatory markers such as hs-CRP. *Tsimikas S, et al. Statin therapy increases lipoprotein(a) levels [published online ahead of print, 2019 May 20]. Eur Heart J 2019;ehz310.*

15. d. CRP is an acute-phase reactant and nonspecific marker of inflammation, produced predominantly in hepatocytes as a pentamer of identical subunits in response to several cytokines. Standard hsCRP assays suffice in settings of active infection, tissue injury, or acute inflammation, which are known to cause marked elevations. Analyses from large-scale clinical trials have used a hsCRP cut point of 2 mg/L for defining increased CV risk. Although hsCRP is commonly used clinically to raise risk estimates, it cannot be used to rule out disease because of its poor sensitivity and low negative predictive value. *Yousuf O et al. High-sensitivity C-reactive protein and cardiovascular disease: a resolute belief or an elusive link? J Am Coll Cardiol 2103;62(5):397–408.*

16. a. Nuclear magnetic resonance (NMR) spectra of serum obtained under quantitative conditions for *lipoprotein* particle analyses contain additional signals that potentially serve as novel cardiovascular disease biomarkers. One of these signals, GlycA, originates from a subset of glycan *N*-acetylglucosamine residues on enzymatically glycosylated acute-phase proteins. *Otvos JD, et al. GlycA: a composite nuclear magnetic resonance biomarker of systemic inflammation. Clin Chem 2015;61(5):714–23.*

17. b. *Lipoprotein*(a) is a low-density lipoprotein (LDL) particle with an added apolipoprotein(a) (apo[a]) attached to the apolipoprotein(b) (apo[b]) component of the LDL particle via a disulfide bridge. The structure of Lp(a) is highly heterogeneous secondary to many different apo(a) isoforms within the population. An individual's Lp(a) level is 80%–90% genetically determined in an autosomal codominant inheritance pattern with full expression by 1–2 years of age and adult-like levels achieved by approximately 5 years of age. Outside of acute inflammatory states, the Lp(a) level remains stable through an individual's lifetime regardless of lifestyle. One of the main obstacles to the clinical use of Lp(a) is that its measurement and target levels have not been standardized. Several available assays report results in mass (mg/dL) instead of concentration (nmol/L), the latter of which is preferred. Unlike other lipids and lipoproteins, direct conversion between these two units is not possible because of the variable number of repeated units in different apo(a) isoforms, which leads to over- or underestimation depending on the particle size. Irrespective of the assay or units used, Lp(a) levels vary among ethnic groups and disease states, as well as whether measurements were from fresh or frozen samples, which have made published target levels inconsistent. *Scheel P, et al.*

Lipoprotein(a) in clinical practice. J Am Coll Cardiol 2019. https://www. acc.org/latest-in-cardiology/articles/2019/07/02/08/05/lipoproteina-in-clinical-practice.

18. c. Low-density lipoprotein cholesterol is commonly estimated by the Friedewald equation, which involves the direct measurement of total plasma cholesterol, HDL-C, and triglycerides. Very-low-density lipoprotein cholesterol is estimated by assuming that its concentration *is* equivalent to 20% of the triglyceride value. β-Quantification of LDL with ultracentrifugation to separate lipoprotein classes is the reference method for LDL-C. The Sampson equation relies on β-quantification of LDL-C as the reference method, topped the Friedewald and Martin methods for accuracy—especially for people with high triglycerides. *Sampson M, et al. A new equation for calculation of low-density lipoprotein cholesterol in patients with normolipidemia and/or hypertriglyceridemia. JAMA Cardiol 2020.* https://doi.org/10.1001/jamacardio.2020.0013.

19. d. Elevated triglyceride levels are a biomarker of cardiovascular (CV) risk. Consistent with these findings, recent genetic evidence *from* mutational analyses, genome-wide association studies, and Mendelian randomization studies provides robust evidence that triglycerides and triglyceride-rich lipoproteins are in the causal pathway for atherosclerotic CV disease. Triglycerides are major components of triglyceride-rich lipoproteins, including very-low-density lipoprotein (VLDL, containing apoB-100, apoC-I, apoC-II, apoC-III, and apoE) and chylomicrons (containing apoB-48, apoE, and apoC-II), which are synthesized and secreted from the liver and intestinal enterocytes, respectively. Lipoprotein lipase (LPL) promotes hydrolysis of the core triglycerides in VLDL and chylomicrons, producing VLDL remnants and chylomicron remnants, both of which are enriched in cholesterol relative to triglycerides. *Budoff M. Triglycerides and triglyceride-rich lipoproteins in the causal pathway of cardiovascular disease. Am J Cardiol 2016;118(1):138–45.*

20. e. Saturated and trans *fatty* acids have increased cardiovascular risk in several studies. Both *n*-6 and *n*-3 polyunsaturated fatty acids have been associated with lower cardiovascular risk. Within the *n*-6 series, linoleic acid seems to decrease cardiovascular risk. Within the *n*-3 series the long-chain fatty acids (eicosapentaenoic and docosahexaenoic acids) are associated with decreased risk for especially fatal coronary outcomes, whereas the role of α-linolenic acid is less clear. *Erkkilä A, et al. Dietary fatty acids and cardiovascular disease: an epidemiological approach. Prog Lipid Res 2008;47:172–87.*

Chapter 4

Oncology

Maximo J. Marin[a] and Xander M.R. van Wijk[b,c]
[a]Department of Pathology and Laboratory Medicine, Keck School of Medicine, University of
Southern California, Los Angeles, CA, United States, [b]Department of Pathology, Pritzker School of
Medicine, The University of Chicago, Chicago, IL, United States, [c]The University of Chicago
Medicine and Biological Sciences, Chicago, IL, United States

1. Neoplasm/tumor is a term used for cells that develop excessive and unregulated proliferation through various mechanisms. In general, which of the following is correct when considering benign and malignant tumors?
   a. Both malignant and benign tumors have gross and microscopic characteristics that suggest it will remain localized.
   b. Both malignant and benign tumors have gross and microscopic characteristics that suggest it will spread to other sites.
   c. Both benign and malignant tumors have the potential to destroy/compromise adjacent structures, but only malignant tumors invade and spread to distant sites.
   d. Both benign and malignant tumors have the potential to destroy/compromise adjacent structures and invade, but only malignant tumors have the potential to spread to distant sites.
2. Malignant tumors can arise from specialized cell types in different tissues throughout the body. Which of the following pair is incorrect? (Tissue type: Malignancy)
   a. epithelial tissue: carcinoma
   b. connective tissue: sarcoma
   c. nervous tissue: lymphoma
   d. muscular tissue: rhabdomyosarcoma
3. In order to differentiate benign from malignant tumors, there are specific characteristics that are evaluated by the pathologist under a microscope to help guide the diagnosis of a neoplasm. Which of the following characteristics is incorrect with regard to increasing malignancy potential?
   a. increased mitoses
   b. lack of differentiation
   c. increased pleomorphism
   d. increased polarity

Self-assessment Q&A in Clinical Laboratory Science, III. https://doi.org/10.1016/B978-0-12-822093-1.00004-1

4. Infections with different organisms/agents are common throughout the world. In general, infections illicit an inflammatory response, some will be completely cleared without treatment and others will not be cleared even with treatment. Which of the following infectious agent is not commonly associated with an increased cancer risk?
   a. syphilis (*Treponema pallidum*)
   b. human papillomavirus (HPV)
   c. *Helicobacter pylori*
   d. hepatitis B virus (HBV)

5. It is well established that viruses can induce tumorigenesis by several mechanisms. Which of the following virus to tumor type pair association is incorrect?
   a. Epstein-Barr virus (EBV): Burkitt's lymphoma
   b. human herpes virus-8 (HHV-8): rhabdomyosarcoma
   c. hepatitis B virus (HBV): hepatocellular carcinoma
   d. human papilloma virus (HPV): squamous cell carcinoma

6. Which of the following is most likely an incorrect mechanism by which a virus can induce tumorigenesis?
   a. increasing genomic instability
   b. preventing apoptosis
   c. genome integration inhibiting angiogenesis protein production
   d. inhibiting normal cell signaling that halts proliferation

7. Which of the following statements about oncogenes and tumor suppressor genes is most likely incorrect?
   a. A proto-oncogene is the unmutated cellular counterpart to an oncogene.
   b. Mutation of an oncogene will induce a constitutively active gene.
   c. Mutations of tumor suppressor genes would decrease the effectiveness of their protein product.
   d. During the expression of an oncogene, a tumor suppressor gene would become activated.

8. Which of the following are essential alterations for malignant transformation to occur?
   1. induction of apoptosis
   2. ability to invade
   3. defects in DNA repair
      a. 1, 3
      b. 1, 2
      c. 2, 3
      d. all of the above

9. Which of the following is not a common inherited susceptibility gene for breast, gynecologic, or gastrointestinal cancer?
   a. *BRCA1*
   b. *APC*
   c. *MLH1*
   d. *GAA*

10. Which of the following chromosomal abnormalities is found in almost all chronic myeloid leukemia (CML) cases?
    a. t(9;22)(q34;q11.2); *BCR-ABL1*
    b. del17p13; *TP53* deletion
    c. t(14;18)(q32;q21); *IGH-BCL2*
    d. t(8;21)(q22;q22); *RUNX1-RUNX1T1*
11. Which of the following are known mechanisms of tumor invasion and metastasis?
    a. secretion of matrix metalloproteinases
    b. modifications of cell–cell interactions
    c. reorganization of the cytoskeleton
    d. all of the above
12. Which of the following outlines the correct order of the pathway in which tumors spread?
    a. primary   tumor > intravasation > localized   invasion > circulatory transport > extravasation
    b. localized   invasion > intravasation > circulatory   transport > extravasation > tissue colonization
    c. primary tumor > extravasation > circulatory transport > intravasation >tissue colonization
    d. localized   invasion > arrest   in   micro   vessels > circulatory   transport > micro metastasis > colonization
13. Which of the following is not a known epidemiologic factor in the incidence of cancer?
    a. age
    b. occupation
    c. chronic inflammatory conditions
    d. wind turbines
14. Normally, when cancer cells arise in the normal tissue, the immune system can detect these rogue cells and subsequently eradicate them. Which of the following mechanisms does the immune system use to either detect and/or eradicate cancer cells?
    1. detection of mutated suppressor proteins
    2. recognition of aberrantly expressed self-proteins
    3. identification of non-self oncogenic viral proteins
       a. 1, 3
       b. 1, 2
       c. 2, 3
       d. all of the above
15. The most recent advances in immunotherapy have revolutionized treatment of some cancers. Which of the following mechanism(s) would most likely not enhance the immune system to detect and or eradicate neoplastic cells?
    1. Downregulation of neoplastic cell machinery that functions to present self-proteins.

    2. Blockage of receptors on neoplastic cells that function as inhibitory receptors for immune cells.

    3. Modification of T-cells to target specific antigens of malignant cells.

    a. 1

    b. 2

    c. 2, 3

    d. 1, 3

16. Which of the following is <u>not</u> a potential predictive biomarker for immunotherapy response?

    a. tumor mutational burden (TMB)

    b. tumor PD-1 expression

    c. tumor infiltrating lymphocytes (TILs)

    d. mismatch repair (MMR) deficiency

17. Traditional anticancer agents target tumor cells by several mechanisms, which of the following is <u>not</u> a correct drug-mechanism pair?

    a. All-trans-retinoic-acid (ATRA): binding to its receptor to induce cell differentiation.

    b. Imatinib mesylate (Gleevec): inhibits tyrosine kinase activity to decrease growth promotion.

    c. 5-Fluorouracil (5-FU): inhibits thymidylate synthase, which depletes nucleoside synthesis for proliferation.

    d. Cyclophosphamide: decreases DNA cross-linking triggering signals for cell death.

18. Which of the following is a known mechanism for tumor therapy resistance?

    a. decreased intracellular drug accumulation

    b. increased repair of drug-induced damage

    c. alteration of drug targets (mutation)

    d. all of the above

19. A 44-year-old man has a diagnosis of colon cancer. His family history is significant for gastrointestinal cancer. No one in the family has had previous genetic testing. Which technology is most likely to be used in genetic screening of this patient?

    a. targeted PCR

    b. full-gene Sanger sequencing

    c. massive parallel sequencing

    d. multiplex ligation-dependent probe amplification

20. What is generally the preferred sample type for detection of circulating tumor DNA (ctDNA)?

    a. plasma

    b. serum

    c. urine

    d. CSF

Answers

1. c. A tumor is considered benign when gross and microscopic features are considered innocuous. This suggests that it will remain localized and generally can be surgically removed. However, although benign tumors are generally well contained cohesive expansile masses, the surrounding normal tissue can become damaged by compression in a collateral damage type of process such as compressing blood vessels and causing death to tissue. In contrast, malignant tumors have the potential to invade adjacent tissue by penetrating into the structures by mechanisms such as secreting proteolytic enzymes to breach underlying normal connective tissue. Another feature that distinguishes malignant from benign tumors is the ability to metastasize. Benign tumors, in general, do not spread to distant sites by a process called metastasis. Metastasis is achieved when the tumor has gained the ability to spread to distant sites by using the vascular system as a conduit and continue their unregulated proliferation and invasive process. *Kumar V. et al., eds., Robbins & Cotran Pathologic Basis of Disease, 9th ed., 2014, pp. 271–2. Weinberg RA, ed., The Biology of Cancer, 1st ed., pp. 589–91.*

2. c. There are various ways tissues could be classified. In general, there are four basic types of tissue: epithelial, muscular, connective (includes blood and blood vessels), and nervous. Within each of these tissue types are specialized cells with different functions. a. Malignant tumors/neoplasms of epithelial origins generally fall under the classification of carcinomas which includes hepatocellular (liver) carcinoma, renal (kidney) cell carcinoma, and transitional (bladder) cell carcinoma. b. Malignant neoplasms of connective tissue origins generally fall under the classification of sarcomas, lymphomas, and leukemias which includes neoplasms such as liposarcoma (adipose tissue), chondrosarcoma (cartilage tissue), osteogenic (bone tissue) sarcoma, Burkitt's lymphoma, and acute myeloid leukemia (lymphoid and hematopoietic tissue). d. Muscular tissue neoplasms include leiomyosarcoma (smooth muscle) and rhabdomyosarcoma (striated muscle). c. Nervous tissue neoplasms include malignant peripheral nerve sheath tumors, and neuroblastomas. Lymphoma is a neoplasm originating from lymphoid tissue, which is classified as connective tissue. *Kumar V. et al., eds., Robbins & Cotran Pathologic Basis of Disease, 8th ed., 2014, p. 263. Weinberg RA, ed., The Biology of Cancer, 1st ed., pp. 28–33.*

3. d. For a. Increased mitoses/proliferation of tumor cells can be seen both in benign and malignant tumors. However, malignant tumors often have a significantly greater increase of mitotic activity. b. Tumor cells that show little to no resemblance of cell origin are highly suggestive of increasing aggressiveness in malignant tumors. c. Variation in nuclear and cell size along with shape indicates an increasing potential for malignancy. d. Lack (not increase) of polarity, i.e., a highly disorganized orientation of tumor

cells, shows an increased potential for malignancy. *Kumar V. et al., eds., Robbins & Cotran Pathologic Basis of Disease. 9th ed., 2014, p. 270.*

4. a. For a. *Treponema pallidum* is a spiral shaped gram-negative bacterial organism that is commonly transmitted by sexual contact. There is currently no association with an increased risk for cancer and an infection with *Treponema pallidum*. b. HPV is a DNA virus that infects squamous epithelial cells and is commonly transmitted by sexual contact. For example, the virus can infect the oropharynx, anogenital region, and cervix. These sites are vulnerable to malignant transformation leading to squamous cell carcinoma. c. *Helicobacter pylori* is a gram-negative bacterium that is helically shaped. It infects the gastric mucosa causing chronic gastritis that results in peptic ulcer disease. A high proportion of gastric cancer patients are infected with *Helicobacter pylori* when compared to aged matched controls. d. Hepatitis B virus infects the hepatocytes of the liver and most infected individuals develop a chronic infection. Of the individuals with chronic infection, a small proportion will progress to cirrhosis which leads to a risk for developing hepatocellular carcinoma. *Rifai N, et al., eds., Tietz Textbook of Clinical Chemistry and Molecular Diagnostics, 6th ed., 2017, pp. 1001–2, 1009–11, 1368–9, 1403–4.*

5. b. For a. Epstein-Barr virus belongs to the Herpesviridae family, a DNA virus. It is a common infection in the general population. Epstein-Barr virus is highly associated with Burkitt's lymphoma, infecting the B-lymphocytes in this case. b. HHV-8 also belongs to the Herpesviridae family, and although most infections are asymptomatic, or have minimal symptoms, immunocompromised patients such as HIV-infected patients can develop malignant transformation of the lymphatic endothelial cells leading to Kaposi's sarcoma, not rhabdomyosarcoma (neoplasm of the muscle tissue). c. HBV infects hepatocytes of the liver leading to chronic infection of the liver and hepatitis leading to possible cirrhosis and hepatocellular carcinoma. d. HPV is a DNA virus that infects squamous epithelial cells and is commonly transmitted by sexual contact. The virus can infect the oropharynx, anogenital region, and cervix. These sites are vulnerable to malignant transformation leading to squamous cell carcinoma. *Rifai N, et al., eds., Tietz Textbook of Clinical Chemistry and Molecular Diagnostics, 6th ed., 2017, pp. 1101, 1368–1369, 1751, 1756–7.*

6. c. For d. HPV E7 oncoprotein inhibits cell signaling that maintains the cell in a quiescent state (no replication), specifically it inactivates the retinoblastoma protein (pRb) and accelerates its degradation which ultimately allows for the cell to gain the potential to proliferate. b. HPV E6 viral oncoprotein coordinates the disruption of the tumor suppressor protein p53. The p53 signaling pathway leads to cell apoptosis. Essentially, when the cell triggers danger as in the case of a virally infected cell, p53 is activated to induce self-killing. Both the E7 and E6 coordination allow for cells to replicate unchecked. a. Because both pRb and p53 are disrupted, one

of the consequences is the dysregulation of the centrosome, which plays a central role in organizing the mitotic spindles for the separation of chromosomes during replication. The result of the centrosome dysregulation leads to centrosome duplication, thus, chromosome instability. c. Although tumorigenesis can occur with integration of the viral genome, inhibiting the proteins that induce angiogenesis would be counterproductive for a tumor growth. *Weinberg RA, ed., The Biology of Cancer, 1st ed., pp. 514–5.*

7. b. There are several essential alterations needed for malignant transformation. Two of those key changes are discussed in this question: self-sufficiency in growth signals and insensitivity to growth inhibitory signals. a. Proto-oncogenes, the unmutated cellular counterpart to an oncogene, are involved in signaling pathways that promote cellular growth and replication. Proto-oncogenes may function as growth factors, receptors, signal transducers, transcription factors, and/or cell cycle regulator components. Tumor suppressor genes and their products function to inhibit uncontrolled proliferation and growth by a number of mechanisms, which, for example, may include a protein product that inhibits cell cycle progression. b. Mutation of a proto-oncogene may result in an active oncogene (mutations of the oncogene would most likely inactivate it), which may not be sufficient for malignant transformation if the d. tumor suppressor gene is functional. This is because it would then become activated by the oncogene signaling leading to suppression by cell arrest or even self-destruction via apoptosis. c. Thus, many malignant tumors have oncogenes with mutated tumor suppressor genes (absent or dysfunctional protein product) of that respective signaling pathway to allow for potential malignant transformation. *Kumar V. et al., eds., Robbins & Cotran Pathologic Basis of Disease. 9th ed., 2014, pp. 283–4.*

8. c. The 7 essential alterations for malignant transformation are self-sufficiency in growth signals, insensitivity to growth-inhibitory signals, evasion of apoptosis, limitless replicative potential, sustained angiogenesis, ability to invade and metastasize, and defects in the DNA repair machinery. Comparing the answer choices given, only induction of apoptosis is incorrect because it is the evasion of apoptosis, not induction of it. *Kumar V. et al., eds., Robbins & Cotran Pathologic Basis of Disease, 9th ed., 2014, pp. 271-2.*

9. d. Answers a. through c. are common examples of tumor suppressor and mismatch repair genes. An inherited defective copy significantly increases the susceptibly of an individual to develop a specific type or types of cancer. A "second hit," i.e., the somatic mutation of the one functioning copy, can result in disease. a. Most inherited forms of breast and ovarian cancer are due to inheritance of a defective copy of *BRCA1* or *BRCA2*, both of which are DNA repair genes. Mutant *BRCA1* carriers have a 44%–75% risk of developing breast cancer by the age of 70; for ovarian cancer this is

43%–76%. Slightly lower frequencies are observed for *BRCA2* mutation carriers. b. The adenomatous polyposis coli (*APC*) gene plays a key role in regulating the level of the proto-oncoprotein beta-catenin. Without functional APC, the Wnt signaling pathway is constantly activated, resulting in chromosomal instability. Germline *APC* mutations cause familial adenomatous polyposis (FAP), a disease characterized by 100–1000s colorectal adenomatous polyps. Mutant *APC* carriers have a 100% lifetime risk for colorectal cancer if left untreated. c. Lynch Syndrome, also known as hereditary nonpolyposis colorectal cancer (HNPCC), is the most common susceptibility syndrome for colorectal cancer, but it also increases the risks of many other types of cancer. It is caused by a germline mutation in one of the mismatch repair genes, such as *MLH1* and *MSH2* (both account for approximately 40% of cases each). Patients with Lynch Syndrome have a 50%–70% lifetime risk of developing colorectal cancer. d. *GAA* mutations cause Pompe disease, an inborn error of metabolism. *Rifai N, et al., eds., Tietz Textbook of Clinical Chemistry and Molecular Diagnostics, 6th ed., 2017, pp. 1050–9.*

10. a. The reciprocal translocation between 9q34 and 22q11.2 results in the formation of a novel BCR-ABL fusion gene, leading to a constitutively active ABL1 (a tyrosine kinase) and dysregulated cellular proliferation. The abnormally short chromosome 22 is known as the Philadelphia chromosome. Answers b., c., and d. are chromosomal abnormalities found in multiple myeloma, follicular lymphoma, and diffuse large B cell lymphoma, and acute myeloid leukemia, respectively. *Rifai N, et al., eds., Tietz Textbook of Clinical Chemistry and Molecular Diagnostics, 6th ed., 2017, pp. 1082-92, 6th ed..*

11. d. Tissues are organized into compartments and are separated by basement membranes and interstitial connective tissue. In order to invade adjacent tissue, the malignant cells must be able to breach this biophysical barrier. a. Cancer cells secrete matrix metalloproteinases to degrade the basement membrane and connective tissue barrier. However, in order for the malignant cell move in and out of tissue, enter the vascular system or colonize tissue, they must be able to (b. and c.) modify their cell-cell interactions to "loosen up" and reorganize their cytoskeleton to begin their migration process. *Rifai et al., eds., pp. 1127–1128, 6th ed. Kumar V. et al., eds., Robbins & Cotran Pathologic Basis of Disease, 9th ed., 2014, p. 306. Weinberg RA, ed., The Biology of Cancer, 1st ed., pp. 621–2.*

12. b. Metastasis is a highly complex and coordinated biological process. The process begins with a transformation and clonal expansion of a cell that eventually leads to a metastatic clone. Then there is localized invasion to the adjacent tissue leading to intravasation. At this point, the tumor cell embolus is transported by the circulatory system until it becomes localized in a vessel and begins the extravasation process leading to a micro metastatic deposit. If the metastatic deposit survives, angiogenesis and

proliferation will begin leading to colonization of the new tissue site. *Kumar V. et al., eds., Robbins & Cotran Pathologic Basis of Disease, 9th ed., 2014, p. 306. Weinberg R, pp. 591, 1st ed.*

13. d. For a. In general, the incidence of cancer increases with age. Still, children are not spared. The types of cancers that predominate in children are different from adults. b. Occupation exposure to carcinogenic agents is commonly seen with specific industries such as agriculture (pesticides), mining and quarrying (asbestos, arsenic, etc.), steel/metal smelting (beryllium, cadmium, etc.), and construction (benzene, vinyl chloride). c. Inflammatory conditions such as reflux esophagitis, cystitis (bladder inflammation), other chronic infections, and various autoimmune disorders can also predispose individuals to cancer. d. There is currently insufficient evidence to suggest that living near wind turbines causes cancer. Kumar V. et al., eds., Robbins & Cotran Pathologic Basis of Disease, 9th ed., 2014, pp. 275–80.

14. d. Although the immune system is very effective at detecting foreign antigens, neoplastic cells present difficulty because they are "self" cells. However, there may be subtle differences, "tumor antigens," that allow the immune system to respond. Still, this is a complex process. Products of mutated genes can be recognized as non-self-antigens. Aberrantly expressed self-proteins, either under- or overexpressed, within specific types of tissue can alert the immune surveillance system that these cells need to be eradicated. Of the three mechanisms, production of viral proteins being expressed by the malignant cell is the most straightforward as the immune system will recognize this as nonself and swiftly target these cells for destruction. *Kumar V. et al., eds., Robbins & Cotran Pathologic Basis of Disease, 9th ed., 2014, pp. 310–4.*

15. a. Downregulating the machinery that presents self-proteins would decrease the ability of immune cells to detect malignant cells. This would not be helpful. Blocking receptors on neoplastic cells that inhibit the immune system is the mechanism of immune checkpoint blockade therapy (e.g., anti-PD-1 antibodies nivolumab and pembrolizumab). Modifying T-cells to target neoplastic cells is the mechanism of CAR T cell immunotherapy. Genetically modified T-cells are enhanced to specifically target antigens on neoplastic cells. For example, a chimeric antigen receptor is engineered to target surface antigens of malignant cells. When binding occurs, the T-cell induces cell death of the neoplastic cells. *Kumar V. et al., eds., Robbins & Cotran Pathologic Basis of Disease, 9th ed., 2014, pp. 310–4. Yang Y. Cancer immunotherapy: harnessing the immune system to battle cancer. J Clin Investig 2015125(9):3335–7. https://doi.org/10.1172/JCI83871.*

16. b. For a. A high TMB is associated with a positive response in various cancer types. b. Tumor PD-L1 (not PD-1, which is expressed by T-cells) expression, however, is a commonly used predictive biomarker in anti-PD-1 therapy. Increased expression is associated with a better response. c. The presence of TILs was associated with a better response in, for example, melanoma patients. d. Mutations that lead to a deficiency in DNA mismatch repair often result in a heavy tumor mutational burden. These unresolved mutations are thought to increase the immunogenicity of MMR-deficient tumors. *George AP, et al. The discovery of biomarkers in cancer immunotherapy.. Comput Struct Biotechnol J 2019;17:484–97. Spencer KR et al. Biomarkers for immunotherapy: current developments and challenges. Am Soc Clin Oncol Educ Book. 2016;35:e493–503.*

17. d. For a. ATRA is the definitive treatment for acute promyelocytic leukemia. When ATRA binds to receptors within the immature leukemic cells, this triggers gene expression for maturation. b. Imatinib binds to the catalytic cleft of the Abl tyrosine kinase to inhibit activity of the abnormal fusion protein, Bcr-Abl, seen in chronic myelogenous leukemia (CML). c. 5-FU, used for gastric and colonic cancers, inhibits thymidylate synthase, which is important for folate, purine, and pyrimidine production. If a fast-dividing malignant cell is depleted of nucleosides for DNA replication, the cell will eventually die. d. Cyclophosphamide metabolites. *Rifai N, et al., eds., Tietz Textbook of Clinical Chemistry and Molecular Diagnostics, 6th ed., 2017, pp. 1424–25. McPherson RA, et al., eds., Henry's Clinical Diagnosis and Management by Laboratory Methods, 23rd ed., 2017. 1468–69. Weinberg RA. The Biology of Cancer. 1st ed., pp. 732, 755–62. Fu D, et al. Balancing repair and tolerance of DNA damage caused by alkylating agents Nature Rev Cancer 2012 12(2):104–20. https://doi.org/10.1038/nrc3185.*

18. d. Neoplasms are constantly evolving by mutating as evidenced by their unstable genome. This gives tumors the ability to find configurations that allow for continued survival as with the case of treatment resistance. a. Decreasing intracellular chemotherapeutic drugs can be achieved by, for example, increasing genes that encode drug efflux pumps at the cell membrane. b. Another mechanism that neoplastic cells can develop is increasing biological DNA repair machinery to remove damaged DNA induced by drugs such as cyclophosphamide. Finally, tumor cells can become resistant to tyrosine kinase inhibitors such as imatinib. c. As the neoplastic cells become exposed to imatinib, a subclone can arise with mutation(s) that modify the cleft where imatinib binds to the Bcr-Abl oncoprotein. The result effectively renders imatinib incapacitated or limited as it no longer recognizes, or has decreased binding affinity to, the modified cleft. This gives this tumor subclone the ability to outgrow its descendants and ultimately making the tumor resistant to treatment. *Weinberg RA, ed., The Biology of Cancer, 1st ed., pp. 756–64.*

19. c. For a. Targeted PCR is most useful in the detection of common pathogenic point mutations and would have been the method of choice if a familial sequence variant was identified. b. Full-gene Sanger sequencing is most

useful for variant detection in disorders associated with a single disease-causing gene. c. Massive parallel sequencing (MPS), also known as next-generation sequencing (NGS), has revolutionized the field of molecular diagnostics as it can perform simultaneous sequencing reactions on millions of targets at a fraction of the per base cost compared to Sanger sequencing. In this case, an MPS-based hereditary gastrointestinal cancer panel would likely be used to confirm a diagnosis of hereditary gastrointestinal cancer. d. Multiplex ligation-dependent probe amplification is most useful to detect large deletions/duplication in three or fewer genes. *Rifai N, et al., eds., Tietz Textbook of Clinical Chemistry and Molecular Diagnostics, 6th ed., 2017, pp. 1064–7.*

20. a. Serum usually has more ctDNA than plasma; however, plasma is preferred as trapped leukocytes in serum which can release their DNA during the clotting process, thereby potentially interfering with ctDNA detection. ctDNA is also present in body fluids other than blood, including urine and CSF; however, ctDNA degradation in urine is a concern and ctDNA concentrations in CSF are far less than in blood. Nevertheless, urine and CSF ctDNA can be particularly useful for detection of urogenital and brain cancer mutations, respectively. *Rifai N, et al., eds., Tietz Textbook of Clinical Chemistry and Molecular Diagnostics, 6th ed., 2017, pp. 1136–7.*

Chapter 5

Tumor markers

Qing H. Meng
MD Anderson Cancer Center, Houston, TX, United States

1. What are the tumor markers?
   a. Biological substance synthesized and released by cancer cells or produced by the host in response to the presence of tumor.
   b. Found in a solid tumor, in circulating tumor cells, in peripheral blood, in
   lymph nodes, in bone marrow, or in other body fluid (urine, stool,
   ascites).
   c. Detected qualitatively or quantitatively by biochemical, immunological,
   and molecular methods.
   d. Can be proteins, glycoproteins, hormones, enzymes, antigens, receptor,
   or oncogenes.
   e. All of the above.
2. Tumor markers are best used to:
   a. screen for cancer in asymptomatic patients
   b. make cancer diagnosis
   c. differentiate malignancy from benign state
   d. make the prognosis
   e. monitor treatment response and recurrence or remission
3. Which of the following is the most appropriate description for CA 15-3?
   a. A good screening test for early detection of breast cancer.
   b. A marker indicating advanced stage of breast cancer with metastasis.
   c. The same protein encoded by human *MUC-1* gene as cancer antigen
   (CA) 27.29.
   d. Cannot be used to monitor therapy
   e. Is breast specific and not seen in other malignancies such as ovary, liver,
   or lung cancer?
4. What is the most appropriate use of the CA 15-3 test in breast cancer
   patients?
   a. screening for disease in the asymptomatic population
   b. diagnosis of disease in symptomatic patients
   c. aid in clinical staging

    d. monitoring breast cancer patients with advanced disease

    e. a sensitive marker for early detection of lung cancer

5. Which of the following statements is most appropriate for Cancer antigen 125 (CA 125):

    a. is sensitive for low stage (stage I–II) ovarian cancer

    b. can be positive in other nonovarian malignancies

    c. cannot be used to assess patients for recurrent ovarian cancer

    d. can distinguish women with ovarian cancer from women with normal benign condition

    e. can be used for ovary cancer screening in the normal population

6. The CA-125 tumor marker is more associated with which disease state?

    a. hepatocellular carcinoma

    b. ovarian carcinoma

    c. medullary carcinoma of the thyroid

    d. colorectal carcinoma

    e. breast cancer

7. Women with tenderness and enlargement of the ovary mass suggestive of an ovarian neoplasm. Which of the following tumor marker(s) should be ordered?

    1. hCG

    2. CA125

    3. CEA

    4. HE4

    5. HER2/neu

        a. 1, 2, and 3

        b. 1 and 3

        c. 2 and 4

        d. 4 only

        e. all of the above

8. A 65-year-old woman is admitted to the hospital with presentation of chest pain, fever, fatigue, dry cough, and shortness of breath. A chest X-ray and CT are performed revealing a peripheral middle lobe mass in the right lung and a large right-sided pleural effusion. Pleural fluid was tested showing pink and hazy with highly increased RBC and leukocytes. Cytological examination on biopsy tissue and pleural fluid confirmed lung adenocarcinoma. Which tumor marker is unlikely changed more significantly in non-small cell lung cancer (NSCLC)?

    a. CEA

    b. neuron-specific enolase (NSE)

    c. cytokeratin 19 (CYFRA-21-1)

    d. CA 125

    e. AFP

9. Which is the most appropriate laboratory approach to differentiate malignant pleural effusion from transudative effusion?

a. pleural fluid CEA

b. plural fluid creatinine

c. pleural fluid pH

d. using Light's criteria

e. pleural fluid RBC

10. Which genetic alteration in the following genes is the most common in nonsmall cell lung cancer (NSCLC) patients as the first FDA-approved liquid biopsy testing guiding targeted therapy?

a. EGFR

b. ALK

c. ROS1

d. BRAF

e. MET

11. A 40-year-old man was presented to ED with complaint of abdominal pain, back pain, loss of appetite, fatigue, weight loss, and dark urine. His physical examination is remarkable for jaundice and large left upper abdominal mass. CT scan of his abdomen revealed an irregular infiltrating mass in the body of the pancreas. A group of tumor markers were ordered along with other laboratory tests. Which tumor marker is most likely to be elevated in this patient?

a. CA19-9

b. CA242

c. Calcitonin

d. AFP

e. CA125

12. An elevated serum concentration of calcitonin is usually associated with which of the following diseases?

a. parathyroid gland tumors

b. ovarian carcinoma

c. medullary carcinoma of the thyroid

d. sarcoma

e. carcinoid tumors

13. AFP is a tumor maker for which type of cancer

a. hepatocellular carcinoma

b. ovarian carcinoma

c. medullary carcinoma of the thyroid

d. colorectal carcinoma

e. sarcoma

14. Beta 2 microglobulin is the tumor marker for which disease state?

a. testicular cancer

b. ovarian carcinoma

c. medullary carcinoma of the thyroid

d. multiple myeloma

e. sarcoma

15. Significantly elevated CEA in a breast cancer patient suggests:
    a. nothing relevant since CEA is not specific for breast cancer
    b. a good prognosis
    c. the patient a heavy smoker
    d. cancer metastasis
    e. accompanying with colorectal cancer
16. Which of the following statement about oncofetal proteins is false?
    a. produced during fetal life
    b. disappear or in low levels in adults
    c. AFP and CEA are the only two oncofetal proteins
    d. become elevated in some malignancies
    e. oncofetal antigens are promising targets for vaccination against several types of cancers
17. Which one of the following conditions is unlikely associated with elevated beta-human chorionic gonadotropin (β-hCG)?
    a. nonseminomatous testicular cancer
    b. prostate carcinoma
    c. gestational trophoblastic disease
    d. normal pregnancy
    e. ovarian germ cell tumors
18. Which of the statements for beta-human chorionic gonadotropin (β-hCG) is correct?
    a. has an identical B-subunit to luteinizing hormone (LH), follicle-stimulating hormone (FSH), and thyroid-stimulating hormone (TSH).
    b. it is homogenous and has only one molecular form of hCG
    c. different immunoassays may measure different forms of hCG
    d. is only elevated in pregnancy.
    e. is not elevated in malignancies
19. A 30-year-old woman was admitted with a 9-month history of amenorrhea and an admission serum β-hCG value of 3800 mIU/mL. X-ray studies revealed a lesion in the left lung, and a CT scan demonstrated small lesions in the frontal and left parietal lobes of the brain. Biopsies confirmed the presence of choriocarcinoma. The patient received radiation and chemotherapy treatment, and serum β-hCG values returned to normal. Three months later, a modest increase in the serum β-hCG signaled a possible recurrence of a tumor. To confirm this, a specimen was sent out to another laboratory and the hCG result from an outside laboratory turned to be negative. X-ray studies revealed a tumor in the right lung. Which is the most likely molecule that the outside laboratory measures only?
    a. intact hCG
    b. beta-core fragment
    c. beta-hCG
    d. intact hCG and free beta-hCG
    e. free beta-hCG

20. Which test is unlikely recommended for colorectal cancer screening?
   a. highly sensitive guaiac FOBT
   b. guaiac-based Hemoccult II cards
   c. fecal immunochemical test (FIT)
   d. fecal DNA
   e. epi proColon

21. A 60-year-old man is referred to an urologist with a PSA of 6.2 ng/mL. His DRE is negative. His father was diagnosed prostate cancer 20 years ago. His free PSA is measured and the result is 8%. What is the next practical approach to make the diagnosis?
   a. prostatic acid phosphatase (PAP)
   b. a 12-core transrectal ultrasound-guided prostate needle biopsy (TRUS) biopsy with pathological examination
   c. prostate Cancer Antigen 3 (PCA3)
   d. prostate Health index (PHI)
   e. 4K Score

22. Total PSA assays measure which of the following?
   1. free PSA
   2. prostatic-specific acid phosphatase (PAP)
   3. PSA-ACT (alpha-1-anti-chymotrypsin)
   4. PSA-A2M (alpha-2-macroglobulin)
      a. 1, 2, and 3
      b. 1 and 3
      c. 2 and 4
      d. 4 only
      e. all of the above

23. Which of the following statements is most appropriate?
   a. PSA is not elevated in benign prostate hypertrophy, prostatitis, and urinary tract infection.
   b. PSA is not elevated in neuroendocrine prostate cancer.
   c. PSA is elevated in patients with use of 5-alpha reductase inhibitors.
   d. normal PSA results can eliminate the possibility of prostate cancer.
   e. PSA can discriminate indolent prostate cancer from aggressive prostate cancer.

24. Which of the following tumor markers are mandatory for testicular cancer?
   1. AFP
   2. hCG
   3. LDH
   4. PSA
      a. 1, 2, and 3
      b. 1 and 3
      c. 2 and 4

    d. 4 only

    e. all of the above

25. Which of the following description for thyroglobulin (Tg) is false?

    a. Is used as a tumor marker for follow-up of patients with differentiated thyroid cancers after thyroidectomy and radioactive iodine ablation.

    b. Serum Tg of >1 ng/mL in an athyrotic individual indicates possible residual or recurrent disease.

    c. Current guidelines recommend measurement of thyroglobulin (Tg) with high sensitivity.

    d. There is no need to measure serum thyroglobulin autoantibodies (TgAb).

    e. Tg can be falsely low or high due to interference.

26. Which of the following description for HER-2 is false?

    a. It is a transmembrane protein normally expressed on the epithelia of numerous organs.

    b. Overexpression of the HER-2 gene is used to guide breast cancer treatment.

    c. It is a member of the epidermal growth factor (EGF) family.

    d. The extracellular protein domain is released into and can be measured in the blood.

    e. It is not used for prognosis

27. Which of the following descriptions on genetic and molecular tumor markers is false?

    a. Genetic mutations affect the biological characteristics and transformation of cells and the development of cancer.

    b. tumor-suppressor gene

    c. proto-oncogene

    d. Cancer is not caused by inherited genetic mutations.

    e. Genetic profiles provide better strategies on risk stratification

28. Detection of the *BCR-ABL* fusion gene is most useful for which type of cancer?

    a. acute myelogenous leukemia (AML)

    b. chronic myelogenous leukemia (CML)

    c. acute lymphocytic leukemia (ALL)

    d. chronic lymphocytic leukemia (CLL)

    e. Lymphoma

29. Which of the following description for BRCA gene is false?

    a. *BRCA1* and *BRCA2* are human genes that produce tumor suppressor proteins.

    b. Women with specific inherited BRCA1 or BRCA2 gene mutations are at high risk of developing breast and ovarian cancers.

    c. BRCA1 mutation is more likely associated with triple negative breast cancer.

d. BRCA gene test is routinely performed on general population for risk assessment of breast and ovarian cancers.

e. BRCA gene test is offered to those who are likely to have an inherited mutation based on personal or family history of breast cancer or ovarian cancer.

30. Given a tumor marker with the following test results for a group of patients: 164 true positive, 27 false positive, 245 true negative, 86 false negative. What is the diagnostic sensitivity and specificity, respectively?

a. 100% and 71%

b. 66% and 74%

c. 66% and 90%

d. 75% and 90%

e. 90% and 66%

Answers

1. **e.** Tumor marker is a biological substance synthesized and released by cancer cells or produced by the host in response to the presence of tumor. It can be found in a solid tumor, in circulating tumor cells, in peripheral blood, in lymph nodes, in bone marrow, or in other body fluid (urine, stool, ascites). It can be measured qualitatively or quantitatively by biochemical, immunological, and molecular methods. Tumor markers can be proteins, glycoproteins, hormones, enzymes, antigens, receptor, or oncogenes. *Burtis CA et al., eds., Textbook of Clinical Chemistry and Molecular Diagnostics, 5th ed., 2012, p. 617.*

2. **e.** Ideal tumor markers should have good diagnostic sensitivity, specificity, high efficacy, high positive predictive value and negative predictive value in clinical setting, and high sensitivity and specificity in laboratory setting. However, no tumor marker can meet all the needs. It has limited roles in screening, diagnosis, differentiation, and prognosis. Tumor markers are mostly used to monitor therapy and predict the recurrence and remission of cancer. *Burtis CA et al., eds., Textbook of Clinical Chemistry and Molecular Diagnostics, 5th ed., 2012, pp. 617–20.*

3. **c.** Cancer antigen 15-3 (CA 15-3) is a protein that is produced by normal breast cells but there is an increased production of CA 15-3 in people with breast cancer. CA 15-3 like many other tumor markers is useful for monitoring treatment response and predicting early recurrence of disease in women with treated carcinoma of the breast. It is not breast specific and can be seen in other malignancies such as lung, colon, pancreas, primary liver, ovary, cervix, and endometrium. Tumor-associated antigens encoded by the human MUC-1 gene are known by several names, including MAM6, milk mucin antigen, cancer antigen (CA) 27.29, and CA 15-3. Several immunoassays using different monoclonal antibodies against different epitopes of a breast cancer-associated antigen have been available to quantitate the levels of tumor-associated mucinous antigens in blood. In other words, CA 27.29 test measures the same protein as CA 15-3 and you do not need both tests. *Burtis CA et al., eds., Textbook of Clinical Chemistry and Molecular Diagnostics, 5th ed., 2012, pp. 624–42.*

4. d. CA 15-3 is a glycoprotein secreted from the mammary epithelial cells and is one of the most frequently used serum tumor markers for breast cancer. Elevations of CA 15-3 are seen in breast cancer patients but also seen in many other benign and malignant diseases of the pancreas, lung, ovary, liver, and colon. CA 15-3 is recommended only for monitoring breast cancer patients with advanced disease. *Burtis CA et al., eds., Textbook of Clinical Chemistry and Molecular Diagnostics, 5th ed., 2012, p. 641.*

5. b. Cancer antigen 125 (CA 125) is a glycoprotein antigen normally expressed in tissues derived from coelomic epithelia. Elevated serum CA 125 levels have been reported in individuals with a variety of nonovarian malignancies including cervical, liver, pancreatic, lung, colon, stomach, biliary tract, uterine, fallopian tube, breast, endometrial carcinomas, and even with a variety of benign conditions. It is mainly being used for evaluating patients' response to ovarian cancer therapy and predicting recurrent ovarian cancer. *Burtis CA et al., eds., Textbook of Clinical Chemistry and Molecular Diagnostics, 5th ed., 2012, pp. 642–3.*

6. b. CA 125 is a glycoprotein antigen expressed by epithelial ovarian tumors and other coelomic epithelia. Serum CA 125 is elevated in approximately 80% of women with advanced epithelial ovarian cancer. Elevated serum CA 125 levels have been reported in individuals with a variety of nonovarian malignancies and benign conditions. CA 125 is elevated during the menstrual cycle. CA-125 is used as an aid in the detection of residual or recurrent ovarian carcinoma. *Burtis CA et al., eds., Textbook of Clinical Chemistry and Molecular Diagnostics, 5th ed., 2012, p. 624, 642.*

7. c. Blood levels of HE4 reportedly have potential as biomarker for epithelial ovarian cancer. HE4 improves the utility of CA125 as a tumor marker in ovarian cancer, and using both markers simultaneously increases the tumor marker sensitivity. The combination of CA125 and HE4 seems to predict the presence of a malignant ovarian tumor more accurately in the Risk of Ovarian Malignancy Algorithm (ROMA). *Burtis CA et al., eds., Textbook of Clinical Chemistry and Molecular Diagnostics, 5th ed., 2012, pp. 642–3.*

8. b. There are two primary types of lung cancer: Nonsmall cell lung cancer (NSCLC) and Small cell lung cancer (SCLC). NSCLS is most common, accounting for over 80% of lung cancer diagnoses. Studies have shown that carcinoembryonic antigen (CEA), cytokeratin 19 (CYFRA-21-1), alphafetoprotein (AFP), carbohydrate antigen-125 (CA-125), and carbohydrate antigen-19.9 (CA-19.9) are usually elevated in NSCLC. NSE (neuron-specific enolase) is usually sensitive for SCLC but less specific. *Burtis CA et al., eds., Textbook of Clinical Chemistry and Molecular Diagnostics, 5th ed., 2012, p. 625.*

9. d. Body cavity effusions are classified as transudative or exudative. Malignant pleural effusions are typically exudative, which is characterized with increased red blood cells, and appear pink or frankly bloody on gross examination. Exudative effusions can be identified using Light's criteria which are pleural fluid protein to serum protein ratio > 0.5, pleural fluid lactate dehydrogenase (LDH) to serum LDH ratio >0.6, or the pleural fluid LDH is greater than two thirds of the upper limit of normal serum LDH. After the pleural fluid is identified as exudative additional testing is required including cell count and differential, and cytologic analysis. The most common sources of metastatic malignancy of the pleural are lung adenocarcinoma, lymphoma/leukemia, gastrointestinal adenocarcinoma, sarcoma, and mesothelioma. *Burtis CA et al., eds., Textbook of Clinical Chemistry and Molecular Diagnostics, 5th ed., 2012, Light RW. Clin Chest Med 2013;34:21–26.*

10. a. It is recommended that every NSCLC patient, especially lung adenocarcinoma should have molecular profiling done on their tumor or blood as liquid biopsy. This can help guiding targeted medications that are available. Targeted treatments are currently approved for people with EGFR mutations, ALK rearrangements, ROS1 rearrangements, and a few other mutations. However, liquid biopsy-based blood testing for the EGFR mutations was first approved by FDA in June 2016 when a tumor biopsy cannot be obtained. EGFR gene mutations are more common in certain groups, such as nonsmokers, women, and Asians. *Burtis CA et al., eds., p. 652; Ulrich BC, et al. Cell-free DNA in oncology: Gearing up for clinic. Ann Lab Med 2018;38:1–8.*

11. a. Carbohydrate antigen 19-9 (CA 19-9) is an antigen, usually elevated in patients with gastrointestinal malignancies such as cholangiocarcinoma, pancreatic cancer, or colon cancer. Benign conditions such as cirrhosis, cholestasis, and pancreatitis also result in elevated serum CA 19-9 concentrations. Potentially useful adjunct for diagnosis and monitoring of pancreatic cancer. *Burtis CA et al., eds., Textbook of Clinical Chemistry and Molecular Diagnostics, 5th ed., 2012, pp. 644–5.*

12. c. Calcitonin is a polypeptide hormone secreted by the parafollicular cells (or C cells) of the thyroid gland. Malignant tumors arising from thyroid C cells (medullary thyroid carcinoma) usually produce elevated levels of calcitonin. Serum calcitonin concentrations may be increased in patients with chronic renal failure, and other benign and malignant conditions such as hyperparathyroidism, leukemic and myeloproliferative disorders, autoimmune thyroiditis, lung cancers, breast and prostate cancer, and various neuroendocrine tumors. Calcitonin measurement can be used to aid in the diagnosis and follow-up of medullary thyroid carcinoma. *Burtis CA et al., eds., Textbook of Clinical Chemistry and Molecular Diagnostics, 5th ed., 2012, p. 636.*

13. a. Alpha-fetoprotein (AFP) is a glycoprotein that is produced in early fetal life by the liver and by a variety of tumors including hepatocellular carcinoma, hepatoblastoma, and nonseminomatous germ cell tumors of the ovary and testis. Serum AFP can be elevated during pregnancy and in patients with benign liver disease (e.g., viral hepatitis, cirrhosis), and gastrointestinal tract tumors. AFP is useful for monitoring therapy and determining the prognosis of hepatocellular carcinoma. AFP is used in conjunction with hCG for germ cell tumors. *Burtis CA et al., eds., Textbook of Clinical Chemistry and Molecular Diagnostics, 5th ed., 2012, pp. 637–9.*

14. d. Beta-2 microglobulin (B2M) is a small membrane protein that is found on the surface of almost all cells, particularly by B lymphocytes and tumor cells in the body. B2M is elevated in the blood and body fluids (CSF and Urine) with cancers such as multiple myeloma, lymphoma, leukemia, and with inflammatory disorders and infections. B2M can be useful as a tumor marker though it is used to assess kidney tubular function. *Burtis CA et al., eds., Textbook of Clinical Chemistry and Molecular Diagnostics, 5th ed., 2012, p. 646; Garewal H et al. Serum beta 2-microglobulin in the initial staging and subsequent monitoring of monoclonal palama cell disorders. J Clin Oncol 1984;2:51–7.*

15. d. Carcinoembryonic antigen (CEA) is a glycoprotein normally found in embryonic endodermal epithelium. Increased levels may be found in patients with primary colorectal cancer or other malignancies including medullary thyroid carcinoma and breast, gastrointestinal tract, liver, lung, ovarian, pancreatic, and prostatic cancers. CEA is also a marker to determine breast cancer metastasis. It is useful for monitoring therapeutic response and for evaluating possible recurrence of those malignancies. *Burtis CA et al., eds., Textbook of Clinical Chemistry and Molecular Diagnostics, 5th ed., 2012, pp. 624; pp. 639–40.*

16. c. Oncofetal antigens are proteins which are typically present only during fetal development but are found in adults with certain kinds of cancer. These proteins include AFP, CEA, trophoblast glycoprotein precursor, and immature laminin receptor protein. Normally oncofetal proteins are present during embryogenesis and may increase with certain cancers, making them potentially useful tumor markers. Oncofetal antigens are promising targets for vaccination against several types of cancers. *Burtis CA et al., eds., Textbook of Clinical Chemistry and Molecular Diagnostics, 5th ed., 2012, pp. 637–9.*

17. b. Human chorionic gonadotropin (hCG) is a glycoprotein secreted by the syncytiotrophoblastic cells of the normal placenta. It consists of 2 noncovalently bound α and β subunits. In addition to pregnancy, hCG may be elevated in seminomatous and nonseminomatous testicular tumors, ovarian germ cell tumors, choriocarcinoma, gestational trophoblastic disease, and benign or malignant nontesticular teratomas. hCG is most useful for monitoring treatment of trophoblastic disease. *Burtis CA et al., eds.,*

Textbook of Clinical Chemistry and Molecular Diagnostics, 5th ed., 2012, pp. 636–7.

18. c. Human chorionic gonadotropin (hCG) is a glycoprotein secreted by the syncytiotrophoblastic cells of the normal placenta. It consists of 2 α and β subunits. In addition to pregnancy, hCG may be elevated in seminomatous and nonseminomatous testicular tumors, ovarian germ cell tumors, choriocarcinoma, gestational trophoblastic disease, and benign or malignant nontesticular teratomas. In addition to secreting intact hCG, tumors may produce numerous heterogeneous molecular forms of free α-subunit, free β-subunit, nicked hCG, nicked free β-subunit, and β core fragment. Assays that detect both intact hCG and free beta-hCG tend to be more sensitive in detecting hCG-producing tumors. *Burtis CA et al., eds., Textbook of Clinical Chemistry and Molecular Diagnostics, 5th ed., 2012, pp. 637–9.*

19. a. Human chorionic gonadotropin (hCG) is a glycoprotein secreted by the syncytiotrophoblastic cells of the normal placenta. It consists of 2 α and β subunits. In addition to secreting intact hCG, tumors may produce numerous heterogeneous molecular forms of free α-subunit, free β-subunit, nicked hCG, nicked free β-subunit, and β core fragment. The whole (intact) molecule hCG assay is less sensitive and less specific as a tumor marker than beta-hCG assay since the alpha subunit is identical for LH, FSH, and TSH. Assays that detect both intact hCG and free beta-hCG tend to be more sensitive in detecting hCG-producing tumors. *Burtis CA et al., eds., Textbook of Clinical Chemistry and Molecular Diagnostics, 5th ed., 2012, pp. 637–9.*

20. b. Guaiac FOBT has its limitations with poor sensitivity and patient compliance. Fecal immunochemical test (FIT) has superior performance characteristics when compared with older guaiac-based Hemoccult II cards. The American College of Gastroenterology (ACG) recommends FIT replaces the older guaiac-based fecal occult blood test. Fecal (or stool) DNA examination is a noninvasive strategy recommended by several medical professional societies for colorectal cancer (CRC) screening in average-risk individuals. The Epi proColon test is a qualitative detection of the plasma methylated septin 9 gene (SEPT9 DNA) using real-time PCR. *Burtis CA et al., eds., Textbook of Clinical Chemistry and Molecular Diagnostics, 5th ed., 2012, p. 1399; Rex DK et al. Am J Gastro 2009; 104:739–50; Lieberman D et al. Screening for colorectal cancer and evolving issues for physicians and patients. a review. JAMA 2016;316: 2135–45.*

21. b. Elevated levels of PAP in the blood may indicate prostate cancer but can also be seen in other conditions. PCA3, PHI, and 4K score are used in differentiation of prostate malignancy from benign conditions and guiding prostate biopsy. None of the above biomarkers can make the diagnosis. A transrectal ultrasound-guided biopsy is the most common approach prostate cancer is diagnosed in the US. *Roeyhrborn CG. Using biopsy to detect prostate cancer. Rev Urol. 2008;10:262–80.*

22. b. PSA exists in serum in multiple forms: complexed to alpha-1-anti-chymotrypsin (PSA-ACT complex), unbound (free PSA), and enveloped by alpha-2-macroglobulin (PSA-A2M) (not detected by immunoassays). *Burtis CA et al., eds., Textbook of Clinical Chemistry and Molecular Diagnostics, 5th ed., 2012, pp. 628–32.*

23. b. PSA is prostate-specific but not prostate-cancer specific. Elevated PSA levels can be seen in benign conditions such as benign prostate hypertrophy, prostatitis, urinary tract infection, and DRE. PSA is not produced by neuroendocrine (small cell) type prostate cancer. Negative PSA can be seen in patients using antiandrogen and 5-alpha reductase inhibitors. Normal results do not eliminate the possibility of prostate cancer. *Burtis CA et al., eds., Textbook of Clinical Chemistry and Molecular Diagnostics, 5th ed., 2012, pp. 628–32.*

24. a. Of those listed, a prostate-specific antigen is not required for testicular cancer. *Burtis CA et al., eds., Textbook of Clinical Chemistry and Molecular Diagnostics, 5th ed., 2012, pp. 628–32.*

25. d. Serum Thyroglobulin (Tg) is mainly used in the follow-up of patients with differentiated thyroid cancers after thyroidectomy and radioactive iodine ablation. Serum Tg concentrations should be undetectable, or very low, after the thyroid gland is removed during treatment for thyroid cancer. Current clinical guidelines consider a serum Tg of >1 ng/mL in an athyrotic individual as suspicious of possible residual or recurrent disease. The presence of antithyroglobulin autoantibodies (TgAb) could lead to misleading Tg results. In immunometric assays, the presence of TgAb can lead to false-low results, whereas it might lead to false-high results in competitive assays. Therefore, both thyroglobulin and suspected antithyroglobulin autoantibody should be measured on the same instrument to eliminate potential interference. Because of the false low interference, measurement of Tg by LC-MS/MS is the preferred method in TgAb positive patients. *Burtis CA et al., eds., Textbook of Clinical Chemistry and Molecular Diagnostics, 5th ed., 2012, pp.649–50.*

26. e. HER-2 is a transmembrane protein expressed on epithelial cells. HER-2 amplification is found in breast and other tissues. Gene overexpression, measured by fluorescence in situ hybridization (FISH) in breast tissue, or protein expression, measured by immunohistochemistry, is used to qualify patients for Herceptin (trastuzumab) treatment. It appears to be as a useful prognostic indicator of overall survival and tumor size. *Burtis CA et al., eds., Textbook of Clinical Chemistry and Molecular Diagnostics, 5th ed., 2012, pp. 654–5.*

27. d. Classes of genes are implicated in cell growth and the development of cancer. Certain types of cancer have seen in a family and are strongly linked to an inherited mutation. Molecular genetic tumor markers can be used in risk stratification and guiding therapy. Genetic mutation can be profiled and liquid biopsy is a potential tool. *Burtis CA et al., eds., Textbook of Clinical Chemistry and Molecular Diagnostics, 5th ed., 2012, pp. 652–8.*

28. b. In the majority of patients with CML, transformation of a hematopoietic stem cell results from the formation of the Philadelphia chromosome, a translocation between chromosomes 9 and 22 creating the BCR-ABL fusion gene. The protein derived from this fusion gene activates signaling pathway resulting in cell growth and proliferation. Detection of the *BCR-ABL* gene is useful in diagnosing CML and in directing treatment. *Burtis CA et al., eds., Textbook of Clinical Chemistry and Molecular Diagnostics, 5th ed., 2012, p. 655.*

29. d. BRCA1 and BRCA2 are human genes that produce tumor suppressor proteins to repair damaged DNA. Women with specific inherited BRCA1 or BRCA2 mutation are at an increased risk of developing breast cancer and ovarian cancer compared with the general population. BRCA1 mutation is also more likely to be triple negative breast cancer. The BRCA gene test is offered to those who are likely to have an inherited mutation based on personal or family history of breast cancer or ovarian cancer. The BRCA gene test is not routinely performed on people at average risk of breast and ovarian cancers. *Burtis CA et al., eds., Textbook of Clinical Chemistry and Molecular Diagnostics, 5th ed., 2012, pp. 657–8.*

30. c. Sensitivity refers to the test's ability to correctly detect ill patients who do have the condition. Mathematically, this can be expressed as:

Diagnostic sensitivity = positivity in disease, expressed as $= TP/(TP+FN) \times 100\%$.

Specificity relates to the test's ability to correctly reject healthy patients without a condition. Mathematically, this can also be written as:

Diagnostic specificity = absence of a particular disease, expressed as $= TN/(TN+FP) \times 100\%$.

	Cancer	No Cancer	Total
Test (+)	164 TP	27 FP	191
Test (−)	86 FN	245 TN	331
Total	250 (TP + FN)	272 (TN + FP)	522 Grand Total

Readers should extend to positive predictive values (PPVs) and negative predictive values (NPVs) with regard to the description and use of screening and diagnosis of the tests as well as the diagnostic efficacy. Another point readers need to know is how to construct and interpret the AUC under ROC curve for a test performance. *Kaplan LA, et al. eds. Clinical Chemistry. Theory, Analysis, Correlation, 3rd ed., 2010. pp. 362–78; Burtis CA et al., eds., Textbook of Clinical Chemistry and Molecular Diagnostics, 5th ed., 2012, pp. 3–58.*

Chapter 6

Sepsis tests

Octavia M. Peck Palmer
University of Pittsburgh School of Medicine, Pittsburgh, PA, United States

1. Which of the following cellular parameter(s) is reportedly a useful parameter to detect sepsis in patients presenting to the emergency department based on either Sepsis-2 criteria or Sepsis-3 criteria?
   1. mean corpuscular volume of 71 fL
   2. neutrophil count greater than 12,000/mm
   3. white blood cell count of 4.5 cells per liter
   4. monocyte distribution width greater than 20.0 U
      a. 1
      b. 1,2,3
      c. 2,4
      d. 4
      e. all of the above
2. In 2015 the United States Centers for Medicare and Medicaid Services defined severe sepsis as having which of the following lactate concentrations?
   a. >2 mmol/L
   b. >4 mmol/L
   c. 2–4 mmol/L
   d. >10 mmol/L
3. In 2015 the United States Centers for Medicare and Medicaid Services defined septic shock as which of the following?
   a. sepsis with hypoperfusion following fluid resuscitation or a lactate >2
   b. organ dysfunction or a lactate >4 mmol/L
   c. severe sepsis with hypoperfusion or a lactate >4 mmol/L
   d. system inflammatory response syndrome or lactate 1.5 or greater
   e. sepsis not responsive to IV antibiotic administration
4. The Third International Consensus Definitions for Sepsis and Septic Shock (Sepsis-3) defined sepsis as
   a. Life-threatening organ dysfunction caused by a dysregulated host response to infection.

Self-assessment Q&A in Clinical Laboratory Science, III. https://doi.org/10.1016/B978-0-12-822093-1.00006-5

   b. Systemic inflammatory response associated with fever, tachycardia, and elevated white blood cells.

   c. Circulatory failure, causing inadequate oxygen delivery to meet cellular metabolic needs and oxygen consumption requirements, producing cellular and tissue hypoxia.

   d. Elevation of white blood cells and the accumulation of pericardial fluid under pressure.

5. In 2013 (revised in 2018), the New York Department of Health mandated hospitals develop sepsis protocols that allow early diagnosis and treatment of sepsis. Which of the following correctly describes the mandate?

   1. Develop/adopt screening measures for early recognition of patients with sepsis, severe sepsis, and septic shock.

   2. Activate a 1-h care bundle within sepsis recognition (includes blood cultures before administration of antibiotics, lactate, 20-mL/kg crystalloid fluid administration).

   3. Hospitals are required to monitor and measure adherence with established sepsis protocols.

   4. Develop and implement sepsis informed protocols for the treatment of septic patients.

   a. 1

   b. 1,2,3

   c. 2,4

   d. 4

   e. all of the above

6. The New York State Sepsis Initiative is also known as which of the following?

   1. Public Health Law

   2. Think Katie First

   3. Strengthening treatment and outcomes for patients sepsis collaborative

   4. Rory's Regulations

   a. 1

   b. 1,2,3

   c. 2,4

   d. 4

   e. all of the above

7. As the Medical Director of the clinical laboratory that performs lactate testing, you are asked to align the hospital's lactate testing practices with the 2018 Surviving Sepsis Campaign (SSC) guidelines. Which of the following accurately depicts the SSCC guidelines for lactate testing in the workup of sepsis?

   a. Order second lactate if the patient's initial lactate concentration was 4 mmol/L at least 2 h postadmission

   b. Interpret a lactate concentration of 1.5 mmol/L to be consistent with sepsis

    c. Remeasure lactate if the initial lactate concentration is >2mmol/L

    d. Lactate concentrations between (2 and 10mmol/L) should be remeasured 12h later to confirm the elevation

8. Sepsis survivors are at a higher risk than the general population for which of the following?

    1. depression

    2. cognitive impairment

    3. cardiovascular disease

    4. death

        a. 1

        b. 1,2,3

        c. 2,4

        d. 4

        e. all of the above

9. What percentage of patients hospitalized with sepsis have blood culture results that are negative for bacteria?

    a. $\leq 1\%$

    b. $\leq 10\%$

    c. $\leq 30\%-50\%$

    d. 0%

10. Since there is no cure for sepsis, which of the following components of the sepsis care bundles is a cornerstone in directing targeted treatment?

    a. lactate

    b. blood cultures

    c. procalcitonin

    d. c-reactive protein

    e. preseptin

11. Procalcitonin is elevated in which of the following?

    a. sepsis

    b. plasmodium falciparum malaria

    c. surgical intervention

    d. heatstroke

    e. all of the above

12. Which of the following analytes comprises 99% of the cluster of differentiation 14 (CD14) within the body and is detectable in both serum (2–6 µg/mL) and urine?

    a. procalcitonin

    b. c-reactive protein

    c. interleukin-6

    d. presepsin

    e. interleukin-10

13. What components comprise the quick Sequential [Sepsis-related] Organ Failure Assessment (qSOFA)?

    1. hypotension (≤ 100mmHg)

   2. tachypnea ($\geq$22/min)
   3. altered mentation
   4. decreased urine output
     a. 1
     b. 1,2,3
     c. 2,4
     d. 4
     e. all of the above

14. Select from following the statement(s) that best reflect the "Surviving Sepsis Campaign International Guidelines for the Management of Septic Shock and Sepsis-Associated Organ Dysfunction in Children."
   1. If no pathogen is identified, narrow or discontinue empiric antimicrobial therapy in accordance with clinical presentation, site of infection, host risk factors, and adequacy of clinical improvement in discussion with an infectious disease or microbiological expert advice.
   2. Clinicians can apply the guidelines to individuals that are 37 weeks' gestation at birth to 18 years of age with severe sepsis or septic shock.
   3. In children with septic shock or sepsis-associated organ dysfunction who are receiving antimicrobials, we recommend daily assessment (e.g., clinical, laboratory assessment) for deescalation of antimicrobial therapy.
   4. No recommendation for utilizing blood lactate concentrations to stratify children with suspected septic shock or other sepsis-associated organ dysfunction into low and high risk of having septic shock or sepsis.
     a. 1
     b. 1,2,3
     c. 2,4
     d. 4
     e. all of the above

15. What clinical laboratory analyte(s) is the Sequential Organ Failure Assessment (SOFA) Score derived?
   1. creatinine ($<$1.2 mg/dL [20 μmol/L] = 0 points)
   2. bilirubin (1.2 mg/dL [20 μmol/L] = 0 points)
   3. platelets ($>$/=150 $\times$ 10/μL = 0 points)
   4. lactate ($<$/=1 mmol/L = 0 points)
     a. 1
     b. 1,2,3
     c. 2,4
     d. 4
     e. all of the above

16. A 73-year-old woman with a past medical history of uncontrolled diabetes and knee replacement (10 weeks ago) presents to the emergency department with complaints of fatigue, night sweats, productive cough, chest pain, and fever (103.2°C). Which of the following interpretations are appropriate for procalcitonin results of 2 μg/L (ED admission) and 0.13 μg/L (3 days postadmission)?

1. patient result of 0.1 μg/L, antibiotics strongly discouraged
2. patient result of 0.1 μg/L, antibiotics discouraged
3. patient result of 2 μg/L, antibiotics recommended
4. patient result of 2 μg/L, antibiotics strongly recommended

   a. 1
   b. 1,2,3
   c. 2,4
   d. 4
   e. all of the above

17. Which of the following is a common and deadly pathogen(s) that causes sepsis

1. *Staphylococcus aureus*
2. *Klebsiella pneumoniae*
3. *Pseudomonas aeruginosa*
4. *Escherichia coli*

   a. 1
   b. 1,2,3
   c. 2,4
   d. 4
   e. all of the above

18. Which population(s) is at a higher risk for sepsis?

1. pregnant individuals
2. elderly individuals
3. neonates
4. adolescent

   a. 1
   b. 1,2,3
   c. 2,4
   d. 4
   e. all of the above

19. Sepsis is a global public health problem. Which ethnic group reportedly exhibits a higher incidence of hospitalizations for infection and a higher risk of severe sepsis?

1. Black/African Americans
2. White/Caucasians
3. Asian Americans
4. all ethnic groups exhibit similar risk of sepsis

   a. 1
   b. 1,2,3
   c. 2,4

d. 4

e. all of the above

20. The most common source of sepsis is?

    a. cancer

    b. knee infection

    c. ear infection

    d. pneumococcal pneumonia

    e. organ transplant

21. A blood sample is collected in a 7-mL K$_2$EDTA blood collection tube and submitted to the clinical laboratory for *Candida* detection (*C. albicans/C. tropicalis, C. glabrata/C. krusei,* and *C. parapsilosis*). The required tube size and type is a 4-mL K$_2$EDTA blood collection tube. Should the clinical lab analyze the sample or contact the healthcare provider and request the required sample type?

    a. Yes, the clinical lab should analyze the sample. The current lab requirement (4 mL) is only to minimize blood collections.

    b. Yes, the clinical lab should analyze the sample. The 7-mL tube will provide a larger volume to ensure Candida is detected.

    c. No, the clinical should not analyze the sample because a 7-mL tube contains a high concentration of EDTA that may destroy the Candida in the blood sample.

    d. No, the clinical lab should not analyze the sample because it will take longer to analyze due to the increased volume.

    e. No, the clinical lab should not analyze the sample because the required sample type should only be analyzed.

22. Which correctly describes MALDI-TOF MS performance in microbial identification

    1. The use of MALDI-TOF MS does not require sample preparation.

    2. MALDI-TOF MS yields significantly discordant antimicrobial susceptibility testing profiles.

    3. MALDI-TOF MS is not a viable tool for microbial identification.

    4. MALDI-TOF MS can identify organisms grown for 4 hours on solid media.

    a. 1

    b. 1,2,3

    c. 2,4

    d. 4

    e. all of the above

23. Which statement accurately describes the peptide nucleic acid fluorescent in situ hybridization (FISH) method?

    1. Labeled DNA mimic-molecules are employed to detect pathogen ribosomal RNA, and FISH is performed on positive blood cultures.

    2. The pathogen's microbial fingerprint is identified via absorbing and scattering incident light.

3. The identification of the pathogen is directly proportional to the binding of the pathogen with an antibody.
4. FISH performed on all blood cultures.
   a. 1
   b. 1,2,3
   c. 2,4
   d. 4
   e. all of the above
24. Which of the following analytes increase during sepsis?
   1. pro-adrenomedullin
   2. procalcitonin
   3. D-dimer
   4. C-reactive protein
   a. 1
   b. 1,2,3
   c. 2,4
   d. 4
   e. all of the above
25. Septic shock is?
   1. a subset of sepsis in which underlying circulatory and cellular metabolism abnormalities are profound enough to increase mortality
   2. sepsis complicated by organ dysfunction
   3. a subclass of sepsis in which the patient exhibits temperature $>38°C$ or $<36°C$, heart rate $>90/min$, and respiratory rate $>20/min$
   4. sepsis accompanied by persisting vasodilation and fever
   a. 1
   b. 1,2,3
   c. 2,4
   d. 4
   e. all of the above

Answers

1. d. The monocyte distribution width value of greater than 20.0 U, has been reported in being effective in detecting sepsis based on either Sepsis-2 criteria or Sepsis-3 criteria, during the initial emergency department presentation. *Crouser ED, et al. Monocyte distribution width: a novel indicator of sepsis-2 and sepsis-3 in high-risk emergency department patients. Crit Care Med 2019; 47:1018–25.*
2. a. In 2015 the Centers for Medicare & Medicaid Services (CMS) released the Severe Sepsis and Septic Shock Early Management Bundle (SEP-1) guidelines, which defined severe sepsis as a lactate concentration of >2 mmol/L or evidence of organ dysfunction. The Sepsis bundle project (SEP) national hospital inpatient quality measures. *Specifications Manual for National Hospital Inpatient Quality Measures Discharges 01-01-18*

(1Q18) through 06-30-18 (2Q18) Version 5.3 [Accessed December 2019].
https://www.sepsiscoordinatornetwork.org/wp-content/uploads/2018/05/
Sepsis-Alliance-SEP-1-Core-Measure.pdf.

3. c. In 2015 the Centers for Medicare & Medicaid Services (CMS) released the Severe Sepsis and Septic Shock Early Management Bundle (SEP-1) guidelines, which defined septic shock as severe sepsis with hypoperfusion despite adequate fluid resuscitation or a lactate > 4. The Sepsis bundle project (SEP) national hospital inpatient quality measures. *Specifications Manual for National Hospital Inpatient Quality Measures Discharges 01-01-18 (1Q18) through 06-30-18 (2Q18) Version 5.3 [Accessed December 2019]. https://www.sepsiscoordinatornetwork.org/wp-content/uploads/2018/05/ Sepsis-Alliance-SEP-1-Core-Measure.pdf.*

4. a. The Third International Consensus Definitions were derived by 19 experts and reviewed by 31 societies. The consensus revised the 1992 definition of sepsis. The systemic inflammatory response syndrome (SIRS) criteria were removed and are no longer to be used to identify sepsis. Sepsis differs from infection in that sepsis is a dysregulated host immune response in the presence of organ dysfunction. *Singer M, et al. The Third International Consensus Definitions for Sepsis and Septic Shock (Sepsis-3). JAMA 2016;315(8):801–10. https://doi.org/10.1001/ jama.2016.0287.*

5. e. In response, the death of a pediatric patient in New York and the rising sepsis-associated deaths, the New York State Sepsis Initiative was released, which mandated hospitals caring for septic patients to adhere to several guidelines. The hospitals were required by Sections 405.2 and 405.4 of the amendment of Title 10 of the New York State Codes, Rules, and Regulations to develop customized sepsis recognition tools that are evidenced based, monitor their adherence to the sepsis protocols, deliver a 1-h care bundle to pediatric patients, and to report to New York state to comply. New York State Department of Health. New York State report on sepsis care improvement initiative: hospital quality performance. *https://www.health. ny.gov/press/reports/docs/2015_sepsis_care_improvement_initiative.pdf. Published March 2017. Accessed December 22, 2019.*

6. d. The New York state sepsis initiative, which requires hospitals to develop and adhere to evidence-based sepsis protocols and deliver a 1-h sepsis care bundle to pediatric patients, is often referred to as "Rory's Regulations." Rory Staunton was a 12-year-old boy from New York City who died after becoming septic. Studies have demonstrated an association between completing the sepsis bundle within 1-hour of presentation with a lower risk-adjusted in-hospital mortality in pediatric septic and septic shock patients. *Evans IVR, et al. Association between the New York sepsis care mandate and in-hospital mortality for pediatric sepsis. JAMA 2018;320(4):358–367. https://doi.org/10.1001/jama.2018.9071.*

7. c. The Surviving Sepsis campaign is a collaborative initiative between the Society of Critical Care Medicine and the European Society of Intensive Care Medicine focused on reducing global sepsis and septic shock associated mortality and morbidity. In 2018 the Hour-1 Surviving Sepsis Campaign Bundle of Care was introduced. Lactate is to be remeasured if the initial lactate concentration is >2 mmol/L. *Levy MM, et al. The surviving sepsis campaign bundle: 2018 update. Inten Care Med 2018;44:925–8. https://doi.org/10.1007/s00134-018-5085-0. Epub 2018 Apr 19.*

8. e. Sepsis survivors, adult patients diagnosed with sepsis and discharged home, exhibit postsepsis morbidity that includes muscle weakness, fatigue, reinfection, depression, cognitive impairment, cardiovascular disease, and amputations. *Prescott HC, et al. Postsepsis morbidity. JAMA 2018;319(1):91. https://doi.org/10.1001/jama.2017.19809.*

9. c. Thirty to fifty percent of patients with sepsis have no pathogen isolated. It is essential to obtain a blood culture specimen before the administration of antibiotics and to reduce the effects of the antibiotic preventing the detection of the pathogen. *Nannan Panday RS, et al. An overview of positive cultures and clinical outcomes in septic patients: a sub-analysis of the Prehospital antibiotics against sepsis (PHANTASi) trial. Crit Care 2019;23:2431–8. https://doi.org/10.1186/s13054-019-2431-8.*

10. b. The importance of obtaining blood cultures is that detection and isolation of the pathogen(s)/organism(s) may allow targeted treatments such as antimicrobial and antifungal therapies. *Nannan Panday RS, et al. An overview of positive cultures and clinical outcomes in septic patients: a sub-analysis of the Prehospital antibiotics against sepsis (PHANTASi) trial. Crit Care 2019;23:2431–8. https://doi.org/10.1186/s13054-019-2431-8.*

11. e. Procalcitonin is a peptide precursor of the hormone calcitonin. Although up-regulated in infection, it is also up-regulated in several noninfectious inflammatory diseases and syndromes. *Wacker C, et al. Procalcitonin as a diagnostic marker for sepsis: a systematic review and meta-analysis. Lancet Infect Dis 2013;13:426–35. https://doi.org/10.1016/S1473-3099 (12)70323-7.*

12. d. Presepsin (also known as soluble CD14-ST) is a 13 kDa peptide that is produced by proteolytic cleavage of sCD14 forms. CD14 is present on a variety of cells, including monocytes (majority), macrophages, neutrophils, B-lymphocytes, hepatocytes, chondrocytes, dendritic cells, and human epithelial intestinal cells. Presepsin is not FDA approved for the diagnosis of sepsis. Systemic reviews and meta-analysis have reported presepsin to have moderately high diagnostic accuracy (ROC AUC 0.88). *Rogić D, et al. Advances and pitfalls in using laboratory biomarkers for the diagnosis and management of sepsis. eJIFCC 2017;28:114-21.*

13. b. The quickSOFA score (also known as qSOFA) may identify adult patients with suspected infection who are at higher risk for a poor outcome outside the intensive care unit. The qSOFA is comprised of three criteria and includes low blood pressure (SBP ≤ 100 mmHg), high respiratory rate (≥ 22 breaths per min), or altered mentation (Glasgow coma scale < 15). One point is assigned for each criterion, and the score ranges from 0 to 3 points. *Seymour CW, et al. Assessment of clinical criteria for sepsis: For the Third International Consensus Definitions for Sepsis and Septic Shock (Sepsis-3). JAMA 2016;315(8):762–74. https://doi.org/10.1001/jama.2016.0288.*

14. e. In February 2020, the first edition of the "Surviving Sepsis Campaign International Guidelines for the Management of Septic Shock and Sepsis-Associated Organ Dysfunction in Children" was released. The guidelines provide guidance to healthcare professionals who provide care to pediatric patients with sepsis and septic shock. *Weiss et al. Surviving sepsis campaign international guidelines for the management of septic shock and sepsis-associated organ dysfunction in children. Pediatr Crit Care Med 2020;21:e52–e106. https://doi.org/10.1097/PCC. 0000000000002198.*

15. b. The Sequential Organ Failure Assessment (SOFA) score was developed in 1994 by the Working Group on Sepsis-Related Problems of the European Society of Intensive Care Medicine and published in 1996. The SOFA score assesses acute morbidity of critical illness in intensive care patients using the function of six organ systems (scored 0–4 points). Several clinical and laboratory variables are calculated to generate a score and include PaO_2/FiO_2 (mmHg) (respiratory system), Glasgow Coma Scale (nervous system), Mean arterial pressure (MAP) OR administration of vasopressors required (cardiovascular system), bilirubin (liver), platelets (coagulation), and creatinine (kidneys). *Lambden S, et al. The SOFA score—development, utility, and challenges of accurate assessment in clinical trials. Crit Care 2019;23:374.* https://doi.org/10.1186/s13054-019-2663-7.

16. c. In 2017 the FDA approved the use of procalcitonin to aid in antibiotic decision making for lower respiratory tract infections. The following are the procalcitonin cutoff values: <0.1 μg/L, antibiotics strongly discouraged; 0.1–0.25 μg/L, antibiotics discouraged; >0.25–0.5 μg/L, antibiotics recommended; and >0.5 μg/L, antibiotics strongly recommended. *Huang D et al. Procalcitonin-guided use of antibiotics for lower respiratory tract infection. N Engl J Med 2018;379:236–49.*

17. e. Enterococcus faecium, Escherichia coli, Klebsiella pneumoniae, Pseudomonas aeruginosa, and Staphylococcus aureus are common bacteria that cause sepsis. *Mayr FB, et al. Epidemiology of severe sepsis. Virulence 2014;5(1):4–11. https://doi.org/10.4161/viru.27372.*

18. b. It is essential to recognize that sepsis can affect anyone; however, some populations are more vulnerable than others and have a higher risk of sepsis. *World Health Organization Newsroom Fact sheets Detail Sepsis,*

https://www.who.int/news-room/fact-sheets/detail/sepsis (accessed 10 January 2020).

19. a. Ethnic differences reported in severe sepsis were due to a higher infection rate and a higher risk of acute organ dysfunction in self-identified black compared to self-identified white individuals diagnosed with community-acquired pneumonia. *Mayr FB, et al. Infection rate and acute organ dysfunction risk as explanations for racial differences in severe sepsis. JAMA 2010;303(24):2495–503. https://doi.org/10.1001/jama.2010.851.*

20. d. Pneumococcal pneumonia is the most common cause of severe sepsis. Studies show that sepsis can originate from infections in the stomach, kidneys, or bladder, and due to trauma and burn. *https://www.cdc.gov/vitalsigns/sepsis/index.html (Accessed December 18, 2019).*

21. c. EDTA possesses antifungal activity and will destroy the Candida present in the whole blood sample. *Clancy CJ, et al. T2 magnetic resonance for the diagnosis of bloodstream infections: charting a path forward. J Antimicrob Chemother 2018;73:iv2–iv5. https://doi.org/10.1093/jac/dky050.*

22. d. MALDI-TOF MS accurately identified organisms grown for only 4 h on solid media and yield concordant antimicrobial susceptibility testing profiles. *Mitchell SL, et al. Performance of microbial identification by MALDI-TOF MS and susceptibility testing by VITEK 2 from positive blood cultures after minimal incubation on solid media. Eur J Clin Microbiol Infect Dis 2017;36:2201–6. https://doi.org/10.1007/s10096-017-3046-0.*

23. a. This 2.5-h method demonstrated 100% sensitivity and specificity for the identification of *C. albicans* and it uses labeled DNA mimic-molecules to detect pathogen ribosomal RNA. FISH is performed on positive blood cultures prepared from blood smears. *Rigby S, et al. Fluorescence In situ hybridization with peptide nucleic acid probes for rapid identification of Candida albicans directly from blood culture bottles. J Clin Microbio 2002;40(6):2182–6; https://doi.org/10.1128/JCM.40.6.2182-2186.2002.*

24. e. During sepsis, elevated concentrations of acute phase proteins, hematosis markers, and inflammatory cytokines are a fatal consequence of a dysregulated host immune response to infection. *Fan SL, et al. Diagnosing sepsis– The role of laboratory medicine. Clin Chim Acta 2016;460:203–10. https://doi.org/10.1016/j.cca.2016.07.002.*

25. a. The Third International Consensus Definitions defined septic shock as a subset of sepsis in which underlying circulatory and cellular metabolism abnormalities are profound enough to increase mortality. Septic shock presents with hypotension that requires vasopressors to maintain a mean arterial pressure of $>/=65$ mmHg and a lactate concentration >2 mmol/L (18 mg/dL) despite adequate volume resuscitation. Furthermore, septic shock has a reported in-house mortality of $>40\%$. *Singer M, et al. The Third International Consensus Definitions for Sepsis and Septic Shock (Sepsis-3). JAMA 2016;315(8):801–10. https://doi.org/10.1001/jama.2016.0287.*

Chapter 7

Serum proteins and electrophoresis

Alan H.B. Wu[a] and Wieslaw Furmaga[b]

[a]University of California, San Francisco, CA, United States, [b]Department of Pathology, University of Texas Health Science Center at San Antonio, San Antonio, TX, United States

1. What is the best interpretation of gel #2 relative to gel #1 (normal SPE) with reference to the band in the β-1 region for this serum sample?

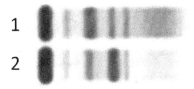

a. monoclonal gammopathy of IgM or IgA
b. increased in transferrin
c. requires confirmatory test such as immunofixation electrophoresis to conclude
d. free light chain disease
e. fibrinogen

2. Which disease is patient #1 likely has or will suffer from (gel #2, normal SPE)?

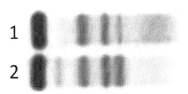

Self-assessment Q&A in Clinical Laboratory Science, III. https://doi.org/10.1016/B978-0-12-822093-1.00007-7

a. nephrotic syndrome
b. α1 antitrypsin deficiency
c. amyloidosis
d. Waldenström macroglobulinemia
e. monoclonal gammopathy of unknown significance

3. What is the best interpretation of gel #2 relative to gel #1 (normal SPE)?

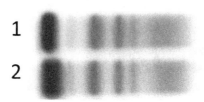

a. Both are typical SPE patterns.
b. The albumin concentration is very high in gel #2.
c. There is a heterozygous genetic variant for albumin in gel #2.
d. This patient will have difficulty transporting drugs and metabolites.
e. This patient may produce an erroneous glycated albumin (fructosamine) level.

4. What is the best explanation for the result in lane 2 with a total protein of 12.0 g/dL?

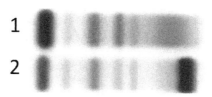

a. The polarity of the gel was switched relative to lane 1.
b. The patient has a very high monoclonal band.
c. This is a fibrinogen artifact (plasma used instead of serum).
d. This patient's protein band reacted more strongly with the Coomassie blue dye.
e. Gel 2 is a concentrated cerebrospinal protein gel.

5. A patient with an increased potassium and lactate dehydrogenase exhibits the following SPE (lane 1). What is a possible explanation for this electrophoretogram?

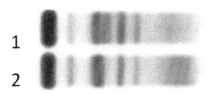

a. liver disease
b. hemolysis
c. kidney failure
d. lipemia
e. hyperbilirubinemia

6. What disease might patient #2 be suffering from (gel #1, normal SPE)?

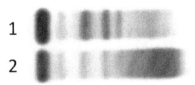

a. human immunodeficiency virus infection
b. alcoholic liver disease
c. multiple myeloma
d. multiple sclerosis
e. nephrotic syndrome

7. The patient in gel 1 has a total protein of 3.8 g/L (gcl #2 normal SPE). Which of the following is true?

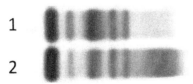

a. The high α2 proteins are a compensatory increase due to renal protein loss.
b. This SPE pattern is indicative of renal tubular damage.
c. This patient is likely to have microalbuminuria.
d. This patient has free light chain disease.
e. This patient has either an IgD or IgE myeloma.

8. The following SPE patterns are from a patient with a confirmed IgM myeloma. Gel 1 is before treatment and gel 2 is after treatment with daratumumab. The arrow indicates a faint band in the gamma region. Which of the following is true?

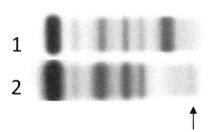

a. The patient has seroconverted from IgM to IgG.
b. The band in the gamma region of gel 2 is an artifact.
c. The second gel is not from the same patient.
d. The patient has recurrent disease.
e. The patient is in remission. The band in the gamma region of gel 2 is daratumumab.

9. Gel 1 is urine and gel 2 is serum from the same patient. The total urine protein was 38 mg/L and the sample was concentrated 200 fold. What is the best interpretation of these findings?

a. A biclonal gammopathy.
b. Urine contains free light chains, serum contains intact immunoglobulins.
c. There was a sample mixup, as these are not from the same patient.
d. This patient has significant renal involvement.
e. Monoclonal gammopathy of undetermined significance (MGUS) can be ruled out.

10. The following figure shows the electrophoresis scan of patient with a large monoclonal band and a high total protein concentration. This serum sample will interfere with what other tests?

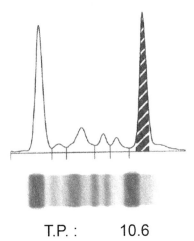

T.P. : 10.6

a. phosphorus
b. direct reading sodium measurement
c. indirect reading potassium measurement
d. glucose
e. none of the above

11. What is the best interpretation of the following serum immunofixation electrophoresis gel?

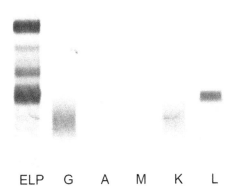

ELP   G   A   M   K   L

a. free light chain disease
b. IgD or IgE myeloma
c. monoclonal gammopathy of undetermined significance
d. pentameric IgM lambda
e. either answer A or B

12. The left immunofixation gel is a serum sample and the right is a urine sample from the same patient. What is the best explanation for these results (K and L antisera are toward bound and free light chains)?

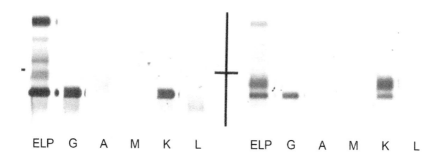

a. biclonal gammopathy
b. Both serum and urine contain intact IgG and free K light chains
c. Serum contains intact IgG and urine contains intact IgG and free K light chains
d. heavy chain disease
e. The gels have been mislabeled. The serum is the right gel and the urine is the left gel

13. What is the best interpretation of the serum immunofixation gel shown as follows?

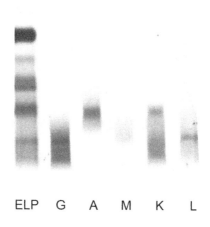

a. One band is a monomer and the other a polymeric complex, resolved through the use of β-mercaptoethanol
b. indicates a more severe clinical course
c. indicative of cryoglobulinemia

    d. indicative of double gammopathy

    e. produced from the same B-cell

14. Which is the best explanation for the following IFE result?

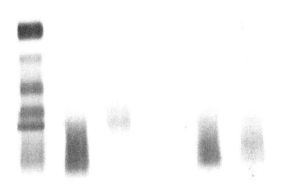

    a. IgD or IgE myeloma

    b. The band between the β2 and ϒ region may be fibrinogen.

    c. hemolysis interference

    d. genetic abnormality of IgM

    e. incorrect antisera used for K and λ light chains

15. Which of the following is most true regarding the immunofixation gel shown as follows?

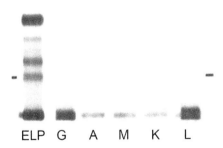

ELP   G   A   M   K   L

    a. monoclonal bands for all immunoglobulins and light chains present

    b. biclone for lambda light chains

    c. prozone (antigen excess) for free lambda light chain

    d. all bands exhibit the prozone effect

    e. this patient has immune complexes

16. Which of the following is an advantage of performing serum protein electrophoresis by mass spectrometry?
    a. Differentiation of MGUS from smoldering myeloma
    b. High sensitivity and specificity for detection of minimal residual disease
    c. Determination of the best treatment approach based on monoclonal antibodies
    d. Determine if stem cell treatment is the best treatment approach
    e. Differentiate between amyloidosis and Waldenstrom's macroglobulinemia

17. Which of the following statements is true regarding capillary electrophoresis (CZE) as a replacement for agarose gel electrophoresis (AGE) for serum protein?
    a. CZE produces a digital signal that may better facilitate archiving
    b. CZE cannot be used on urine
    c. presence of radio-opaque agents can cause an interference AGE but not CZE
    d. CZE provides less resolution for bisalbuminemia
    e. CZE replaces the need for immunofixation electrophoresis

18. Which of the following protein deficiency causes liver cirrhosis and emphysema?
    a. haptoglobin
    b. complement C3 and C4
    c. α1-antitrypsin
    d. transferrin
    e. α2-macroglobulin

19. Which of the following is not an acute phase reactant?
    a. IgG, IgA, IgM
    b. haptoglobin
    c. C-reactive protein
    d. transferrin
    e. ferritin

20. Multiple myeloma is characterized by which of the following EXCEPT?
    a. osteolytic lesions
    b. plasmacytosis in the bone marrow
    c. renal failure
    d. anemia
    e. hypoviscosity

21. Immunofixation electrophoresis:
    1. Antigen excess causes solubility of complexes and loss of visualization
    2. Determines serum viscosity
    3. Can be used to detect oligoclonal banding for cerebrospinal fluid
    4. Provides quantitative light chain analysis

   a. 1

   b. 1,2,3

   c. 2,4

   d. 4

   e. all of the above

22. Which protein is the most common underlying defect for amyloidosis?

   a. light chain

   b. heavy chain

   c. hereditary transthyretin

   d. wild-type transthyretin

   e. apolipoprotein

23. What is true about serum proteins?

   a. α1 acid glycoprotein is exclusively synthesized by the liver.

   b. In massive in vitro hemolysis the major component in α1 region may completely disappear.

   c. The primary physiologic role for ceruloplasmin is transporting the copper from the liver to the brain.

   d. β-Lipoprotein produces "feathery edge" in the β region on the serum electrophoresis.

24. In the immunofixation electrophoresis (IFE) the antibodies precipitate specific proteins in the gel. What is true about IFE?

   a. The analytical sensitivity of IFE is higher than conventional urine electrophoresis but it still requires preconcentration of the urine.

   b. The analytical specificity of IFE is significantly higher than conventional urine electrophoresis.

   c. The main goal to perform IFE on CSF is to detect multiple bands of the IgM in the gamma region.

   d. The antisera used in IFE contain antibody against IgG/kappa, IgG/lambda, IgM/kappa, IgM/lambda, IgA/kappa, and IgA/lambda antibodies.

25. Which protein does not travel with α2 fraction on the agarose gel electrophoresis?

   a. haptoglobin

   b. β lipoprotein

   c. ceruloplasmin

   d. transferrin

26. Prealbumin is a major component of circulated proteins, with molecular weight of 62,000 Da. What is true about this protein?

   a. Prealbumin migrate toward cathode faster than albumin.

   b. Prealbumin has a dimeric structure and it is the main binding protein for thyroxine.

   c. The transporting complex for vitamin A consists of prealbumin and retinol binding protein and they are manufactured in the liver.

d. In the hemodialysis patients, low level of prealbumin associates with lower risk for mortality and hospitalization due to hypovolemia.

27. The serum protein electrophoresis of a 14-year-old patient with pulmonary emphysema showed an absence of α1 fraction. What is true about this patient?

  a. α1 antitrypsin is coded by *aat1* gene located on chromosome 13th.

  b. The patient with homozygous for ZZ allele exhibits normal phenotypic function.

  c. The best diagnosis method for α1 antitrypsin deficiency is measuring AAT concentration, followed by isoelectric-focusing electrophoresis and genotyping.

  d. The liver cirrhosis in AAT-deficient children can be prevented with α1 antitrypsin replacement therapy.

Answers

1. c. Fibrinogen is consumed in the clot and does not appear in serum. Transferrin migrates in this region and can be increased in inflammation and iron deficiency anemia, although this band appears higher than expected. Monoclonal gammopathies such as IgM or IgA can occur in this region as well as free light chains. Therefore immunofixation is needed to determine the etiology of this band. *Morrison T, et al. Laboratory assessment of multiple myeloma. Adv Clin Chem 2019;89:1–58.*

2. b. Absence or significant reduction of the α1-band is indicative of α1-antitrypsin deficiency, a genetic disorder associated with liver and lung disease. Antitrypsin binds and inactivates trypsin, a proteolytic enzyme. *Morrison T, et al. Laboratory assessment of multiple myeloma. Adv Clin Chem 2019;89:1–58.*

3. c. High albumin levels are unusual outside the context of dehydration. A gel that is overloaded due to high albumin concentrations would be spread evenly over the middle of the band. In this case, the band appears to spread more cathodically (toward the globulin region) and not anodically. Therefore this is more likely to be bisalbuminemia, a genetic abnormality that has no medical significance. Glycated albumin is used for short-term glucose control, but there have been no studies to suggest that the presence of this form causes an interference. *Morrison T, et al. Laboratory assessment of multiple myeloma. Adv Clin Chem 2019;89:1–58.*

4. b. This is a case of an extremely high monoclonal gammopathy. *Morrison et al. Laboratory assessment of multiple myeloma. Adv Clin Chem 2019;89:1–58.*

5. b. High potassium and LDH could mean hemolysis which will produce a haptoglobin-hemoglobin complex that migrates between the α2 and β regions. *Morrison T, et al. Laboratory assessment of multiple myeloma. Adv Clin Chem 2019;89:1–58.*

6. b. The SPE pattern demonstrates a low albumin, high ϒ globulins, and a fusion of the β2 and ϒ bands due to an increase in IgA and IgM is due to chronic liver disease such as cirrhosis. *Morrison T, et al. Laboratory assessment of multiple myeloma. Adv Clin Chem 2019;89:1–58.*

7. a. This is a typical pattern for nephrotic syndrome where the total protein is low, and has an increase in high-molecular-weight α2 proteins. They are upregulated to maintain osmotic pressure. The low albumin is due to macroalbuminuria, not microalbuminuria as seen in early diabetic nephropathy. *Morrison T, et al. Laboratory assessment of multiple myeloma. Adv Clin Chem 2019;89:1–58.*

8. e. A faint band that migrates elsewhere from the original site of the monoclonal band on a patient treated with daratumumab (IgG monoclonal antibody) indicates the presence of that therapeutic treatment. There are means to eliminate or move the band by binding it to a receptor altering its electrophoretic migration. *Morrison T, et al. Laboratory assessment of multiple myeloma. Adv Clin Chem 2019;89:1–58.*

9. b. Two or biclonal gammopathy bands appear in serum. Light chains disease can affect renal function as evidenced by light chains in serum and intact immunoglobulins in urine which are likely not seen here. MGUS can express both intact and free light chains. *Dispenzieri A. Prevalence and risk of progression of light-chain monoclonal gammopathy of undetermined significance (LC-MGUS): a newly defined entity. Lancet 2010;375 (9727):1721–8.*

10. a. Pseudohyperphosphatemia is caused by a spectral interference with the molybdate reaction. Pseudohyponatremia occurs with high protein content with indirect reading ion selective electrode readings but not with direct reading assays. *Jelinek AG, et al. Unexpected test results in a patient with multiple myeloma. Clin Chem 2014;60:1375–9. Mirvis E, et al. Hyponatraemia in patients with multiple myeloma. BMJ Case Rep 2015;https:// doi.org/10.1136/bcr-2015-212838.*

11. e. While free light chains are the most likely explanation, this gel did not contain antisera for IgD and IgE, therefore their presence could be the cause for the absent heavy chain could be an explanation for this pattern. This sample should be retested for their potential presence. *Morrison T, et al. Laboratory assessment of multiple myeloma. Adv Clin Chem 2019;89:1–58.*

12. c. Typically the free light chain migrates more anodic (toward albumin) than the intact immunoglobulin, as seen in this case. The normal SPE pattern is seen in the left gel confirming that there was no mislabeling. Heavy chain disease is characterized by the absence of light chains (not seen here). *Morrison T, et al. Laboratory assessment of multiple myeloma. Adv Clin Chem 2019;89:1–58.*

13. d. Biclonal gammopathy, also termed double monoclonal, is the production of two monoclonal bands that are different from separate B-cells. Two bands can also be seen in immune complex formation, but in that case, the subtypes are the same. *Srinivasan VK, et al. Occurrence of double monoclonal bands on protein electrophoresis: an unusual finding. Indian J Hematol Blood Transfus. 2016;32(Suppl. 1):184–18.*

14. b. If plasma is used instead of serum, fibrinogen appears in this region of SPE. Pretreatment with ethanol can be helpful in removing this band. *Davison AS, et al. Can lithium-heparin plasma be used for protein electrophoresis and paraprotein identification? Ann Clin Biochem 2006;43:31–4.*

15. c. Monoclonal bands in each lane are due to an artifact. The prozone effect is when there is antigen excess and has occurred in the lambda region. In this situation, the middle of the thick lambda band has disappeared due to solubility of the complex, giving the appearance of two bands. *Morrison T, et al. Laboratory assessment of multiple myeloma. Adv Clin Chem 2019;89:1–58.*

16. b. MALDI-TOF/MS analysis of tryptic digests can have high sensitivity and specificity when the specific amino acid sequences of the variable portion of antibodies are known in advance. *Barnidge DR, et al. Monitoring M-proteins in patients with multiple myeloma using heavy-chain variable region clonotypic peptides and LC-MS/MS. J Proteom Res 2014;13:1905–10.*

17. a. CZE does improve the resolution for bisalbuminemia, but there is no medical significance of this finding. Radio-opaque agents can interfere with CZE by producing a spike, a disadvantage of this technique. The major advantage is automation and archival. *Bossuyt X, et al. Separation of serum proteins by automated capillary zone electrophoresis. Clin Chem Lab Med 2003;41:762–72.*

18. c. α1-Antitrypsin deficiency is an inherited disorder and can be detected by the absence of an α1 band on serum protein electrophoresis. *Mahadeva R. α1-antitrypsin deficiency, cirrhosis and emphysema. Thorax 1998;53:501–5.*

19. d. Acute phase reactants increase under conditions of inflammation and infection. Transferrin is not upregulated under these conditions and is therefore not an acute phase reactant. *Markanda A. Acute phase reactants in infections: evidence-based review and a guide for clinicians. Acute Phase Reactants Infect 2015;2:1–7.*

20. e. Hyperviscosity not hypoviscosity can occur in multiple myeloma if the total protein is high, but is more common in Waldenstrom's macroglobulinemia, where there is a proliferation of IgM antibodies. *Mehta J, et al. Hyperviscosity syndrome in plasma cell dyscrasias. Semin Thromb Hemost. 2003;29:467–71.*

21. b. Serum protein electrophoresis identifies monoclonal bands but do not determine the heavy or light subtypes. Immunofixation electrophoresis makes use of specific antibodies to determine if IgG, IgA, IgM, IgD, IgE, and kappa and lambda light chains are present. *Csako G. Immunofixation electrophoresis for identification of proteins and specific antibodies. In: Kurien BT, et al., eds. Protein Electrophoresis. Methods in Molecular Biology (Methods and Protocols), vol. 869. Humana Press, Totowa, NJ.*

22. a. Amyloidosis is characterization by deposition of insoluble complexes into organs such as the heart and kidneys. The most common cause is associated with free light chains. Transthyretin is a thyroid binding protein that can also cause a genetic and spontaneous amyloidosis. *Benson MD, et al. Amyloid nomenclature 2018: recommendations by the International Society of Amyloidosis (ISA) nomenclature committee. Amyloid 2018; 25:215–9.*

23. a. α1 Acid glycoprotein is synthesized by the liver, granulocytes, and monocytes. In massive intravascular hemolysis, the major component of α2 region, that is haptoglobin, may completely disappear. The primary physiologic role of ceruloplasmin is to support oxidation-reduction reaction in the plasma. On serum electrophoresis β-lipoprotein produces "feathery edge" in the β region and it is the most anodal protein in this fraction. *Clarke W. Contemporary practice in clinical chemistry. AACC press. 2nd ed., pp. 234–6.*

24. a. The urine IFE requires preconcentration procedure. The analytical sensitivity of IFE is significantly higher than conventional urine electrophoresis. The main goal to perform IFE on CSF is to detect oligoclonal banding which is defined as multiple IgG bands. The antisera used in IFE contain antibody against gamma, α and μ heavy chains, and kappa and lambda light chains. *Clarke W. Contemporary practice in clinical chemistry. AACC press. 2nd ed., pp. 242–43.*

25. d. Transferrin migrates with β fraction. Haptoglobin, β lipoprotein, ceruloplasmin, and α2 macroglobulin migrate with α2 fraction. *McPherson, Pincus; Henry's Clinical Diagnosis and Management by Laboratory Methods; Elsevier: 23rd ed., p. 255.*

26. c. Prealbumin migrates toward anode faster than albumin. It has tetrameric structure and each unit can bind to one molecule of thyroxine. The thyroxine binding globulin has 100 times higher capacity to bind thyroxine than prealbumin. Prealbumin along with retinol binding protein caries vitamin A. This complex is manufactured in the liver. In hemodialysis patients a low level of prealbumin associates with higher risk for mortality and hospitalization due to infection. *McPherson, Pincus; Henry's Clinical Diagnosis and Management by Laboratory Methods; Elsevier: 23rd ed., pp. 257–8.*

27. c. α1 Antitrypsin is coded by *SERPINA1* gene located on chromosome 14th. The homozygous for MM allele patients exhibits normal phenotypic function. An efficient diagnosis of α1 antitrypsin deficiency is done by measuring AAT concentration, followed with isoelectric-focusing electrophoresis and genotyping. The α1 antitrypsin replacement therapy cannot prevent liver cirrhosis in children with AAT deficiency. *McPherson, Pincus; Henry's Clinical Diagnosis and Management by Laboratory Methods; Elsevier: 23rd ed., pp. 259–60.*

Chapter 8

Interference and HIL indices

Y. Victoria Zhang[a], Shu-Ling Fan[b], Michael Karasick[a], and Alan H.B. Wu[c]
[a]University of Rochester, Rochester, NY, United States, [b]UMass Memorial Medical Center, Worcester, MA, United States, [c]University of California, San Francisco, CA, United States

1. Which of the following is the mechanism for biotin interference?
   a. Endogenous biotin binds directly to solid-phase particles.
   b. Dietary biotin binds to both the capture and signal antibodies producing a signal in the absence of the analyte.
   c. Biotin binds to solid-phase particles linked to streptavidin interfering with antibodies linked to biotin.
   d. Biotin quenches both fluorescence and chemiluminescence signals.
   e. Interference occurs in the absence of dietary biotin.
2. An abnormal serum chloride (135 mmol/L) with a normal sodium is reported on a patient who has epilepsy and a normal calcium.
   a. The patient is dehydrated.
   b. Bromide interferes with ion-selective electrodes for chloride.
   c. The sample is contaminated with intravenous normal saline.
   d. The patient has primary hyperparathyroidism.
   e. The patient has pseudohyponatremia due to multiple myeloma.
3. A patient is admitted with ethylene glycol exposure. Which of the following test may be adversely affected?
   a. Serum ethanol with the alcohol dehydrogenase method.
   b. Methanol with the head space gas chromatography method.
   c. Lactate with the lactate oxidase enzymatic assay.
   d. Lactate dehydrogenase utilizing the pyruvate to lactate direction.
   e. Lactate dehydrogenase utilizing the lactate to pyruvate direction.
4. Estimate the glucose concentration, starting from 100 mg/dL after a tube of blood is left uncentrifuged at room temperature for 8 h.
   a. no change
   b. 90 mg/dL
   c. 75 mg/dL
   d. 50 mg/dL
   e. 20 mg/dL

Self-assessment Q&A in Clinical Laboratory Science, III. https://doi.org/10.1016/B978-0-12-822093-1.00008-9

5. A 55-year-old female has a serum hCG concentration of 10 U/L (reference range 0–5 IU/mL) who states that she is not sexually active. Repeat testing on dilution produces the appropriately expected result. Her follicle stimulating hormone was increased. What is the best explanation for this result?

   a. She has a gestational tumor.

   b. She is not truthful about her sex history and is actually pregnant.

   c. She has an ectopic pregnancy.

   d. She has human antimouse antibodies.

   e. The hCG was produced from the anterior pituitary.

6. A patient exposed to sodium nitrite has oxygen saturation measurements by cooximetry and pulse oximetry. Which of the following is likely?

   a. Results will be the same within precision limits.

   b. The apparent oxygen concentration will be falsely high in the pulse oximetry measurement.

   c. The carboxyhemoglobin concentration will be falsely high in the pulse oximetry measurement.

   d. The pulse oximetry measurement is most accurate overall.

   e. Accuracy is dependent on the total hemoglobin concentration.

7. A hemolyzed serum sample is tested to contain 1 g/dL of hemoglobin. What is the approximate contribution of potassium?

   a. negligible

   b. 0.1 mmol/L

   c. 0.4 mmol/L

   d. 1.0 mmol/L

   e. 1.5 mmol/L

8. The following figure represents an immunoassay. The dots indicate the analyte. The top antibody is the capture and the bottom the detection. Which is the best explanation?

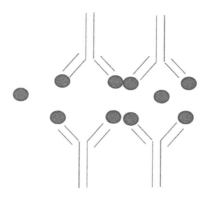

a. Indicates a heterophile interference

b. This is the prozone effect.

c. This is the hook effect.

d. The calibration curve is saturated, producing a result in the plateau region.

e. This is an accurate measurement.

9. Which of the following is not an effective countermeasure to investigate the presence of a human antimouse antibody

   a. add mouse IgG

   b. dilute the sample

   c. check other laboratory tests for medical consistency of the result

   d. use an alternative vendor for the assay

   e. increase the incubation time for the assay

10. What is the source of hemolytic interference?

    a. white blood cells

    b. red blood cells

    c. platelets

    d. plasma

11. Which electrolyte is affected the most with hemolytic interference?

    a. sodium

    b. chloride

    c. potassium

    d. calcium

12. Mechanisms of in vitro hemolysis include (choose all that apply):

    a. mechanical destruction

    b. glucose exhaustion

    c. autoimmune disease

    d. freezing

13. Which enzyme assays are affected by hemolytic interference? (choose two)

    a. lactate dehydrogenase

    b. γ-glutamyl transferase

    c. troponin

    d. creatine kinase

14. Where approximately is the peak spectral absorbance of hemoglobin?

    a. 300–400 nm

    b. 400–500 nm

    c. 500–600 nm

    d. 600–700 nm

15. What is the cause of icteric interference?

    a. urea

    b. albumin

    c. bilirubin

    d. immune globulin

16. Which assays demonstrate inference from icterus? (Choose all that apply)
    a. glucose
    b. triglycerides
    c. uric acid
    d. magnesium

17. Which of the following is a primary cause of lipemic interference?
    a. HDL
    b. IDL
    c. LDL
    d. VLDL

18. Which of the following measurement methods is most affected by turbidity?
    a. ion specific electrodes
    b. nephelometry
    c. spectroscopy
    d. potentiometry

19. When a result with possible serum index interference is obtained, one should:
    a. fail the test
    b. repeat testing on a different platform
    c. centrifuge the sample and repeat testing on the same platform
    d. consult laboratory protocols

20. What combination of tests will get affected the most by hemolysis?
    a. glucose, total calcium, progesterone
    b. TSH, albumin, hCG
    c. lactate dehydrogenase, potassium, AST
    d. cortisol, chloride, PTH

21. Serum indices (hemolysis, icterus, lipemia) are considered errors of which phase of the testing process?
    a. preanalytical phase
    b. analytical phase
    c. postanalytical phase
    d. critical value reporting phase

22. What is not the mechanism that hemolysis interferes with routine chemistry test results?
    a. release of analytes found in high concentrations in red blood cells
    b. hemoglobin color overlaps with the spectrophotometric measurement of testing
    c. disturb the binding of antigen-antibody binding
    d. the oxidation–reduction chemical reactivity of the iron in hemoglobin

23. What is not a common process to eliminate lipemic interference?
    a. ultracentrifugation
    b. extraction

   c. dilution

   d. chromatography separation

24. What condition is not likely to cause lipemic interference?

   a. nonfasting

   b. acute pancreatitis

   c. total parenteral nutrition

   d. adrenal insufficiency

25. What causes icteric samples?

   a. direct bilirubin and indirect bilirubin

   b. delta bilirubin

   c. creatinine

   d. ethylene glycol

26. What is not the mechanism of icterus interference?

   a. spectral interference: absorbance in the bilirubin range (400–540 nm)

   b. chemical reaction with reagent component (e.g., H_2O_2)

   c. turbidity results in light scattering

   d. increase of both conjugated and unconjugated bilirubin

27. Acetaminophen can be measured by the following methods. Which method might produce falsely positive result in a hemolyzed sample with no acetaminophen?

   a. EMIT immunoassay

   b. mass spectrometry

   c. enzymatic/colorimetric assay

   d. thin-layer chromatography

28. What test result might be affected in a severely lipemic sample?

   a. acetaminophen measured by the enzymatic/colorimetric method

   b. sodium measured by the electrochemistry method

   c. gentamicin measured by the EMIT immunoassay

   d. fentanyl measured by the mass spectrometry

29. Which of the following does not describe how serum indices are detected/determined?

   a. visual observation of the sample

   b. automated serum indices inspection

   c. chemically measured by a testing reagent

   d. give a semiquantitative (1^+ to 4^+) estimation

Answers

1. c. Some immunoassays use solid-phase particles that are linked to streptavidin. When biotin-labeled antibodies are added, they form a complex with streptavidin and the solid-phase reagent to capture the analyte. The addition of a second antibody labeled with a fluorophore or chemiluminescent tag is added to complete the sandwich and enable detection of the complex.

Paramagnetic particles linked to streptavidin are not specific to any analyte. *Colon PJ, et al. Biotin interference in clinical immunoassays. J Appl Lab Med 2018;2:941–51.*

2. c. Potassium bromide is used to treat certain types of epileptic seizures especially in children. The ion-selective electrode for chloride binds to bromide and can cause false positive results. *Wang T, et al. Variable selectivity of the Hitachi chemistry analyser chloride ion-selective electrode toward interfering ions. Clin Biochem 1994;27:37–41.*

3. c. The presence of glycolate, an ethylene glycol metabolite and similar in structure to lactate, causes interferences with many lactate assays. *Tintu A, et al. Interference of ethylene glycol with L-lactate measurement is assay-dependent. Ann Clin Biochem 2013;50:70–2.*

4. d. Glucose declines at a rate of 5%–7% per hour at room temperature. *Turchiano M. Impact of sample collection and processing methods on glucose levels in community outreach studies. J Environ Pub Health 2013; Article ID 256151, https://doi.org/10.1155/2013/256151.*

5. e. Perimenopausal and postmenopausal nonpregnant women can have mildly increased hCG that is accompanied by a high FSH. Pregnant women have low FSH and much higher hCG. *Wu AHB, et al. Mild positive hCG in a perimenopausal female: normal, malignancy, or phantom? Lab Med 2009;40:463–6.*

6. b. Pulse oximetry only measures oxygenated and reduced hemoglobin while cooximetry will also measure methemoglobin, which would be expected to be high in this case, and carboxyhemoglobin, expected to be normal in this case. Pulse oximetry will produce a falsely high value of methemoglobin level. *Jurban A. Pulse oximetry. Crit Care 2015;19:272. https://doi.org/10.1186/s13054-015-0984-8.*

7. c. Studies conducted by Dimeski showed that for every 1 g/dL of hemoglobin, under conditions of normal white cell count, added 0.29–0.53 mmol/L of potassium. Because of the large variability, the authors do not recommend testing for free plasma hemoglobin and institute a potassium correction factor to be reported for hemolyzed specimens. *Dimeski G. Correction and reporting of potassium results in haemolysedsamples. Ann Clin Biochem 2005;42:119–23.*

8. c. An excess of the antigen produces the hook effect, a decreasing of the signal. The calibration curve does not plateau but folds back onto itself. *Namburi RP et al. High dose hook effect. J NTR Univ Health Sci 2014;3:5–7. http://www.jdrntruhs.org/text.asp?2014/3/1/5/128412.*

9. e. Increasing the incubation time will not be useful in investigation the presence of or eliminating an interference from the presence of human antimouse antibody. Using other tests for internal consistency may be possible, e.g., a falsely high TSH will not produce a low free T4. *Klee GG. Human anti-mouse antibodies. Arch Pathol Lab Med 20009;126:921–3.*

10. b. Hemolysis is the destruction of red blood cells. The cell contents being released into the specimen are the basis of associated interference. Hemolysis is the destruction of red blood cells. The cell contents being released into the specimen are the basis of associated interference. *CLSI C56-A: Hemolysis, Icterus, and Lipemia/Turbidity Indices as Indicators of Interference in Clinical Laboratory Analysis. p. 8.*

11. c. Potassium is found in greater concentrations within RBCs than the surrounding plasma. Hemolysis results in, sometimes greatly, increased potassium concentrations. Calcium is found in significant amounts in certain cells but not RBCs. Sodium and chloride have a higher concentration in plasma than in cells. *Yucel D, et al. Effect of in vitro hemolysis on 25 common biochemical tests. Clin Chem 1992;38:575–7.*

12. a,b,d. Mechanisms of in vitro destruction include destruction from mechanical causes during preanalytical processes, cell lysis due to cell stress from freezing or lack of glucose, or other inherent red cell abnormalities. Autoimmune-mediated RBC destruction occurs in vivo, requiring RBC interaction with cells of the immune system. *CLSI C56-A: Hemolysis, icterus, and lipemia/turbidity indices as indicators of interference in clinical laboratory analysis. p. 8.*

13. a,d. Lactate dehydrogenase is found in RBCs so hemolysis results in increased sample concentrations. RBCs also contain adenylate kinase which catalyzes the removal of the ADP substrate of some creatine kinase assays, leading to falsely elevated results. γ-Glutamyl transferase and troponin assays are unaffected by hemolysis. *CLSI C56-A: Hemolysis, icterus, and lipemia/turbidity indices as indicators of interference in clinical laboratory analysis. p. 9.*

14. b. The peak absorbance of hemoglobin is at 420 nm. *CLSI C56-A: Hemolysis, icterus, and lipemia/turbidity indices as indicators of interference in clinical laboratory analysis. p. 9.*

15. c. Elevated bilirubin is the cause of icterus. *CLSI C56-A: Hemolysis, icterus, and lipemia/turbidity indices as indicators of interference in clinical laboratory analysis. p. 9.*

16. a,b,c. Explanation: Bilirubin can interfere with assays utilizing H_2O_2 as an intermediary including cholesterol, glucose, triglycerides, and uric acid. Interference is also possible due to bilirubin's spectral range (400–540 nm). *CLSI C56-A: Hemolysis, icterus, and lipemia/turbidity indices as indicators of interference in clinical laboratory analysis. pp. 9–10.*

17. d. VLDLs and chylomicrons are the primary cause of lipemic interference, thought to be due to elevated triglycerides and/or the large particle size. *CLSI C56-A: Hemolysis, icterus, and lipemia/turbidity indices as indicators of interference in clinical laboratory analysis. p. 10.*

18. b. Turbidity affects nephelometry, which measures light scatter by sample. Increased turbidity results in increased scattering, affecting measurement and interpretation. *CLSI C56-A: Hemolysis, icterus, and lipemia/turbidity indices as indicators of interference in clinical laboratory analysis. p. 11.*

19. d. When faced with a result that may be affected by increased serum indices, one should consult the protocol. Different platforms and analytes have different established cutoffs for interference with testing and there is no blanket approach to cover the multitude of situations that can arise. *CLSI C56-A: Hemolysis, icterus, and lipemia/turbidity indices as indicators of interference in clinical laboratory analysis.*

20. c. In red blood cells, the concentration of lactate dehydrogenase is about 160-fold greater, potassium is 22-fold greater, and magnesium is 3-fold greater than the normal concentrations in serum or plasma. *CLSI C56-A: Hemolysis, icterus, and lipemia/turbidity indices as indicators of interference in clinical laboratory analysis. p. 19.*

21. a. The majority of detected clinical laboratory errors originate in the preanalytical phase of testing. Hemolysis, icterus, and lipemia are prevalent preanalytical errors, often leading to rejected samples because of interference. *Clinical Laboratory News. March 2016. Detecting and handling hemolysis using serum indices. https://www.aacc.org/publications/cln/articles/2016/march/detecting-and-handling-hemolysis-using-serum-indices.*

22. c. Hemolysis influences the accuracy and reliability of routine chemistry testing by two different mechanisms. The first is through the release of analytes found in high concentrations in erythrocytes. The second mechanism by which hemolysis affects test accuracy and reliability is through interference from hemoglobin itself. Due to its red color, hemoglobin absorbs light between 340–440 nm and 540–580 nm, allowing it to interfere with assays across a broad range of wavelengths. *Clinical Laboratory News. March 2016. Detecting and handling hemolysis using serum indices. https://www.aacc.org/publications/cln/articles/2016/march/detecting-and-handling-hemolysis-using-serum-indices.*

23. d. In most cases, lipemia can be removed from the sample and measurement can be done in a clear sample without interferences. There are several ways of removing lipids, such as centrifuge, extraction, or dilution, and laboratory experts should carefully choose which one to use depending on the tests that have to be measured in the sample. *Nikolac N. Lipemia: causes, interference mechanisms, detection and management. Biochem Med (Zagreb) 2014;24:57–67.*

24. d. The most common preanalytical cause of lipemia is inadequate time of blood sampling after the meal. In the hospital patients, lipemia can also be caused by sampling too soon after administration of parenteral lipid emulsions. In patients with acute pancreatitis, the triglycerides level is usually significantly elevated. *Nikolac N. Lipemia: causes, interference mechanisms, detection and management. Biochem Med (Zagreb) 2014; 24):57–67.*

25. a. The bilirubin in the sample should be unconjugated and/or conjugated. *CLSIC56-A: Hemolysis, icterus, and lipemia/turbidity indices as indicators of interference in clinical laboratory analysis. p. 20.*

26. c. Bilirubin may interfere by two mechanisms: spectral interference due to its strong absorbance between 400 and 540 nm, and chemical interference due to its chemical reactivity. For example, bilirubin has been shown to produce a negative bias on assays that involves H_2O_2 as an intermediate reaction. *CLSI C56-A: Hemolysis, icterus, and lipemia/turbidity indices as indicators of interference in clinical laboratory analysis. p. 19.*

27. c. Hemolysis and icterus had insignificant interference on the Syva EMIT and the DRI immunoassays for the analysis of acetaminophen, but significant interference effect on the Roche enzymatic/colorimetric assay. *Zhang YV. Effect of hemolysis, icterus, and lipemia on three acetaminophen assays: Potential medical consequences of false positive results. Clin Chim Acta 2018;487:287–92.*

28. b. This mechanism of volume displacement strongly affects concentration of electrolytes. The normal plasma consists of approximately 92% of water and 8% of lipids. In the lipemic sample, the proportion of lipid phase increases and can be up to 25%. Analytes that are not distributed in the lipid phase (i.e., electrolytes) are distributed in the aqueous part of the sample, which now accounts for only 75% of the sample. Methods that measure concentration of electrolytes in the total plasma volume (including the lipid phase), like indirect potentiometry, result with falsely decreased concentration of electrolytes because of the high dilution prior to analysis. Multiplying obtained result after the measurement to the full plasma volume results with an error in electrolyte concentration. *Nikolac N. Lipemia: causes, interference mechanisms, detection and management. Biochem Med (Zagreb) 2014;24:57–67.*

29. c. Initial visual observation of sample has been used in the clinical laboratory for decades. The use of automated HIL indices overcomes the inherent limitations. This is particularly applicable in the highly automated laboratory where visual inspection is difficult due to the high volume of samples and the required speed for evaluating sample quality. *CLSI C56-A: Hemolysis, icterus, and lipemia/turbidity indices as indicators of interference in clinical laboratory analysis. p. 12.*

Chapter 9

Clinical cases

Wieslaw Furmaga
Department of Pathology, University of Texas Health Science Center at San Antonio, San Antonio, TX, United States

1. Nephrologist treats diabetes insipidus patient with hypertonic solutions. He asks you what is the relation between monitoring instrument and number of molecules in the solution. You answer that number of molecules in the solution affects its colligative properties such as freezing point, boiling point, and vapor pressure. Which statement about properties of solutions is true?
   a. Osmolality refers to the number of moles of solute in a liter of solution.
   b. Osmolarity refers to the number of moles of solute in a kg of water (solvent).
   c. In clinical practice both terms osmolality and osmolarity can be used interchangeably, because the changes in water volume over the range of human body temperature over is negligible.
   d. Osmolarity is preferred in clinical usage, because the molecular mass of water does not change with temperature as water volume does.

2. An 85-year-old male was admitted to the hospital by psychiatry service due to suicidal attempt. He was found unresponsive in his room with the empty aspirin bottle on his chest. May the osmotic gap be used to estimate how much salicylic acid is still in his system?
   a. The measured osmolality is always larger than calculated one, when anions, such as salicylate, formate, lactate, β-hydroxybutyrate, accumulate in the body.
   b. The accumulation of nonelectrolyte solutes such as ethanol, ethylene glycol, methanol, and mannitol do not produce osmotic gap because their presence is balanced by sodium.
   c. Accumulation of neutral and cationic amino acids leads to production of osmolal gap.
   d. Osmolal gap may be used as a test for the presence of toxic substances such as salicylate and formate, but not mannitol, neutral and cationic amino acids.

Self-assessment Q&A in Clinical Laboratory Science, III. https://doi.org/10.1016/B978-0-12-822093-1.00009-0

3. Internal medicine resident asked why anuric patients must be treated with the fluids. What is true about this issue?

a. Because sweat does not contain electrolytes, the amount of water eliminated from the skin is affected by extracellular electrolyte balance.

b. Respiratory loss of water is constant and is included in water balance calculation as a constant number.

c. In absence of diarrhea and constipation the amount of water loosing through GI tract is constant.

d. Colon is the place where most of the water is excreted to keep stool soft.

4. The prevalence of serious infections in the hospitals ranges from 4.5% to 29.3%. What is true about laboratory markers for the serious infections?

a. The best markers for diagnosis of a severe infection are interleukins and white blood cell count.

b. Procalcitonin and C reactive protein have different cutoff values for "ruling in" and for "ruling out" serious infections.

c. White blood cell indicators have a great value to ruling out and are no valuable for "ruling in" serious infections.

d. In contrast to outpatient settings, in emergency departments procalcitonin and CRP have the same cutoffs to rule out and rule in serious infections.

5. There are diagnostic and management uncertainties over mild thyroid dysfunction. What is true about subclinical thyroid dysfunction?

a. Physiological and within reference ranges fluctuation of the free T4, may cause clinically significant alteration of TSH concentration, which may be outside normal values.

b. The prevalence of subclinical hypothyroidism is higher in iodine-deficient populations.

c. In subclinical hyperthyroidism a serum TSH concentration may be normal, but free T4 and T3 concentrations are usually above the normal ranges.

d. It is important to aggressively treat patients with TSH concentration even slightly below reference limits for hyperthyroidism, because they are at high risk for permanent damage.

6. The normal concentration of free thyroxin (free T4) in the serum of pregnant women is essential for normal psychomotor development of the child. A free thyroxin index (FT4I) has been developed to allow for better interpretation of thyroid function during pregnancy. What is true about thyroid testing in pregnant women?

a. The pregnancy-related changes in protein composition affect performance of antibody-based methods for the free T4 testing.

b. The concentration of free T4 during the third trimester is significantly lower than in nonpregnant women.

    c. Currently available antibody-based methods detect effectively the first trimester hypothyroidism, but they are not able to identify gestational hyperthyroidism.

    d. When compared to the reference ranges from equilibrium dialysis and tandem mass spectrometry, the thyroxin level by FT4I is abnormally low during the first trimester, but it is abnormally high during the second and third trimester.

7. The most common consequences of prolonged cocaine abuse are myocardial ischemia and true myocardial infarct. What is true about the pathophysiology of these conditions?

    a. Cocaine induces tachycardia that is due to its direct action on the myocardium.

    b. Cocaine increases oxygen demand by myocardium by its chronotropic stimulation and an increase in the afterload.

    c. Direct stimulation of adrenal cortex is responsible for cocaine-related vasoconstriction.

    d. A number of studies showed that a chronic cocaine usage protects against arteriosclerosis.

8. An 83-year-old male received morphine for pain control after colectomy done 6 weeks ago. He found that opioids stimulate different receptors and asked you to explain what it means. What would you tell him?

    a. Morphine analgesia is exclusively mediated through sigma receptors.

    b. Opioid-related sedation and inhibition of respiratory function and gastrointestinal transit attribute to MOR (mu) receptor.

    c. DOR and KOR (delta and kappa) receptors are responsible for respiratory depression.

    d. Stimulation of sigma receptors produces similar reaction to stimulation of MOR (mu) receptors.

9. The mother of your friend has had long time suspicion that her son takes "some drugs." She hears the conversation between him and man about designer drugs. Now she asks you about these drugs. What would you tell her?

    a. The synthetic cocaine such as 3,4-methylenedioxymethamphetamine is so-called designer drug and it is intended to produce sedation.

    b. The assays for each of the designed drugs are easy to develop and are widely available.

    c. These drugs can be administered orally, inhaled, intravenously, or rectally.

    d. Death from overdose of these designer drugs is very common.

10. Restrained by police, 45-year-old female was brought to ED. She was running away from officer with no reason. She incoherently speaks about someone she met in police car. Physical examination was significant for pulse of 154 with blood pressure 176/123. The officer asks you how cocaine is detected. What you would tell him?

a. Benzoylecgonine (BE) is an inactive product of only nonenzymatic cocaine metabolism.

b. BE is a final metabolic product and it is excreted unchanged in urine.

c. The presence of M-hydroxybenzoylecgonine (m-HOBE) in the urine confirms Cocaine use.

d. Norcocaine may be detected in urine in high concentration in nonalcoholics.

11. During the processing, the blood tube of a 28-year-old man was accidently dropped to the garbage and was found after 2 h. Because patient was difficult to draw, the specimen was considered as irretrievable, and it is testing which showed hyperpotassemia. Clinician who received this report calls you with a question about Pseudohyperkalemia. What is certain information about Pseudohyperkalemia you can tell to the clinician?

a. Rhabdomyolysis and intravascular hemolysis are the most common cause of pseudohyperkalemia.

b. The absence of electrocardiographic (ECG) abnormalities rule out both pseudohyperkalemia and true hyperkalemia.

c. 24-h urine K^+ does not distinguish hyperkalemia due to impaired renal excretion from the excessive shift of an intracellular K^+ to the extracellular space.

d. In cases of chronic hyperkalemia due to renal insufficiency the serum creatinine and BUN is better diagnostic test than K^+ secretion in 24-h urine.

12. The testing of a 46-year-old healthy female showed mild hyperkalemia. Clinician called you with a question what could cause an increased potassium in this patient. The investigation of preanalytical possibilities such as prolonged tunicae or hemolysis gave no explanation. What are the other possibilities of hyperkalemia?

a. Heparin therapy causes direct aldosterone digestion and reduction its concentration.

b. Treatment with nonsteroidal antiinflammatory agents or beta blockers may provoke hyperkalemia by stimulating negative feedback on adrenal gland.

c. Renal tubes may lose their response for aldosterone. This defect may involve only potassium secretion (pseudohypoaldosteronism type II), or sodium reabsorption and potassium secretion (pseudohypoaldosteronism type I).

d. On the day of testing patient might drink a lot of water what produced secondary hyperaldosteronism.

13. A 65-year-old postmenopausal woman who works in your laboratory sent you an email with question about effectiveness of strontium ranelate (SrR) as a protective supplement against osteoporosis. Based on information published in Clinical Chemistry journal, what would you say about stramonium in treating of osteoporosis?

   a. An intake of SrR at bedtime by postmenopausal women with osteoporosis reduces both bone resorption and a risk of bone fracture.

   b. Either calcium or SrR administered to the postmenopausal women inhibits a bone resorption.

   c. Anti-PTH effect of SrR in the postmenopausal women is significantly higher than the same effect induced by calcium.

   d. Similarly to calcium, strontium activates CaR and directly inhibits the formation of mature osteoclasts.

14. Despite numerous problems with methodology and interpretation from obtained results, laboratory tests detecting cardiac infarction become widely used in clinical settings. What is true about these assays?

   a. Ischemia-modified albumin (IMA) is formed during interaction between albumins and damaged myocytes and it is a specific marker for cardiac injury.

   b. One of many limitations for troponin testing is its homogeneity and its degradation into homogenous cTnT-I-C complex.

   c. The problem with cTn assays relates to its relatively high concentration what leads to interference with heterophilic antibodies and prezonal effect.

   d. The cTn 99th percentile depends on assay accuracy at the low-end and on chosen reference population.

15. Despite a new multidisciplinary treatment and advances in diagnosis, the cancer-related mortality has not decreased. As today, there is no tumor marker that detects early stages of cancers with clinically useful sensitivity and specificity. What is true about cancer detection by laboratory methods?

   a. Tumor markers share numerous tumor-associated epitopes such as C19-9, CA 15-3. CA 125.

   b. The oncogenic p53 protein is coded by p53 oncogene and it is responsible for stimulation of cell growth and proliferation.

   c. Significantly increased serum level of adhesion molecules such as E-selectin, intercellular adhesion molecule (ICAM), and vascular cell adhesion molecule (VCAM) has been observed in breast cancer patients who respond well to chemotherapy.

   d. Dimerization of HER2/Neu oncoprotein causes cellular apoptosis.

16. There are three classes of tumor markers used for diagnosis, monitoring, and prognosis of human malignancies. What is true about these classes?

    a. Oncofetal antigens such as CA 19-9, CA 125, and CA 15-3 are normally expressed during fetal development but not in children and adults.

    b. AFP does not appear in the normal tissue and it is marker for primary hepatocellular carcinoma.

    c. Enzymes (alkaline phosphatase) and hormone-like peptides (parathyroid hormone-like protein) may be elevated in the serum of the patients with specific tumors.

    d. The concentration of acute phase reactants as well as CRP increases significantly in most of the occult neoplasms.

17. You were asked to give the general overview to the medical students about basic concepts of human proteins with clinical examples. What would you tell to the medical students?

    a. Determined by amino acid sequence, a three-dimensional structure of the protein is termed *quaternary structure.*

    b. Because it is not a quantitative test, electrophoresis cannot be used to monitor patients with myeloma, nephrotic syndrome, or cirrhosis.

    c. Abnormal fragments of some proteins (e.g., serum amyloid-associated protein, immunoglobulin light chains, prealbumin) may precipitate into *quaternary structure* named multiple myeloma.

    d. The serum albumin concentration can be used for a nutritional status evaluation, for assessment of the liver synthetic capacity, and degree of nephropathy or enteropathy.

18. A 65-year-old male suffers from severe back pain which does not respond to NSAD. The diagnostic tests exclude both occult neoplasm and autoimmune diseases. The serum protein electrophoresis has been signed out as a normal but the primary care physician who ordered this test wants more information. What is true about serum protein electrophoresis?

    a. Patients with β-fibrillar amyloid component may have an abnormal prealbumin band detected by standard electrophoresis.

    b. If the serum for electrophoresis is collected on heparin as an anticoagulant, a distinct prealbumin band can be seen.

    c. The level of alfa1 antitrypsin (AAT) in heterozygous (MZ) patients is not sufficient for antiprotease activity and for protection against pulmonary or hepatic diseases.

    d. Serum haptoglobin rises in response to stress, infection, or acute inflammation, but falls following transfusion reaction, burns, or intravascular hemolysis.

19. Transferrin (siderophilin) is the major β-globulin component, which carries ferric ions from the intracellular ferritin to bone marrow. What is true about transferrin?

    a. The response to increased iron concentration in the plasma, the synthesis of transferrin increases through an accelerated transcription of mRNA.

  b. In case of iron overload transferrin forms an abnormal paraprotein band on serum electrophoresis.

  c. A sequencing of hemochromatosis gene is currently the best screening method for hemochromatosis.

  d. Double glycosylated sialic form of transferrin is characteristic of CSF and it is used as an evidence of CSF leaking in cases of skull fracture or other head trauma.

20. A classification of metabolic acidosis may be based on anion gap calculation. Which of the following conditions listed associates with an increased anion gap?

  a. renal tubular acidosis

  b. uremic acidosis

  c. intestinal loss of bicarbonate

  d. acidosis following respiratory alkalosis

21. The metabolic acidosis can be classified as a renal and extra renal acidosis. The extra renal acidosis is caused by organic acidosis, diarrheal loss of bicarbonate, and acidosis due to exogenous toxins. What statement about extra renal acidosis is true?

  a. In case of severe diarrhea urine anion gap increases and it is calculated as: urine (Cl^-) – urine $(Na^+ + K^+)$.

  b. Because diarrheal fluid contains more Cl^- than $Na^+ + K^+$, the high urine anion gap in diarrhea is explained by the larger loss of Cl^- than $Na^+ + K^+$.

  c. In cases of metabolic acidosis with loss of bicarbonate other than diarrhea, urine anion gap does not change as long as no electrolytes are lost by extra renal path.

  d. The diagnosis of organic acidosis is difficult because tissue hypoxia and lactic acidosis are difficult to detect with laboratory testing.

22. An 18-year-old girl has been evaluated for her delayed puberty, hypertension 155/90, lack of breast, and ambiguous genitalia. A provisional diagnosis made by PCP has been deficiency of 17-hydroxylase. What is true about clinical presentation of this deficiency?

  a. increase synthesis of cortisol, androgens, and estrogens

  b. increased synthesis of progesterone, corticosterone, and DOC

  c. hypotension at birth

  d. male pseudo-hermaphroditism

23. The excess of terminal hair in the androgen-dependent areas, such as inner thigh, chest, face, etc., in women, may be a symptom of endocrine pathology. What is true about this condition?

  a. Idiopathic hirsutism is only a working diagnosis, and it is always a symptom of an underlying pathology.

  b. Hyperprolactinemia-related hirsutism may be accompanied by amenorrhea and/or galactorrhea.

    c. Hirsutism is always the first symptom of early onset of the congenital adrenal hyperplasia.

    d. hCG-secreting ovarian tumors are the main cause of hirsutism.

24. What is true about male hypogonadism?

    a. The primary hypogonadism due to testicular damage leads to elevation of LH and suppression of FSH.

    b. Free testosterone in the primary testicular male hypogonadism is low but total serum testosterone is normal.

    c. A decreased or low normal concentration of LH and FSH indicates pituitary or hypothalamic disease (secondary hypogonadism).

    d. Sperm production is low in primary testicular hypogonadism, but it is normal in the secondary pituitary or hypothalamic hypogonadism.

25. Most pituitary tumors are benign and associate with characteristic neurological symptoms. What is true about the laboratory testing for the pituitary tumors?

    a. Low grade of hyperprolactinemia associated with a large pituitary tumor is most likely due to prolactinoma.

    b. The best screening test for Cushing disease is a cortisol urine concentration.

    c. The quantitation of the serum insulin-like growth factor I is a marker for central diabetes mellitus.

    d. For diagnosis of LH- or FSH-secreting tumors, both hormones along with testosterone for male and estradiol for female must be measured.

Answers

1. c. Osmolarity refers to the number of moles of solute in a liter of solution. Osmolality refers to the number of moles of solute in a kg of water (solvent). Osmolality is preferred in clinical usage, because the molecular mass of water does not change with temperature as water volume does. *McPherson RA, et al. Henry's Clinical Diagnosis and Management by Laboratory Methods; 22nd ed., 2011, p.170.*

2. c. The accumulation of anions such as salicylate, formate, lactate, and β-hydroxybutyrate do not produce osmotic gap because in healthy body they are balanced by sodium. The measured osmolality is larger than calculated when nonelectrolyte solutes such as ethanol, ethylene glycol, methanol, and mannitol are accumulated in the body. Osmolal gap may be used as a test for the presence of toxic substances such as ethanol, ethylene glycol, and methanol as well as mannitol, neutral and cationic amino acids. *McPherson RA, et al. Henry's Clinical Diagnosis and Management by Laboratory Methods; 22nd ed., 2011, p. 171.*

3. c. Majority of the fluid entering GI tract is absorbed in the small intestine. Small amount is absorbed in the colon and only about 100 mL of water is excreted in the feces. The sweat is not electrolyte free. It contains about 50 mmol/L of sodium and 5 mmol/L of potassium. The amount of the water lost from the skin is proportional to the heat producing (30 mL of water by 100 calories) and radiating from the body. The water loss though respiration depends on calorie production, that is 13 mL per 100 calories produced. This amount increases to the clinically significant amounts during hyperventilation. *McPherson RA, et al. Henry's Clinical Diagnosis and Management by Laboratory Methods; 22nd ed., 2011, p. 173.*

4. b. A cutoff to "ruling in" a serious infection for procalcitonin is 2 ng/mL and for C reactive protein is 80 mg/L. A cutoff to "ruling out" a serious infection for procalcitonin is 0.5 ng/mL and 20 mg/L for C reactive protein. The best markers for diagnosis of severe infection are procalcitonin and C reactive protein. White blood cell count has no value for "ruling out" and it is less valuable than inflammatory markers for "ruling in" serious infections. In both, outpatient clinics and emergency departments, procalcitonin and C reactive protein have different cutoffs to "rule in" or "rule out" serious infection. *Van den Bruel A, et al. Diagnostic value of laboratory tests in identifying serious infections in febrile children: systematic review. BMJ. 342:d3082, 2011.*

5. a. In the physiological situations, slight changes in free T4 concentration (within reference ranges) lead to significant alterations (outside reference ranges) of the TSH production. The prevalence of subclinical hyperthyroidism seems to be far higher in iodine-deficient populations. In the subclinical hyperthyroidism serum TSH may be below the reference ranges with normal concentrations of free T4 and T3. In a study from Israel, among the patients with serum TSH slightly lower than reference ranges, 51.5% return to the normal ranges when retested over the period of 5-year follow-up. *Cooper DS, et al. Subclinical thyroid disease. Lancet 2012;379 (9821):1142–54.*

6. a. Measured by immunological methods, a free T4 concentration is affected by pregnancy-related changes in albumin and thyroxin binding globulin levels. If the third trimester free T4 is measured by equilibrium dialysis/ GS/MS, its concentration differs less than 10% from nonpregnant women. Currently available immune method may not detect first trimester hypothyroidism, important for normal child development. When compared to the reference ranges from a gold standard equilibrium dialysis/GS/MS, the thyroxin level by FT4I is high during the first trimester and normalized to the nonpregnant level during the second and third trimester. *Lee RH, et al. Free T4 immunoassays are flawed during pregnancy. Am J Obstet Gynecol 2009;200:260.e1–260.e6.*

7. b. Cocaine increases myocardial oxygen demand by increasing chronotropy and afterload. Cocaine induces tachycardia by its sympathomimetic effects. Another sympathomimetic effect of cocaine action leads to peripheral vasoconstriction. Cocaine stimulates platelets aggregation what accelerates atherosclerotic process. *McPherson RA, et al. Henry's Clinical Diagnosis and Management by Laboratory Methods; 22nd ed., 2011; chapter 23.*

8. b. Sedation and inhibition of respiratory function and gastrointestinal peristaltic function are controlled through MOR (mu) receptor. Although morphine analgesia is mainly mediated through MOR (mu) receptors, the same effect can be achieved by stimulation of DOR or KOR (delta or kappa) receptors. Stimulation of DOR or KOR (delta and kappa) receptors does not affect respiration. Sigma are completely distinct from the classic opioid receptors and their stimulation produces tachycardia and tachypnea. *Burtis CA, et al. Tietz Textbook of Clinical Chemistry and Molecular Diagnostics Fifth Edition 2012, p. 1146.*

9. c. The road of administration of these drugs varies and can be orally, inhaled, intravenously, or rectally. So-called designer drugs are the synthetic phenylethylamines such as 3,4-methylenedioxymethamphetamine (MDMA or ecstasy) that are intended to produce euphoria. There are no specific assays available for most of these drugs in urine. Death from overdose of these designer drugs has been reported but it is uncommon. *McPherson RA, et al. Henry's Clinical Diagnosis and Management by Laboratory Methods; 22nd ed., 2011, chapter 23.*

10. c. The benzoylecgonine (BE) is a marker for cocaine use but its presence in the urine can be challenged in legal proceedings by the argument that cocaine was added to the urine and was transformed to BE by nonenzymatic metabolism. *M*-hydroxybenzoylecgonine (*m*-HOBE) arises exclusively via in vivo metabolism, therefore its presence in urine confirms Cocaine use or cocaine expose if present in the meconium. Cocaine is metabolized to inactive BE via both nonenzymatic hydrolysis and enzymatic transformation. BE is further transformed to *m*-hydroxybenzoylecgonine (*m*-HOBE) and *p*-hydroxybenzoylecgonine (*p*-HOBE). In cholinesterase-deficient patients, as well as in cocaine and ethanol users, the cocaine is metabolized to hepatotoxic Norcocaine. *Rifai N, et al. Tietz Textbook of Clinical Chemistry and Molecular Diagnostics, 6th ed., 2018, pp.866–869.*

11. d. Renal insufficiency is the most common cause of hyperkalemia. In these situations, the serum creatinine and BUN is better test than measuring of K^+ in 24 h urine. A prolonged use of a tourniquet, thrombocytosis, severe leukocytosis, and in vitro hemolysis are the most common causes of hyperkalemia. In pseudohyperkalemia there are no characteristic for hyperkalemia EKG changes. The absence of EKG changes does not rule out true hyperkalemia because such changes are rare in chronic hyperkalemia. Potassium excretion in 24 h urine distinguishes hyperkalemia due to excessive intake from true hyperkalemia caused by the shift of K^+ from the cell to the intracellular space or due to impaired renal excretion. *McPherson RA, et al. Henry's Clinical Diagnosis and Management by Laboratory Methods; 22nd ed., 2011, pp. 177–8.*

12. c. There are three major mechanisms leading to decreased renal excretion of potassium: 1. reduced aldosterone secretion or aldosterone

responsiveness, 2. renal failure, and 3. low concentration of sodium in the distal tubes. These situations may affect potassium secretion only (pseudohypoaldosteronism type II), or both sodium reabsorption and potassium secretion (pseudohypoaldosteronism type I). Heparin therapy inhibits production of steroid hormones and aldosterone in the zona glomerulosa. Treatment with angiotensin-converting enzyme inhibitors, nonsteroidal antiinflammatory agents, and beta blockers suppresses production of renin or angiotensin II, what leads to hyperkalemia. The beta-blockers decrease potassium transport into the cells. Dehydration produces secondary hyperaldosteronism due to reduction of sodium delivery to the cortical collecting ducts. *McPherson RA, et al. Henry's Clinical Diagnosis and Management by Laboratory Methods; 22nd ed., 2011, pp. 176–7.*

13. d. Similarly to calcium, strontium activates CaR and directly inhibits the formation of mature osteoclasts. The high calcium concentration inhibits the bone resorption by osteoclasts through direct action on the CaR. It is still not proven that treatment with calcium or SrR administered at bedtime to the women with postmenopausal osteoporosis reduces bone resorption and actual risk of bone fracture. An acute inhibitory effect on the bone resorption has been showed only in fasting healthy young women receiving either calcium or SrR in the morning. The effect on postmenopausal women waits to be investigated. In young adults, SrR effect on decreasing PTH and the other markers of bone resorption is similar to the effects observed after administration of the calcium. *Maresova KB, et al. A comparison of the acute effects of calcium and strontium ranelate on the serum marker of bone resorption. Clin Chem Lab Med 2012;50(2):333–5.*

14. d. Because the accuracy at the low end of cTn assays improves, the 99th percentile changes and depends on the reference population chosen. Ischemia-modified albumin (IMA) is a variant of albumin formed during interactions between albumin and free radicals at sites of tissue ischemia. Formation of IMA is not specific for cardiac damage. One of the limitations for troponins testing is their heterogeneity that originates from cTn complex which degrades into ternary cTnT-I-C complex, binary cTnI-C, free cTnT, oxidized forms, phosphorylated forms, and degraded forms. Another problem with cTn assays relates to the low level of normal concentration of cTn in the plasma. This creates the situation in which even small interferences (heterophilic antibodies, fibrin, and other substances) lead to the abnormally high and false results. *McPherson RA, et al., ed., Henry's Clinical Diagnosis and Management by Laboratory Methods; 22nd ed., 2011, pp. 247–8.*

15. a. Various tumor markers share tumor-associated epitopes such as C19-9, CA 15-3. CA 125 is expressed in different degree in all carcinomas. The antioncogenic p53 protein is coded by p53 tumor suppressor genes

and it is responsible for suppressing cell growth and stimulate apoptosis. Deletions or mutations of the p53 gene greatly predispose cells to undergoing malignant transformation. An increased serum levels of adhesion molecules such as E-selectin, intercellular adhesion molecule (ICAM), and vascular cell adhesion molecule (VCAM) have been observed in late stage of breast cancer and these markers are used to predict shorter survival. Oncoprotein HER2/Neu is a transmembrane receptor with intracytoplasmic tyrosine kinase domain. The effect of HER2/Neu dimerization is activation of cell division and proliferation. *McPherson RA, et al., ed., Henry's Clinical Diagnosis and Management by Laboratory Methods; 22nd ed., 2011, pp. 1433–4.*

16. c. Oncofetal antigens are normally expressed during fetal development but do not normally occur in the tissues or sera of children and adults. The examples of these substances are α-fetoprotein (AFP) and carcinoembryonic antigen (CEA). During fetal period, AFP is synthesized by normal, nonneoplastic tissues such as yolk sac and hepatocytes. It is a tumor marker for hepatocellular carcinoma and yolk sac germ cell tumors. There is no evidence that occult neoplasms lead to elevation of the acute phase reactants or CRP. *McPherson RA, et al., ed., Henry's Clinical Diagnosis and Management by Laboratory Methods; 22nd ed., 2011, p. 1433.*

17. d. The quantitation of albumin can be used for the nutritional status evaluation, degree of nephropathy or enteropathy, liver synthetic function, as well as for interpretation of calcium and magnesium levels. The folding pattern or three-dimensional structure of the protein, which is determined by its amino acid sequence, is termed its *tertiary structure*. More complex structures of the proteins, such as dimers, trimers, and tetramers, are termed *quaternary structure*. Clinical conditions (myeloma, nephrotic syndrome, cirrhosis, or extensive body burn) with an alteration in the level of particular proteins can be monitored by protein electrophoresis. The fragments rich in β-regions and release from some proteins, such as serum amyloid-associated protein, immunoglobulin light chains, or prealbumins, may precipitate in the tissue and form amyloid deposits. *McPherson RA, et al., ed., Henry's Clinical Diagnosis and Management by Laboratory Methods; 22nd ed., 2011, p. 254.*

18. d. Serum haptoglobin is an acute phase reactant and rises in response to stress, infection, and acute inflammation. The intravascular hemolysis due to transfusion reaction, thermal burns, or autoimmune hemolytic anemia causes haptoglobin concentrations to decrease, as its complex with hemoglobin is removed from the circulation. The analytical sensitivity of standard serum electrophoresis is too low to detect prealbumin β-fibrillar amyloid component which is present in the type I familial amyloidotic polyneuropathy. This condition is caused by precipitated β-structured fragments which are derived from mutated prealbumins

and precipitate in the nerve fibers. Anticoagulation treatment with heparin activates in vivo lipoprotein lipase, which digests β-lipoprotein into smaller fragments. The size of these fragments is similar to the prealbumin and they precipitate in the prealbumin position. This is in vivo effect and it does not occur if heparin is added to samples already collected. Alfa 1 antitrypsin is coded by the *SERPINA1* gene located on chromosome 14. Patients who exhibit the heterozygous phenotype (MZ) may have reduced antitrypsin inhibitory capacity in serum, but they are generally asymptomatic; however, *homozygous ZZ individuals suffer from pulmonary or hepatic disease. McPherson RA, et al., ed., Henry's Clinical Diagnosis and Management by Laboratory Methods; 22nd ed., 2011, pp. 254–62.*

19. a. The level of transferrin is regulated by an increase in mRNA transcription in response to the rising concentration of iron in the circulation and around hepatocytes. In severe iron deficiency, transferrin forms abnormal paraprotein (pseudoparaproteinemia) band. The calculation of percent of saturation from the serum iron and transferrin levels is considered as the best index screen for hemochromatosis and for identifying previously unrecognized cases. A transferrin variant (called asialotransferrin, τ protein, or β_2-transferrin) does not have a sialic acid and it is present in CSF, aqueous and vitreous humor of the eye, and perilymph of the ear. The asialotransferrin is used as a marker for detection of CSF in fluid coming from a fistula or nasal drainage in cases of skull fracture or other head trauma. *McPherson RA, et al., ed., Henry's Clinical Diagnosis and Management by Laboratory Methods; 22nd ed., 2011, pp. 260–1.*

20. b. Metabolic acidosis with increased anion gap (normochloremic acidosis) includes ketoacidosis, D- and L-lactic acidosis, β-hydroxybutyric acidosis, uremic acidosis, and ingestion of toxins (salicylate, methanol, ethylene glycol, toluene, and acetaminophen). Metabolic acidosis with normal anion gap (hyperchloremic acidosis) includes renal tubular acidosis, early uremic acidosis, acidosis following respiratory alkalosis, intestinal loss of bicarbonate, administration of chloride-containing acid (HCl, NH_4Cl), and kctoacidosis during the recovery phase. *McPherson RA, et al., ed., Henry's Clinical Diagnosis and Management by Laboratory Methods; 22nd ed., 2011, p. 185.*

21. c. In other than diarrheal loss of bicarbonate types of metabolic acidosis, urine anion gap is not altered as long as there is no extra renal loss of electrolytes. The urine anion gap, which is calculated as urine $(Na^+ + K^+)$ – urine Cl^-, is reduced or negative in cases of severe diarrhea. The low urine anion gap observed in patients with severe diarrhea is explained by the larger loss of $Na^+ + K^+$ than Cl^-. It is because diarrheal fluid contains more $Na^+ + K^+$ than Cl^-. The presence of organic acidosis is usually obvious by laboratory detection of lactic acid, hyperglycemia, and ketone bodies.

Mcpherson RA, et al., ed., Henry's Clinical Diagnosis and Management by Laboratory Methods; 22nd ed., 2011, pp. 182–3.

22. b. 17-Hydroxylase deficiency causes a decrease production of DHEA and androstenedione, which causes and increased synthesis of progesterone, corticosterone, and DOC. 17-Hydroxylase deficiency leads to decreasing synthesis of cortisol, androgens, and estrogens resulting in pubertal failure. 17-Hydroxylase deficiency is rarely diagnosed at birth. These girls get medical attention later due to hypertension or failure to develop secondary sexual characteristics at puberty. Male patients may have ambiguous or female genitalia, and seek medical attention later in their life due to hypertension or a lack of breast development. http://emedicine.medscape.com/article/919218-clinical. *Rifai N, et al., eds., Tietz Textbook of Clinical Chemistry and Molecular Diagnostics; 6th ed., 2018, pp. 107–10.*

23. b. Idiopathic hirsutism is a condition which cannot be explained by biochemical abnormalities. Hirsutism can be a symptom of the late onset of the congenital adrenal hyperplasia, as well as testosterone-secreting ovarian tumors. Wallach's interpretation of diagnostic tests. *Willamson MA et al., eds., Wallach's interpretation of diagnostic tests. 10th ed. 2015, pp. 263–5.*

24. c. The primary hypogonadism due to testicular damage leads to elevation of both LH and FSH, and low concentration of testosterone. Sperm production is low in both primary and secondary hypogonadism. Male hypogonadism is defined as an impairment of sperm and/or testosterone production. *Williamson MA, et al., eds., Wallach's interpretation of diagnostic tests. Willamson MA et al. 10th ed., 2015, pp. 272–4.*

25. d. Low-grade hyperprolactinemia and a large pituitary tumor are most likely due to loss of dopamine inhibitory action on prolactin secretion but not to prolactin-secreting adenoma. The best screening test for Cushing disease is 24 h cortisol urine excretion. The best diagnostic test for acromegaly due to growth hormone-secreting tumors is measuring serum concentration of the insulin-like growth factor I. *Williamson MA, et al., eds., Wallach's interpretation of diagnostic tests. Willamson MA et al. 10th ed., 2015, pp. 277–9.*

Chapter 10

Blood gases, electrolytes, and iron

John Toffaletti[a], Anthony Okorodudu[b], Neil Harris[c], and Alan H.B. Wu[d]

[a]Duke University, Durham, NC, United States, [b]University of Texas Medical Branch, Galveston, TX, United States, [c]University of Florida, Gainesville, FL, United States, [d]University of California, San Francisco, CA, United States

1. Which of the following will increase the pH?
a. $HCO_3 \downarrow$, acid $CO_2 \downarrow$
b. $HCO_3 \downarrow$, acid $CO_2 \uparrow$
c. $HCO_3 \uparrow$, acid $CO_2 \uparrow$
d. $HCO_3 \uparrow$, acid $CO_2 \downarrow$

2–7. Match the clinical condition (a–f) to the laboratory result (9–14):

Laboratory Result	Condition
2. $HCO_3 = 16\,mmol/L$	a. respiratory acidosis
3. $pCO_2 = 28\,mmHg$	b. respiratory alkalosis
4. $pO_2 = 88\,mmHg$	c. metabolic acidosis
5. $pCO_2 = 55\,mmHg$	d. metabolic alkalosis
6. $pO_2 = 60\,mmHg$	e. apparently normal status
7. $HCO_3 = 34\,mmol/L$	f. hypoxemia

8. In a person who has had an acid-base disorder, which set of bicarbonate and pCO_2 results most likely indicate a compensated acid/base condition?

HCO3 (mmol/L)	PCO$_2$ (mmHg)
a. 35	35
b. 16	40
c. 16	25
d. 24	40
e. 24	50

Self-assessment Q&A in Clinical Laboratory Science, III. https://doi.org/10.1016/B978-0-12-822093-1.00010-7
© 2021 Elsevier Inc. All rights reserved.

9–11. Match the condition (a–c) to the set of results:

HCO$_3$/pCO$_2$ results	Condition
9. 18/50	a. chronic respiratory acidosis
10. 26/50	b. mixed acid-base disorder
11. 32/50	c. acute respiratory acidosis

12–14. For each set of information (19–21), choose the most likely mixed acid-base disorder from the four choices (a–d):

12. low blood potassium, vomiting, and hyperventilation
13. pulmonary arrest with poor perfusion
14. high anion gap, diuretic therapy
    a. metabolic acidosis + respiratory acidosis
    b. metabolic alkalosis + respiratory alkalosis
    c. metabolic acidosis + respiratory alkalosis
    d. metabolic acidosis + metabolic alkalosis

15. A patient is brought to your emergency room in shock by EMS following an automobile accident. You are given the following arterial blood gases. The patient is being given oxygen by nasal cannula.

	Result	Reference intervals
pH	7.29	7.35–7.45
pO$_2$	90	90–105 mmHg (Torr)
pCO$_2$	55	35–45 mmHg
HCO$_3^-$	27.5	21–27 meq/L
% Hgb saturation	95%	95%–99%

    What is the most likely diagnosis?
    a. respiratory acidosis
    b. respiratory alkalosis
    c. metabolic acidosis
    d. metabolic alkalosis

16. Which condition (a–c) would most likely decrease the pO$_2$ in a sample for blood gas analysis?
    a. A pea-sized air bubble in the syringe.
    b. The sample is analyzed 45 min after collection at room temperature.
    c. The sample is stored on ice for 30 min.
    d. Exposure of the sample to the atmosphere.

17. Calcium homeostasis is most directly maintained by which of the following?
    a. thyroxine

b. PTH
c. 1, 25(OH)$_2$D
d. 25(OH)D
e. b & C

18. Which of the following are most consistent with hypoparathyroidism?
   a. hyperphosphatemia
   b. hypophosphatemia
   c. hypocalcemia
   d. hypercalcemia
   e. a & c

19. All of the following cause hypercalcemia except:
   a. hyperparathyroidism
   b. malignancy
   c. 1, 25(OH)$_2$D deficiency
   d. endocrine disorders
   e. drugs

20. Which of the following is the most frequent cause of hypermagnesemia?
   a. increased intake
   b. hypoaldosteronism
   c. acidosis
   d. renal failure
   e. none of the above

21. Which of the following may raise the set point in the parathyroid gland for negative feedback, eliciting hypercalcemia?
   a. vitamin D deficiency
   b. hypomagnesemia
   c. chronic lithium therapy
   d. acute pancreatitis
   e. renal disease

22. Which of the following is incorrect about phosphorus homeostasis?
   a. it is not absorbed by passive transport
   b. it is freely filtered in the glomerulus
   c. PTH induces phosphaturia by inhibition of Na-P cotransport

d. vitamin D increases intestinal absorption and renal reabsorption of phosphorus

e. administration of growth hormone increases phosphate levels in the serum.

23. Hyperphosphatemia can be caused by which of the following disease states?
    a. renal failure, chronic, and acute
    b. decreased glomerular filtration rate
    c. leukemia
    d. rhabdomyolysis
    e. all of the above

24. Which of the following does not enhance urinary excretion of calcium?
    a. hypercalcemia
    b. phosphate deprivation
    c. acidosis
    d. glucocorticoids
    e. parathyroid hormone

25. Most kidney stones are made up of which of the following as the major constituent?
    a. calcium oxalate monohydrate; (Whewellite); $CaC_2O_4 \cdot H_2O$
    b. uric acid; $C_5H_4N_4O_3$
    c. cystine; $(SCH_2CH(NH_2) \cdot COOH)_2$
    d. magnesium ammonium phosphate hexahydrate (struvite); $MgNH_4PO_4 \cdot 2H_2O$
    e. ammonium acid urate; $NH_4H \cdot C_5H_2O_3N_4 \cdot H_2O$

26. Hypercalcemia is caused by all of the following except:
    a. milk-alkali syndrome
    b. decreased solar exposure with resultant decrease in vitamin D synthesis
    c. sarcoidosis
    d. thyrotoxicosis
    e. thiazide diuretics

27. The following differential diagnoses are consistent with high calcium and low PTH.
    a. bone metastasis
    b. multiple myeloma
    c. sarcoidosis

d. PTH-related peptide
e. all of the above

28. Patients with a buildup of serum phosphate (i.e., due to renal failure), high PTH, and chronically low serum calcium are classified as having the follows:
    a. secondary hyperparathyroidism
    b. vitamin D intoxication
    c. tertiary hyperparathyroidism
    d. primary hyperparathyroidism
    e. ectopic PTH-producing tumor
29. Secondary osteoporosis results from all of the following conditions except:
    a. vitamin D overdose
    b. Cushing syndrome
    c. hyperthyroidism
    d. acromegaly
    e. chronic heparin administration

30. Hypophosphatemia is not found in which condition?
    a. acute alcoholism
    b. malabsorption syndrome
    c. osteomalacia
    d. chronic renal failure
    e. steatorrhea
31. A thin, elderly African American woman has various complaints of aches and pains including focal bone pain in her back and a history of a recent kidney stone. She was treated for breast cancer 3 years earlier but has not had any follow-up since completing her treatment 2 years ago. The following labs were obtained:

	Result	Reference interval
Ca^{+2}	13.9 mg/dL	8.4–10.4
PO_4^{-3}	4.4 mg/dL	2.5–4.5
Intact PTH	<2 pg/mL	10–65
Albumin	3.9 g/dL	3.5–5
Creat	1.0 mg/dL	0.4–1.2
BUN	20 mg/dL	5–20

What is the most likely diagnosis?
a. hyperparathyroidism
b. milk-alkali syndrome

c. sarcoidosis

d. hypercalcemia of malignancy

e. renal osteodystrophy

32. A young adult male presents to the ER with a history of new-onset seizures. In fact, a grand mal seizure was witnessed by the ER physician. The glucose at the time of a seizure that occurred in the ER was 109 mg/dL (normal fasting glucose 65–99 mg/dL) and the calcium was 6.0 mg/dL (Reference interval 8.4–10.2 mg/dL). The albumin was normal. Further studies show: $PO_4^{-3} = 7.0$ mg/dL (Reference interval = 3.0–4.5 mg/dL). The alkaline phosphatase is normal. What test should next be ordered to determine the cause of the patient's seizure?

   a. 25 (OH)D

   b. 1,25 (OH)$_2$D

   c. intact PTH

   d. calcitonin

   e. insulin

33. With respect to nonheme iron absorption from the gastrointestinal tract:

   a. Only heme iron can be absorbed from the diet. Nonheme iron is NOT absorbed.

   b. Transferrin is required for iron uptake into the enterocyte.

   c. Iron enters the enterocyte (from the lumen. via passive diffusion.

   d. Iron is absorbed in the ferric (Fe^{3+}) state.

   e. Iron is absorbed in the ferrous (Fe^{2+}) state.

34. With respect to iron absorption from the gastrointestinal tract:

   a. Between 1 and 2 mg is typically absorbed per day in normal individuals.

   b. Between 15 and 25 mg is typically absorbed per day in normal individuals.

   c. The colon is the main site of absorption.

   d. Vitamin C inhibits iron absorption.

   e. GI bleeding uncommonly causes iron deficiency.

35. Which of the following transporters is responsible for transporting iron out of the enterocyte to the circulatory system?

   a. DMT-1: the divalent metal transporter

   b. ferroportin

   c. ferritin

   d. heme

   e. transferrin

36. Which iron profile is seen frequently in iron deficiency?
   a. ↓ Fe, N TIBC, N % Sat, N ferritin
   b. ↓ Fe, N TIBC, ↓ % Sat, N ferritin
   c. ↓ Fe, ↑ TIBC, ↓↓ % Sat, ↓↓ ferritin
   d. ↑ Fe, ↓ TIBC, ↑% Sat, ↑↑ ferritin
   e. N Fe, N TIBC, N % Sat , ↑↑ ferritin

37. Which iron profile is may be seen in thalassemia major or megaloblastic anemias?
   a. ↓ Fe, N TIBC, N % Sat, N ferritin
   b. ↓ Fe, N TIBC, ↓ % Sat, N ferritin
   c. ↓ Fe, ↑ TIBC, ↓↓ % Sat, ↓↓ ferritin
   d. ↑ Fe, ↓ TIBC, ↑% Sat, ↑↑ ferritin
   e. N Fe, N TIBC, N % Sat, ↓↓ ferritin

38. Transferrin
   a. is an α_1 globulin
   b. increases with acute inflammation
   c. binds 4 ferric (Fe^{3+}. ions per molecule)
   d. binds 2 ferric (Fe^{3+}. ions per molecule)
   e. binds 20% of serum iron. The remainder is free iron.

39. With respect to cellular iron uptake
   a. The iron-transferrin complex is taken into the cell by receptor-mediated endocytosis.
   b. The iron-ferritin complex is taken into the cell by receptor-mediated endocytosis.
   c. Cells absorb free iron from solution.
   d. Iron is released from transferrin at high (i.e., alkaline pH).
   e. Transferrin receptor expression is decreased in iron deficiency.

Answers
1. d. Decreasing bicarbonate and increasing acidic CO_2 are associated with a low pH and increasing bicarbonate and decreasing acidic CO_2 are associated with a high pH. The other two combinations tend to cancel each other. *McPherson RA, et al. eds., Henry's Clinical Diagnosis and Management by Laboratory Methods, 22nd ed., 2011, pp. 100–2.*
2. c. A HCO_3 of 16 mmol/L is low, suggesting a metabolic acidosis.
3. b. A pCO_2 of 28 mmHg is low, suggesting a respiratory alkalosis.
4. e. A pO_2 of 88 mmHg is normal status.
5. a. A pCO_2 of 55 mmHg is high, suggesting a respiratory acidosis.
6. f. A pO_2 of 60 mmHg is low, which is hypoxemic.
7. d. A HCO_3 of 34 mmol/L is elevated, suggesting a metabolic alkalosis.

McPherson RA, et al. eds., Henry's Clinical Diagnosis and Management by Laboratory Methods, 22nd ed., 2011, pp. 100–2.

8. c. Metabolic compensation will cause the HCO_3 to follow a rise or fall in PCO_2 (respiratory disorder). Likewise, respiratory compensation will cause the PCO_2 to follow a rise or fall in HCO_3 (metabolic disorder). Example c is the only one with both the HCO_3 and PCO_2 altered in the same direction. *McPherson RA, et al. eds., Henry's Clinical Diagnosis and Management by Laboratory Methods, 22nd ed., 2011, pp. 100–2.*

9. b. The HCO_3 is decreased while the PCO_2 is elevated. Since these are in opposite directions, this appears to be a mixed disorder.

10. c. The HCO_3 is normal, while the PCO_2 is elevated. Since there is no evidence for compensation, this appears to be an early, or acute, respiratory acidosis.

11. a. Both HCO_3 and PCO_2 are elevated, suggesting a chronic (compensated) respiratory *acidosis. While a compensated metabolic alkalosis is also possible, that is not a choice.*

   McPherson RA, et al. eds., Henry's Clinical Diagnosis and Management by Laboratory Methods, 22nd ed., 2011, pp. 100–2.

12. b. A low potassium is associated with a metabolic alkalosis, vomiting is associated with metabolic alkalosis, and hyperventilation is associated with respiratory alkalosis.

13. a. Pulmonary arrest should cause a respiratory acidosis and poor perfusion should lead to metabolic acidosis.

14. d. A high anion gap suggests metabolic acidosis and diuretic therapy could cause *metabolic alkalosis. McPherson RA, et al. eds., Henry's Clinical Diagnosis and Management by Laboratory Methods, 22nd ed., 2011, pp. 100–2.*

15. a. The combination of a low pH and a high pCO_2 establishes this case as a respiratory acidosis. The near normal bicarbonate indicates that this is early and uncompensated.

   McPherson RA, et al. eds., Henry's Clinical Diagnosis and Management by Laboratory Methods, 22nd ed., 2011, pp. 100–2.

16. b. Whole blood samples standing at room temperature will metabolize oxygen and decrease their pO_2. While it is possible that a sample with a very high pO_2 can decrease on exposure to atmosphere, there is no mention of the sample having a very high pO_2. *Liss HP, et al. Stability of blood gases in ice and at room temperature. Chest, 1993;103:1120–22.*

17. e. Free calcium is tightly regulated by PTH and 1, 25(OH)$_2$D. *McPherson RA, et al. eds., Henry's Clinical Diagnosis and Management by Laboratory Methods, 22nd ed., 2011, pp. 194–5.*

18. e. Hypocalcemia and hyperphosphatemia are most consistent with hypoparathyroidism. *McPherson RA, et al. eds., Henry's Clinical Diagnosis and Management by Laboratory Methods, 22nd ed., 2011, pp. 195–6.*

19. c. 1, 25(OH)₂D deficiency will lead to hypocalcemia because of decreased absorption. *McPherson RA, et al. eds., Henry's Clinical Diagnosis and Management by Laboratory Methods, 22nd ed., 2011, pp. 199–200.*

20. d. The most common cause is renal failure. *McPherson RA, et al. eds., Henry's Clinical Diagnosis and Management by Laboratory Methods, 22nd ed., 2011, p. 196–7.*

21. c. Lithium can induce hyperparathyroidism in some patients treated treat bipolar disorder. *Medsace. Hyperparathyroidism and Hypercalcaemia with Lithium Treatment. https://www.medsafe.govt.nz/profs/PUArticles/September2014LithiumTreatment.htm.*

22. a. Phosphorus is mainly absorbed by passive transport. There is also an active energy-dependent process, which is stimulated by vitamin D. *McPherson RA, et al. eds., Henry's Clinical Diagnosis and Management by Laboratory Methods, 22nd ed., 2011, pp. 195–6.*

23. e. All of the others are common causes of hyperphosphatemia. *McPherson RA, et al. eds., Henry's Clinical Diagnosis and Management by Laboratory Methods, 22nd ed., 2011, pp. 195–6.*

24. e. PTH diminishes urinary calcium excretion. *McPherson RA, et al. eds., Henry's Clinical Diagnosis and Management by Laboratory Methods, 22nd ed., 2011, pp. 197–9.*

25. a. Calcium constitutes the majority of kidney stones; Magnesium Ammonium Phosphate Hexahydrate (Struvite) is the 2nd most common constituent of kidney stone. *McPherson RA, et al. eds., Henry's Clinical Diagnosis and Management by Laboratory Methods, 22nd ed., 2011, pp. 477–8.*

26. b. Vitamin D is a major factor in calcium homeostasis. It requires sunlight irradiation of the skin for its production. *McPherson RA, et al. eds., Henry's Clinical Diagnosis and Management by Laboratory Methods, 22nd ed., 2011, p. 199–200.*

27. e. Some tumors can secrete PTH-rp which causes increase in calcium levels.
McPherson RA, et al. eds., Henry's Clinical Diagnosis and Management by Laboratory Methods, 22nd ed., 2011, pp. 198–9.

28. a. Due to renal failure, the phosphate builds up and this phosphate concentration is inversely proportional to the calcium level. The hypocalcemia stimulates PTH secretion and causes the parathyroid gland to synthesize more PTH as a compensation with resultant parathyroid hyperplasia. The tertiary hyperparathyroidism is associated with conditions where the renal failure has been treated (i.e., renal transplant, dialysis, or healing of the primary injury/disease)—the parathyroid glands are occasionally not able to return to normal with resultant hypercalcemia and high PTH. *McPherson RA, et al. eds., Henry's Clinical Diagnosis and Management by Laboratory Methods, 22nd ed., 2011, pp. 197–9.*

29. a. An adequate vitamin D intake (400–800 U/day) is recommended for prevention but high levels themselves do not cause osteoporosis. *McPherson RA, et al. eds., Henry's Clinical Diagnosis and Management by Laboratory Methods, 22nd ed., 2011, pp. 199–200.*

30. d. Hyperphosphatemia and hyperkalemia are expected in chronic renal failure. Oral phosphate supplements are used to treat these patients. *McPherson RA, et al. eds., Henry's Clinical Diagnosis and Management by Laboratory Methods, 22nd ed., 2011, pp. 195–6.*

31. d. An increased total calcium due to malignancy causes a suppression of the parathyroid hormone levels and is the most common cause of hypocalcemia. *McPherson RA, et al. eds., Henry's Clinical Diagnosis and Management by Laboratory Methods, 22nd ed., 2011, pp. 197–9.*

32. c. A low total and ionized calcium together with a high phosphorus, in the absence of release from bone (normal alkaline phosphatase), suggests a primary hyperparathyroidism and testing for PTH. The intact assay is the preferred method today. *Rifai N, et al. eds., Tietz Textbook of Clinical Chemistry and Molecular Diagnostics, 6th ed., 2018, pp. 604–625.*

33. e. Nonheme iron in the gastrointestinal lumen has to be reduced to the ferrous (Fe^{2+}. form prior to absorption. This is accomplished by brush-border Duodenal Cytochrome *b* (Dcytb.. Vitamin C also may assist in this process. Iron absorption is via a protein symporter called DMT1. *Andrews NC, et al. Iron homeostasis. Annu Rev Physiol 2007;69:69–85.*

34. a. Iron absorption occurs in the duodenum. Dietary iron is approximately 10–15 mg per day. Intestinal absorption is ~1.0 mg per day in males and ~1.4 mg in females (to balance iron loss. Ascorbic acid enhances iron uptake. *Andrews NC, et al. Iron homeostasis. Annu Rev Physiol 2007;69:69–85.*

35. b. Ferroportin is a basolateral membrane iron transporter. Also called "iron regulated transporter 1 (Ireg1)" or "metal transporter protein 1" (MTP1). The gene family is termed *SLC40*. Ferroportin exports absorbed iron from the enterocyte to the circulation. On the other hand, DMT1 (divalent metal-ion transporter-1. is on the brush-border of the enterocyte and it is responsible for iron entry from the gastrointestinal lumen. It is also known as NRAMP2. *Andrews NC. Molecular control of iron metabolism. Best Practice Res Clin Haematol 2005;18:159–69.*

36. c. In iron deficiency, the ferritin is low, reflecting a depletion of iron stores. Transferrin (and therefore the TIBC. is typically elevated. Serum iron is low, but this is the least helpful of the measurements, since low serum iron is seen in the anemia of chronic inflammation. *Kaplan LA, et al., eds. Clinical chemistry. Theory, Analysis, Correlation. 5th ed., 2010, pp. 755–70.*

37. d. In other nutritional anemias (such as B_{12} or folic acid deficiency) or in the more severe thalassemias, there is ineffective erythropoesis. The latter is associated with a high rate of red cell production, but the majority of the

cells do not survive to enter the circulation. Iron absorption is increased, leading to iron overload-especially if the patient receives a red cell transfusions. Iron overload produces an elevated ferritin and a normal or suppressed transferrin/TIBC. The % saturation will be high. *Kaplan LA, et al., eds. Clinical chemistry. Theory, Analysis, Correlation. 5th ed., 2010, pp. 755–70.*

38. d. Transferrin, a plasma beta globulin (molecular weight 79.5 kDa. binds up to two ferric (Fe^{3+}) ions). Ferrous iron will not bind to transferrin. Transferrin actually *decreases* with inflammation and is sometimes termed a *negative* acute-phase reactant. There is no free iron in serum. *Rifai N, et al. eds., Tietz Textbook of Clinical Chemistry and Molecular Diagnostics, 6th ed., 2018, pp. 604–25.*

39. a. Cells do not take up free iron. Transferrin (carrying ferric iron) binds to a specific transmembrane receptor (transferrin receptor) and the complex is internalized via receptor-mediated endocytosis. The endosome is then acidified, This causes the release of iron into the endosome, The transferrin, now free of iron, is recycled to the surface together with the transferrin receptor. At the surface the iron-free transferrin (apotransferrin. is released). Iron deficiency increases cell surface expression of the transferrin receptor. *Andrews NC. Molecular control of iron metabolism. Best Practice Res Clin Haematol 2005;18:159–69.*

Chapter 11

Pancreatic markers, GI absorption, vitamins

Kent Lewandrowski[a] and Alan H.B. Wu[b]

[a]Massachusettes General Hospital, Boston, MA, United States, [b]University of California, San Francisco, CA, United States

1. All of the following concerning cystic fibrosis are true except:
   a. An autosomal recessive disorder characterized by a nonfunctional transmembrane conductance regulator protein.
   b. The majority of cases are due to a three base pair deletion at position 508 of the CFTR protein ($\triangle$F508).
   c. Typically shows low levels of chloride in the sweat chloride test.
   d. Cystic fibrosis affects multiple types of exocrine glands.
   e. Cystic fibrosis is the most common lethal genetic disorder in Caucasian populations.

2. All of the following are true concerning macroamylase except:
   a. Macroamylasemia occurs in approximately 1% of patients and in 2.5% of subjects with hyperamylasemia.
   b. In hyperamylasemia due to macroamylase, the urine amylase to creatinine clearance ratio is typically low.
   c. Macroamylasemia can occur in patients with a normal serum amylase.
   d. Macroamylasemia usually results from amylase-immunoglobulin complexes.
   e. Approximately 20% of cases of macroamylasemia are associated with adenocarcinoma of the pancreas.

3–7. Match the following gastrointestinal endocrine tumors with their corresponding associated syndrome or presentation:

3. glucagonoma	a. Zollinger-Ellison syndrome
4. somatostatinoma	b. Verner-Morrison WDHA syndrome
5. gastrinoma	c. Whipples triad
6. VIPoma	d. migratory necrolytic erythema, mild diabetes mellitus
7. insulinoma	e. mild diabetes mellitus, diarrhea, steatorrhea

Self-assessment Q&A in Clinical Laboratory Science, III. https://doi.org/10.1016/B978-0-12-822093-1.00011-9
125

8. All of the following laboratory tests are useful in determining the severity of acute pancreatitis except:
   a. elevated BUN and creatinine
   b. $PO_2 < 60$ mmHG
   c. LDH > 500 U/L
   d. amylase > 5 times the upper reference limit
   e. hematocrit > 44%

9. All of the following are true of gastrin except:
   a. stimulates secretion of gastric acid and pepsinogen
   b. molecular forms include big gastrin, little gastrin, and mini gastrin
   c. produced and stored by G cells in the gastric antrum
   d. gastrin secretion is inhibited by direct action of acid G cells
   e. gastrin is very stable at 4°C for up to a week or more

10. All of the following are true concerning D-xylose and the D-xylose absorption test except:
    a. D-xylose is absorbed mainly in the gastrin antrum
    b. D-xylose is excreted by the kidney into the urine
    c. D-xylose absorption can be evaluated using either a blood or urine sample
    d. low absorption of D-xylose is observed in some cases of malabsorption
    e. D-xylose absorption is typically normal in patients with pancreatic insufficiency

11. All of the following statements concerning steatorrhea are true except:
    a. steatorrhea is defined as more than 5 g of lipids in feces over 24 h
    b. normal subjects excrete essentially no (i.e., <0.5 g) of fecal fat over 24 h
    c. microscopic examination of feces for fat globules is a useful screening test
    d. in patients with steatorrhea, the serum carotenoid level is low
    e. fecal lipid levels may be determined by a titrimetric method by converting lipids to fatty acids followed by titration with sodium hydroxide in neutral alcohol

12. All of the following statements concerning celiac disease are true except:
    a. synonyms include celiac sprue, nontropical sprue, and gluten-induced enteropathy
    b. patients may develop neurological manifestations due to vitamin deficiencies
    c. the definitive diagnostic test is a small intestinal biopsy showing marked hyperplasia of intestinal villi
    d. serologic testing includes antibodies to gliadin (IgA, IgG), endomysium (IgA) reticulin (IgA), and IgA-antitissue transglutaminase
    e. antiendomysial antibody testing performed by immunofluorescence has high sensitivity and specificity but is difficult to perform and is subject to intraobserver variation in interpretation

13. All of the following statements concerning the secretin-cholecystokinin (CCK) test are true except:

  a. The test can assess the volume and bicarbonate secretion of the pancreas and secretion of enzymes.

  b. Cerulein may be used in place of CCK.

  c. Patients with pancreatic malabsorption typically exhibit a low bicarbonate concentration less than 80 mmol/L.

  d. Improved radiological imaging techniques have decreased use of the secretin-CCK test.

  e. all of the above

14. All of the following are true concerning intestinal disaccharidase deficiency except:

  a. The disorder may be congenital or acquired.

  b. Congenital lactase deficiency is common in African Americans and Asians.

  c. Nonhydrolyzed disaccharides are osmotically active resulting in secretion of water and electrolytes into the intestine with diarrhea.

  d. Stools are typically watery and acidic with a positive reducing substances test.

  e. Lactose tolerance testing using an oral lactose challenge is useful to test for lactase deficiency but establishing a definitive diagnosis requires measurement of enzyme activity in an intestinal biopsy.

15. Which of the following statements concerning vitamin D are true (K format)?

  1. Cholecalciferol (vitamin D_3) is produced in the skin whereas ergocalciferol (vitamin D_2) is obtained in the diet.

  2. 25 (OH) is the most active form of the vitamin.

  3. 25 (OH)D is produced in the liver and $1,25(OH)_2D$ in the kidney.

  4. Vitamin D status is best assessed by measurement of $1,25(OH)_2D$.

  5. Measurement of $24,25(OH)_2D$ is useful to assess vitamin D status in patients with renal failure.

  a. 1,2,3

  b. 1,3

  c. 2,4

  d. 4 only

  e. all

16–20. Match the common name of the following vitamins with their corresponding trivial chemical name:

16. vitamin B_1	a. pteroylglutamic acid
17. vitamin B_2	b. pyridoxine
18. vitamin B_6	c. thiamine
19. folic Acid	d. riboflavin
20. vitamin B_{12}	e. cyanocobalamin

21. All of the following concerning vitamin K are true except:
   a. vitamin K includes phylloquinone (K type) synthesized in plants and menaquinones (K_2 type) of bacterial origin.
   b. the most important function of vitamin K is as a cofactor to vitamin K-dependent carboxylase.
   c. vitamin K toxicity typically manifests with stomatitis, rash, and right ventricular heart failure.
   d. vitamin K status is usually assessed by functional methods such as the prothrombin time.
   e. warfarin conversion of the 2,3-exposide of vitamin K to the vitamin K quinines.

22–26. Match the following vitamins with the corresponding deficiency syndrome:

22. vitamin B_6	a. angular stomatitis, dermatitis
23. thiamine	b. pellagra
24. vitamin B_2	c. seizures
25. niacin	d. beriberi
26. pantothenic acid	e. burning feet syndrome

27. All of the following concerning CA 19-9 are true except:
   a. CA19-9 is a carbohydrate antigen glycolipid.
   b. the antigen is only expressed in subjects that are blood group antigen Kell positive.
   c. the antigen is a sialylated derivative of the Le^a blood group antigen.
   d. may be elevated in pancreatitis.
   e. CA19-9 levels correlate with pancreatic cancer staging.

28. Concerning testing for *H. pylori*, which of the following statements are true (K-format)
   1. Recent administration of proton pump inhibitors may lead to false negative urea breath testing.
   2. Urea breath testing may be performed using either ^{14}C or ^{13}C labeled urea.
   3. The stool test detects *H. pylori* antigens using immunoassay techniques.
   4. Serology is useful to confirm eradication following treatment.
   5. Infection by *H. pylori* stimulates both gastric and pancreatic urease production.
   a. 1,2,3
   b. 1,3
   c. 2,4
   d. 4 only
   e. all of the above

29. All of the following concerning assays for amylase are true except

  a. Chromolytic starch methods are largely obsolete.

  b. Amylase assays should only be performed on serum or heparinized samples because other anticoagulants chelate calcium and inhibit amylase activity.

  c. Amylase activity is stable for several days at room temperature and up to 2 weeks at 4°C.

  d. Substrates such as maltotetrose and 4-NP-glycosides should be avoided in total amylase assays due to a 4-fold increase in activity arising from the salivary type isoenzyme.

  e. most assays measure both salivary and pancreatic isoenzymes simultaneously.

Answers

  1. c. The sweat chloride level in cystic fibrosis is elevated not low. *McPherson RA, et al. eds., Henry's Clinical Diagnosis and Management by Laboratory Methods, 22nd ed., 2011, p. 313.*

  2. e. Pancreatic cancer is not specifically associated with macroamylase. *McPherson RA, et al. eds., Henry's Clinical Diagnosis and Management by Laboratory Methods, 22nd ed., 2011, pp. 312–3.*

  3. d.

  4. e.

  5. a.

  6. b.

  7. c. Many of these gastrointestinal tumors present with specific symptoms that constitute a syndrome while others present nonspecifically. *McPherson RA, et al. eds., Henry's Clinical Diagnosis and Management by Laboratory Methods, 22nd ed., 2011, pp. 316–25.*

  8. d. The amylase level is not useful in predicting prognosis in acute pancreatitis. Evidence of renal failure, respiratory failure, a markedly elevated LDH, and hemoconcentration are all associated with a poor prognosis. *McPherson RA, et al. eds., Henry's Clinical Diagnosis and Management by Laboratory Methods, 22nd ed., 2011, pp. 438–9.*

  9. e. Gastrin is unstable due to the action of proteolytic enzymes in serum or plasma. *Warburton R, et al. The in vitro stability of gastrin in serum and whole blood. Ann Clin Biochem1987; 24: 320–31.*

  10. a. D-xylose is absorbed in the intestine. *McPherson RA, et al. eds., Henry's Clinical Diagnosis and Management by Laboratory Methods, 22nd ed., 2011, p. 122.*

  11. b. Normal fecal fat levels are <5 g/day and are typically around 2 gm/day. *Healthline. Fecal fat testing. https://www.healthline.com/health/fecal-fat*

  12. c. Intestinal biopsy shows villous atrophy not hyperplasia. *McPherson RA, et al. eds., Henry's Clinical Diagnosis and Management by Laboratory Methods, 22nd ed., 2011, pp 322–3.*

13. e. Trypsin is the most commonly measured enzyme. *Lieb JG, et al. Pancreatic function testing: Here to stay for the 21st century. World J Gastroenterol. 2008;14(20):3149–58.*

14. b. Congenital lactase deficiency is rare. In contrast, acquired lactase deficiency is common. *Medscape. What causes lactase deficiency? https:// www.medscape.com/answers/187249-159229/what-causes-secondary-lactase-deficiency.*

15. b. $1,25(OH)_2D$ is the most active form and $25(OH)D$ is generally the best test to assess vitamin D status. Measurement of $24,25(OH)_2D$ is not used clinically. *McPherson RA, et al. eds., Henry's Clinical Diagnosis and Management by Laboratory Methods, 22nd ed., 2011, pp. 199–200.*

16. c

17. d

18. b

19. a

20. e. Dual scientific and trivial names add to confusion. *McPherson RA, et al. eds., Henry's Clinical Diagnosis and Management by Laboratory Methods, 22nd ed., 2011, pp. 418–23.*

21. c. *Some* cases of vitamin K toxicity due to K_3 exhibit Heinz bodies and hemolytic anemia. K_1 and K_2 do not have a defined toxicity syndrome. *McPherson RA, et al. eds., Henry's Clinical Diagnosis and Management by Laboratory Methods, 22nd ed., 2011, pp. 831–4.*

22. c

23. d

24. a

25. b

26. e. Vitamins and minerals are necessary for maintenance of health. *McPherson RA, et al. eds., Henry's Clinical Diagnosis and Management by Laboratory Methods, 22nd ed., 2011, pp. 418–23.*

27. b. CA19-9 expression is not related to the Kell blood group. *McPherson RA, et al. eds., Henry's Clinical Diagnosis and Management by Laboratory Methods, 22nd ed., 2011, p. 1393.*

28. a. Serology is not useful to confirm eradication because of a slow decline in antibody titers. *H. pylori* produces urease. The stomach and pancreas do not. *McPherson RA, et al. eds., Henry's Clinical Diagnosis and Management by Laboratory Methods, 22nd ed., 2011, p. 1103.*

29. d. Maltotetrose and 4-NP-glycosides are commonly used as substrates and do not show a strong preference for either isoenzyme. *McPherson RA, et al. eds., Henry's Clinical Diagnosis and Management by Laboratory Methods, 22nd ed., 2011, p. 438.*

Chapter 12

Endocrinology

Ibrahim Hashim[a], Carole A. Spencer[b], James H. Nichols[c], and
Alan H.B. Wu[d]

[a]University of Texas Southwestern, Dallas, TX, United States, [b]University of Southern California,
Los Angeles, CA, United States, [c]Vanderbilt University, Nashville, TN, United States, [d]University of
California, San Francisco, CA, United States

1. Patients with Sheehan syndrome will show biochemical evidence of
   a. hypothyroidism
   b. failure to lactate
   c. hypopituitarism
   d. all of the above
2. The pituitary gland secretes all of the following except
   a. TSH
   b. cortisol
   c. prolactin
   d. β-lipotropin
3. The following hormone acts on several target tissues rather than on a specific organ
   a. GH
   b. ACTH
   c. TSH
   d. FSH
4. Growth hormone action on target tissues is mediated through
   a. GFHRH
   b. IGF-1
   c. ghrelin
   d. somatostatin
5. IGF-I is produced by
   a. adenohypophysis
   b. neurohypophysis
   c. hepatocytes
   d. adipocytes

6. The following is true about IGFs except
   a. IGF I levels are not affected by age
   b. IGF-I is >75% bound to IGF-BP-3
   c. random IGF-I has a better diagnostic test than random GH
   d. there are six IGF-Binding Proteins
7. Ghrelin
   a. inhibits GH secretion
   b. inhibits GHRH secretion
   c. stimulates GH secretion
   d. stimulates somatostatin secretion
8. Growth hormone secretion is not
   a. stimulated by exercise
   b. stimulated by hypoglycemia
   c. stimulated by glucocorticoids
   d. inhibited by somatostatin
9. Assessment of excess GH status in early stages is best by
   a. measuring serum GH levels at random
   b. monitoring serum GH levels during dynamic testing stimulation
   c. monitoring serum GH levels during a dynamic testing following suppression
   d. observing physical changes
10. Follow-up of patients following surgical treatment for acromegaly is best by
    a. observing physical changes
    b. measurement of GH at random
    c. measurement of IGF-1 at random
    d. measurement of somatostatin at random
11. Investigation of GH deficiency is usually by
    a. stimulation of GH production by exercise alone
    b. stimulation of GH production by exercise and pharmacological studies alone
    c. stimulation of GH production by pharmacological studies alone
    d. none of the above
12. IGF binding proteins
    a. bind GF
    b. control GH secretion
    c. IGFBP-3 binds >75% of circulating IGF-1
    d. all of the above
13. In patients being investigated for GH deficiency. A serum GH response following stimulation that is considered normal is
    a. >7–10 ng/mL
    b. 2–5 ng/mL
    c. <5 ng/mL
    d. no change from basal

14. Vasopressin and oxytocin are
  a. secreted by the adenohypophysis
  b. secreted by the neurohypophysis
  c. stored in the adenohypophysis
  d. stored in the neurohypophysis
15. Control of prolactin release differs from that of other pituitary hormones in that
  a. it is under negative feedback inhibition
  b. it is under positive feedback inhibition
  c. it is secreted in pulsatile fashion from pituitary
  d. none of the above
16. Hyperprolactinemia could be due to all except
  a. macroprolactinemia
  b. prolactinoma
  c. use of dopamine agonists
  d. renal failure
17. Macroprolactin interference in immunoassays is due to
  a. aggregates of monomeric prolactin and immunoglobulins
  b. monomeric prolactin alone
  c. dimeric prolactin alone
  d. cleaved prolactin fragments
18. The most common pituitary tumors are
  a. prolactinomas
  b. GH-secreting tumors
  c. ACTH-secreting tumors
  d. TSH-secreting tumors
19. In patients with secondary hypothyroidism
  a. TSH is elevated
  b. TSH is reduced
  c. FT4 is elevated
  d. FT3 is elevated
20. Hyperprolactinemia causes
  a. inhibition of GnRII
  b. increased TSH levels
  c. increased testosterone levels
  d. hypergonadism
21. Prolactin levels greater than 200 µg/L are suggestive of
  a. prolactinoma
  b. pituitary stalk compression
  c. hypothyroidism
  d. stress
22. Hyperprolactinemia due to macroprolactin can be investigated by
  a. PEG precipitation
  b. gel filtration chromatography

    c. repeat analysis using different analyzer

    d. all of the above

23. In patients suspected with gonadal failure as a cause of infertility measure of best

    a. FSH/LH

    b. TSH

    c. testosterone

    d. all of the above

24. The hormone most responsive to changes in blood osmolality is

    a. ADH

    b. cortisol

    c. prolactin

    d. renin

25. The following is NOT a cause of acquired diabetes insipidus

    a. hypocalcaemia

    b. hypokalemia

    c. amyloidosis

    d. lithium

26. Normal response following a water deprivation test in the investigation of diabetes insipidus is

    a. increased urine and plasma osmolality

    b. increased urine osmolality and reduced serum osmolality

    c. increased urine osmolality and no change in plasma osmolality

    d. reduced urine osmolality and reduced serum osmolality

27. During a water deprivation tests, patients with hypothalamic or nephrogenic diabetes insipidus will show

    a. weight increase

    b. hyponatremia

    c. increased plasma osmolality

    d. increased urine osmolality

28. In patients with hypothalamic diabetes insipidus

    a. circulating levels of ADH are elevated

    b. circulating ADH is not functional

    c. circulating levels of ADH are low

    d. serum and urine osmolality are normal

29. Syndrome of antidiuretic hormone secretion (SIADH) is characterized by

    a. hyponatremia

    b. hypovolemia

    c. urine sodium concentration <20 mmol/L

    d. urine osmolality lower than serum osmolality

30. ADH measurement is not required during

    a. water deprivation test

    b. water loading test

c. hypertonic solution infusion

d. investigation of SIADH

31. Oxytocin is

a. stored in the posterior pituitary

b. stored in the anterior pituitary

c. inhibitor of milk production

d. inhibited by estrogens

32. The following is seen in patient suspected of acromegaly

a. elevated serum phosphate concentration

b. decreased serum growth hormone releasing factor concentration

c. increase in serum IGF-1 levels

d. increased serum somatostatin concentration

Answers

1. d. Sheehan's syndrome occurs when there is postpartum pituitary necrosis affecting the pituitary hormones such as TSH and prolactin. *Shivaprasad C. Sheehan's syndrome: Newer advances. Indian J Endocrinol Metab. 2011; S203–7.*

2. b. Cortisol is produced by the adrenal fasciculata and reticularis zones. *McPherson RA, et al., eds. Henry's Clinical Diagnosis and Management by Laboratory Methods, 22nd ed., 2011, pp. 390–9.*

3. a. Growth hormones promote growth in soft tissues, cartilage, and bone by promoting protein synthesis. *McPherson RA, et al., eds. Henry's Clinical Diagnosis and Management by Laboratory Methods, 22nd ed., 2011, pp. 368–70.*

4. b. Insulin-like growth factors, particularly IGF-1, act to incorporate sulfate into cartilage and stimulate DNA and RNA synthesis and collagen formation. *McPherson RA, et al., eds. Henry's Clinical Diagnosis and Management by Laboratory Methods, 22nd ed., 2011, pp. 1404–5.*

5. c. IGF is produced by the liver but small amounts are produced in the bone. *McPherson RA, et al., eds. Henry's Clinical Diagnosis and Management by Laboratory Methods, 22nd ed., 2011, pp. 1404–5.*

6. a. IGF-1 levels are lowest in the neonates and progressively increase as the subject ages. *McPherson RA, et al., eds. Henry's Clinical Diagnosis and Management by Laboratory Methods, 22nd ed., 2011, pp. 1404–5.*

7. c. Ghrelin is a 28 amino acid peptide that is released from neuroendocrine cells and stimulates release of growth hormone and growth hormone releasing hormone. *Pradhan G, et al. Ghrelin: Much more than a hunger hormone. Curr Opin Clin Nutr Metab Care 2013;16:619–24.*

8. c. Growth hormones are released in exercise, stress, low glucose, increased amino acids, thyroxine, and sex hormones. *McPherson RA, et al., eds. Henry's Clinical Diagnosis and Management by Laboratory Methods, 22nd ed., 2011, pp. 368–70.*

9. c. There are a number of physiologic and pharmacologic conditions that can increase growth hormone release, therefore, suppression testing with glucose may be the most effective manner to access excess release. *McPherson RA, et al., eds. Henry's Clinical Diagnosis and Management by Laboratory Methods, 22nd ed., 2011, pp. 368–70.*

10. b. Quarterly or semiannual random growth hormone levels are recommended after surgery for many years to determine recurrence. *McPherson RA, et al., eds. Henry's Clinical Diagnosis and Management by Laboratory Methods, 22nd ed., 2011, pp. 368–70.*

11. b. Growth hormone deficiency cannot be accurately diagnosed by random growth hormone levels. Two provocative tests are recommended, exercise and pharmacologic. *McPherson RA, et al., eds. Henry's Clinical Diagnosis and Management by Laboratory Methods, 22nd ed., 2011, pp. 368–70.*

12. d. There are six major IGF binding proteins that bind growth hormone, with the insulin-like growth factor binding protein being the most important in plasma. *McPherson RA, et al., eds. Henry's Clinical Diagnosis and Management by Laboratory Methods, 22nd ed., 2011, pp. 1404–5.*

13. a. Growth hormone deficiency is unlikely with a GH concentration of >7 ng/mL. The intermediate response is between 3 and 6 ng/mL. *McPherson RA, et al., eds. Henry's Clinical Diagnosis and Management by Laboratory Methods, 22nd ed., 2011, pp. 368–70.*

14. d. Both vasopressin and oxytocin are produced by the magnocellular neurons of the hypothalamus and stored in the neurohypophysis. *McPherson RA, et al., eds. Henry's Clinical Diagnosis and Management by Laboratory Methods, 22nd ed., 2011, p. 370.*

15. c. Among the metabolites that inhibit the secretion of prolactin include dopamine, endothelin-1, calcitonin, and transforming growth factor-β. *McPherson RA, et al., eds. Henry's Clinical Diagnosis and Management by Laboratory Methods, 22nd ed., 2011, pp. 367–8.*

16. c. There are a number of pharmacologic agents that can increase prolactin, including haloperidol, risperidone, antihypertensive agents, beta blockers, calcium channel blockers, and antihistamines, but not dopamine agonists. *McPherson RA, et al., eds. Henry's Clinical Diagnosis and Management by Laboratory Methods, 22nd ed., 2011, pp. 367–8.*

17. a. Macroprolactinemia is an aggregate of prolactin with immunoglobulins and is not biologically active and is not associated with any disease. *McPherson RA, et al., eds. Henry's Clinical Diagnosis and Management by Laboratory Methods, 22nd ed., 2011, pp. 367–8.*

18. a. Prolactinomas can be microadenoma, macroadenoma, or a nonsecretory tumor or pseudoprolactinoma that do not produce excess prolactin. *McPherson RA, et al., eds. Henry's Clinical Diagnosis and Management by Laboratory Methods, 22nd ed., 2011, pp. 367–8.*

19. b. Secondary hypothyroidism is caused by a disease in the pituitary or hypothalamus and results in a low TSH and circulating thyroid hormone levels. *McPherson RA, et al., eds. Henry's Clinical Diagnosis and Management by Laboratory Methods, 22nd ed., 2011, pp. 376–81.*

20. a. Prolactin inhibits the release of gonadotropin-releasing hormone resulting in a functional state of hypogonadotropism. Other pituitary hormones are usually normal. *McPherson RA, et al., eds. Henry's Clinical Diagnosis and Management by Laboratory Methods, 22nd ed., 2011, pp. 376–81.*

21. a. Large prolactin tumors can produce prolactin levels that exceed 1000 μg/L. If a pituitary tumor is present, values below 200 μg/L usually indicate a pseudoprolactinoma. *McPherson RA, et al., eds. Henry's Clinical Diagnosis and Management by Laboratory Methods, 22nd ed., 2011, pp. 367–81.*

22. d. The most common procedure is use of polyethylene glycol precipitation, but all of the answers can be used. *McPherson RA, et al., eds. Henry's Clinical Diagnosis and Management by Laboratory Methods, 22nd ed., 2011, pp. 367–81.*

23. a. In addition to low FSH and LH, female hypogonadism is also associated with low serum estradiol. *Eaglin AR. Male and female hypogonadism. Eaglin AR. Male and female gonadism. Nurs Clin N Am 2018;53:395–405.*

24. a. Vasopressin (antidiuretic hormone) regulates water conservation and is directly linked to serum osmolality, the number of dissolved solutes in blood. *McPherson RA, et al., eds. Henry's Clinical Diagnosis and Management by Laboratory Methods, 22nd ed., 2011, pp. 371–2.*

25. a. Diabetes insipidus is a deficiency of antidiuretic hormone secretion (primary) or renal responsiveness to ADH (secondary). *McPherson RA, et al., eds. Henry's Clinical Diagnosis and Management by Laboratory Methods, 22nd ed., 2011, pp. 371–2.*

26. c. This test involves deprivation of water overnight and measurement of urine and serum osmolality. Normal individuals will lose 3% of body weight resulting in a dilute urine but a normal serum sodium. *McPherson RA, et al., eds. Henry's Clinical Diagnosis and Management by Laboratory Methods, 22nd ed., 2011, pp. 371–2.*

27. c. Patients with nephrogenic diabetes insipidus will have increases in both plasma osmolality and serum sodium with a less concentrated urine than normal. *McPherson RA, et al., eds. Henry's Clinical Diagnosis and Management by Laboratory Methods, 22nd ed., 2011, pp. 371–2.*

28. c. Central or hypothalamic diabetes insipidus is characterized by the inability to produce ADH resulting in dehydration and hypernatremia. *McPherson RA, et al., eds. Henry's Clinical Diagnosis and Management by Laboratory Methods, 22nd ed., 2011, pp. 371–2.*

29. a. The syndrome of antidiuretic hormone secretion is characterized by excessive retention of water resulting in a dilutional hyponatremia.

McPherson RA, et al., eds. Henry's Clinical Diagnosis and Management by Laboratory Methods, 22nd ed., 2011, pp. 371–2.

30. a. ADH measurements are not necessary during a water deprivation test, only serum and urine osmolality and electrolytes. *McPherson RA, et al., eds. Henry's Clinical Diagnosis and Management by Laboratory Methods, 22nd ed., 2011, pp. 371–2.*

31. a. Oxytocin stimulates uterine contraction and can activate milk letdown with nursing. It originates from the posterior pituitary. *McPherson RA, et al., eds. Henry's Clinical Diagnosis and Management by Laboratory Methods, 22nd ed., 2011, p. 370.*

32. d. Acromegaly is caused by excessive growth hormone in adults and an increase in IFG-1. *McPherson RA, et al., eds. Henry's Clinical Diagnosis and Management by Laboratory Methods, 22nd ed., 2011, pp. 368–70.*

Chapter 13

Adrenal, thyroid hormones, and pregnancy

Janine D. Cook[a], Carole A. Spencer[b], Alison Woodworth[c], and Alan H.B. Wu[d]

[a]Substance Abuse Mental Health Services Administration, Rockville, MD, United States, [b]University of Southern California, Los Angeles, CA, United States, [c]University of Kentucky, Lexington, KY, United States, [d]University of California, San Francisco, CA, United States

1. Aldosterone's effect on the kidneys is manifested as
   a. potassium reabsorption.
   b. passive water excretion.
   c. hydrogen ion reabsorption.
   d. sodium reabsorption.
   e. passive chloride excretion.
2. Conn's syndrome, or primary hyperaldosteronism, is clinical recognized by
   a. decreased plasma aldosterone/renin ratio.
   b. increased plasma aldosterone and decreased renin.
   c. increased serum cortisol.
   d. decreased serum cortisol.
   e. decreased plasma aldosterone and increased renin.
3. Cortisol
   a. is water soluble.
   b. is unessential for human life.
   c. is transported in circulation unbound to carrier proteins.
   d. peaks in the late afternoon in those with a normal sleep/wake cycle.
   e. is derived from cholesterol and classified as a steroid.
4. The most diagnostically sensitive screening test for Cushing's syndrome is
   a. 24-h urinary free cortisol
   b. salivary cortisol
   c. serum cortisol
   d. plasma ACTH
   e. dexamethasone suppression test

Self-assessment Q&A in Clinical Laboratory Science, III. https://doi.org/10.1016/B978-0-12-822093-1.00013-2

5. The section of the adrenals that is responsible for the production of catecholamines is the
   a. cortical zona glomerulosa
   b. medulla chromaffin
   c. cortical zona reticularis
   d. cortical zona fasciculata

6. The enzyme deficiency of the adrenal steroid synthesis pathway that is the most common cause of congenital adrenal hyperplasia is
   a. 3-β-hydroxysteroid dehydrogenase
   b. 11-β-hydroxylase
   c. 17-α-hydroxylase
   d. 18-hydroxylase
   e. 21-hydroxylase

7. Pheochromocytoma is characterized by decreased levels of cortisol.
   a. the cause of potentially lethal secondary hypotension.
   b. associated with a slow heartbeat and decreased body temperature.
   c. a tumor associated with the adrenal cortex.
   d. diagnosed through fractionated catecholamines and/or their metabolites.

8. In Cushing's syndrome,
   a. the ectopic form is primarily caused by a cortisol-producing tumor, usually located in the lungs, pancreas, or thymus.
   b. the primary form is related to an ACTH-secreting pituitary tumor.
   c. the primary form results in bilateral adrenal hyperplasia because of excessive ACTH stimulation.
   d. the secondary form is referred to as Cushing's disease.
   e. differentiation between primary and secondary Cushing's is through serum cortisol determination.

9. The classic features of Cushing's syndrome include
   a. amenorrhea and hirsutism in women because of excess aldosterone production.
   b. hypertension and hypokalemia because of excess sex steroid production.
   c. the cushingoid appearance (truncal obesity, moon facies, and buffalo hump) resulting from excess cortisol production.
   d. glucose intolerance due to decreased cortisol concentrations.
   e. hypoglycemia, hyponatremia, and hyperkalemia due to decreased cortisol production.

10. Addison's disease
    a. is associated with the rare, autoimmune-mediated destruction of the adrenal cortex.
    b. has symptoms that are related to the progressive increase in adrenocortical hormones, especially cortisol, aldosterone, and the adrenal sex steroids.
    c. is an acute form of primary adrenal insufficiency.

d. is characterized by an acute presentation of symptoms.

e. is associated with increased serum cortisol and aldosterone and decreased plasma ACTH.

11. Catecholamines

    a. specifically, epinephrine, function solely as CNS neurotransmitters.

    b. are inhibited by stressors such as fear and pain.

    c. are fractionated into epinephrine, norepinephrine, and dopamine.

    d. have epinephrine, norepinephrine, and dopamine as metabolites.

    e. have norepinephrine as the major component.

12. A familial form of Cushing's syndrome is

    a. adrenal carcinoma.

    b. multiple endocrine neoplasia type 1.

    c. ACTH-secreting pituitary tumor.

    d. ectopic ACTH-producing tumor.

    e. chronic exogenous glucocorticoid therapy.

13. The dexamethasone suppression test is:

    a. Low dose protocol uses 0.5 mg every 6 h for 2 days with a urine cortisol of <20 μg/d as a normal result.

    b. Overnight protocol is considered a confirmatory test.

    c. Involves the administration of either a 250 or 1 μg dose of cosyntropin, both of which are supraphysiological.

    d. Involves the administration of metyrapone, an inhibitor of 11 β-hydroxylase, the last enzyme in the cortisol biosynthetic pathway.

    e. Utilizes a cortisol analog that stimulates pituitary ACTH secretion.

14. Pseudo Cushing's syndrome is

    a. characterized by the classical signs and symptoms of full-blown Cushing's syndrome.

    b. differentiated from true Cushing's syndrome by metyrapone dynamic testing.

    c. characterized by cortisol resistance.

    d. characterized by decreased serum cortisol levels.

    e. common in depressed and alcohol-abusing individuals.

15. Hyperandrogenism

    a. has the clinical features of androgen insufficiency.

    b. presents as ambiguous genitalia in the male infant.

    c. presentation depends upon the age of onset and the degree of androgen excess.

    d. is associated with increased levels of estrogens.

    e. causes hirsutism in the adult male and gynecomastia in the adult female.

16. In the evaluation of suspected adrenal insufficiency in patients on long-term glucocorticoid therapy,

    a. suppression testing is recommended to assess pituitary-adrenal function.

    b. removal of exogenous glucocorticoids returns the hypothalamic-pituitary-adrenal axis to normal within hours.

    c. the ACTH stimulation test accurately reflects adrenal reserve.

    d. the dose of glucocorticoids and the duration of therapy predict the degree of secondary adrenal insufficiency.

    e. the basal serum cortisol level is an accurate assessment of pituitary-adrenal function.

17. An acquired form of adrenal insufficiency is

    a. congenital adrenal hyperplasia.

    b. chronic glucocorticoid administration.

    c. mutations of P_{450} oxidoreductase.

    d. tuberculosis or HIV infection.

    e. ACTH unresponsiveness.

18. Autoimmune Addison's disease

    a. is associated with autoimmune polyendocrine syndrome (APS) Type 1 and the AIRE gene mutation.

    b. is always associated with autoantibodies.

    c. related to APS types 1 and 2 exhibits similar genetics, age of onset, prevalence, and associated endocrine abnormalities.

    d. exhibits autoantibodies to enzymes involved in adrenal steroid catabolism.

    e. is associated with autoimmune polyendocrine syndrome types 3 and 4.

19. Adrenal insufficiency associated with AIDS

    a. is a rare finding.

    b. is caused by infectious agents destroying the adrenal gland, pituitary, or hypothalamus.

    c. is best diagnosed with a basal serum cortisol.

    d. is diagnosed by peak serum cortisol levels >20 µg/dL in the ACTH stimulation test.

    e. is classified as secondary if antiadrenal cell antibodies are detected.

20. In the septic patient, adrenal insufficiency

    a. is easily diagnosed in those who are critically ill.

    b. is a rare occurrence.

    c. is associated with a better prognosis.

    d. glucocorticoid replacement therapy is guided by baseline serum cortisol levels.

    e. and its associated mortality risk are best assessed with the dexamethasone suppression test.

21. In patients suspected with Cushing's syndrome, the following is a reliable initial screening test

    a. 24-h urine free cortisol measurement

    b. 24-h urinary 17-ketosteroids

    c. 8 am cortisol concentration

    d. 11 pm cortisol concentration

22. Highly sensitive TSH assays (3rd. generation functional sensitivity ~0.01 mIU/L) are necessary for evaluating thyroid status in which conditions (select all answers)?
   a. central hypothyroidism (hypopituitarism)
   b. hospitalized patients with nonthyroidal illnesses (NTI)
   c. pregnancy
   d. differentiated thyroid cancers (DTC)
   e. Hashimoto's thyroiditis

23. Which combination of tests would be most sensitive for thyroid case finding?
   a. free T4 + free T3 concentrations
   b. free T3 + TSH concentrations
   c. TSH + thyroid peroxidase antibody (TPOAb) concentrations
   d. TSH + free T4 concentrations
   e. thyroglobulin (Tg) + Tg antibody (TgAb) concentrations

24. Given the following results in a clinically euthyroid patient, the most likely diagnosis is:

T4:	14.1 (4.5–12.0 μg/dL)
T3:	243 (80–200 ng/dL)
TSH:	4.9 (0.1–5.0 μIU/mL)
T-uptake:	1.01 (0.74–1.24)

   a. nonthyroidal illness
   b. human chorionic gonadotropin (hcG)-secreting tumor
   c. medullary thyroid cancer
   d. thyroid hormone resistance
   e. increased affinity of thyroxine-binding proteins (TBPs) for T4

25. A male patient is taking exogenous androgens to increase his muscle mass. His thyroid gland is enlarged. His pulse is 88, but he is taking propranolol for hypertension. His hypertension is due to salt retention from taking androgens. What is the explanation for the patient's TSH?

T4:	9.2 (4.5–12.0 μg/dL)
T3:	109 (80–200 ng/dL)
TSH:	<0.1 (0.1–5.0 μIU/mL)
T-uptake:	0.54 (0.74–1.24)

   a. hyperthyroidism
   b. presence of excess TBF
   c. androgen-lowered TSH level
   d. central hypothyroidism
   e. euthyroid state

26. A middle-aged adult male presents without goiter but with cold intolerance, constipation, dry skin, and hypercholesterolemia. He was in a major traffic accident 6 months ago, after which he was unconscious for 96 h. Which of the following choices best fits the data and the clinical history?

    FT4:   0.55 (0.7–1.48 ng/dL)
    TSH: 0.250 (0.35–5000 mIU/mL)
    T3:   45 (80–200 ng/dL)

    a. Hashimoto thyroiditis
    b. central hypothyroidism
    c. inborn error in thyroid hormone synthesis
    d. congenital thyroid gland hypoplasia
    e. atrophic thyroiditis

27. All of the following clinical or laboratory findings can contribute to the diagnosis of preeclampsia except:
    a. Two incidents of Systolic blood pressure of >140 mmHg and/or Diastolic pressure of >90 mmHg at least 4 h apart.
    b. Increased circulating antiangiogenic factors such as sFlt-1.
    c. Proteinuria of >100 mg/24 h on one occasion.
    d. Proteinuria of >300 mg/L in a spot urine sample or $>1+$ on a urine dipstick on at least two occasions.
    e. Elevated serum AST and ALT.

28. A 23-year-old female presents to the ED with an intrauterine pregnancy at 27 weeks 3 days gestation. The patient complains of abdominal pain, contractions, and mild vaginal spotting for the past 24 h. The patient has a history of chlamydia 3 months ago and 3 previous preterm deliveries. Cervicovaginal fetal fibronectin (fFN) testing was negative. How should this patient be treated?
    a. Give patient antenatal corticosteroids to induce fetal lung maturity.
    b. Admit the patient and treat with a tocolytic agent, like terbutaline.
    c. Discharge patient with recommendation of pelvic rest and follow-up in one to two weeks.
    d. Induce delivery immediately.
    e. fFN testing is inaccurate in the presence of a chlamydia infection.

29. All of the following are true for fFN measurements except:
    a. A positive test equates to an fFN concentration >50 ng/mL in cervicovaginal fluid.
    b. High-risk asymptomatic patients should have testing done between 22^0 and 30^6 weeks' gestation.
    c. In symptomatic patients, fFN should be measured between 24^0 and 34^6 weeks' gestation.
    d. Results are reliable when sample is collected from a patient whose amniotic membrane is ruptured.

e. Results are not reliable when sample is collected from a patient experiencing gross or moderate vaginal bleeding.

30. The following are indicative of normal maternal adaptation during pregnancy.

1. decreased hemoglobin and hematocrit
2. elevated BUN and creatinine
3. increased hormone binding proteins
4. decreased free and total cortisol
   a. 1, 2, and 3
   b. 1 and 3
   c. 2 and 4
   d. 4 only
   e. all of the above

31. All of the following describe hyaline membrane disease except:

a. also known as respiratory distress syndrome (RDS)
b. reduced incidence with injection of corticosteroids at least 48 h prior to delivery
c. greater prevalence in infants born at 29 weeks' gestation than 34 weeks
d. caused by deficiency of pulmonary surfactant
e. lower surface tension at the alveolar air-liquid interface

32. All of the following are true of thyroid function and thyroid function tests during pregnancy except:

a. Decreased maternal thyroid hormone in the first trimester impairs neurologic development of the fetus.
b. All pregnant women should be screened for thyroid dysfunction in the first trimester.
c. TSH decreases in response to increasing hCG in the first trimester.
d. Thyroid binding globulin increases, causing an approximate 1.5 fold increase in Total T3 and Total T4.
e. The presence of thyroperoxidase (TPO) antibodies in pregnancy indicates a higher risk for postpartum thyroiditis.

33. When establishing the Medians used to calculate Multiples of the Median (MoM) for prenatal screening markers, which are valid statements?

a. Medians for each specific marker are the same regardless of assay manufacturer.
b. Medians must be established for each relevant gestational week for each specific population being evaluated.
c. Medians can be shared among different laboratories performing prenatal screening.
d. There is no need to reestablish medians once they are determined.
e. An MoM is calculated by dividing the individual test result by a nationally published median for that gestational week.

34. A patient who is 17 weeks pregnant has a maternal serum (MS) α-fetoprotein (AFP) value of 2.5 MoM (multiples of the median). Which of the following is inconsistent with these results?

a. The patient has a folic acid deficiency.

b. The patient is pregnant with healthy twins.

c. Repeat testing is recommended prior to confirmatory testing.

d. The fetus is at high risk for a closed neural tube defect.

e. The patient is 19 weeks pregnant and at low risk for a neural tube defect.

35. Which is (are) true of a positive Amniotic Fluid (AF)-AFP result for diagnosis of an open neural tube defect (ONTD)?

1. The MoM is generally >2.0 or 2.5 (depending on the laboratory).

2. The presence of fetal blood can cause a false positive result.

3. A measurement of amniotic fluid acetylcholinesterase activity is the recommended confirmatory test.

4. AF and MS have similar concentrations of AFP throughout pregnancy.

a. 1, 2, and 3

b. 1 and 3

c. 2 and 4

d. 4 only

e. all of the above

36. Inhibin

a. inhibits FSH secretion by adenohypophysis

b. produced by Sertoli cells

c. inhibits LH secretion by adenohypophysis

d. 1,2 only

37. Which is true regarding the biochemistry of inhibin molecules?

a. members of the IgG superfamily of proteins

b. covalent, homodimeric proteins

c. inhibin A and B are both increased in maternal serum (MS) in pregnancy

d. dimeric inhibin A (DIA) shows the weakest association with gestational age of all of the MS Down syndrome screening markers

e. MoM values for MS-DIA are not affected by maternal weight

38. Which is true regarding the biochemistry of unconjugated estriol (uE₃)?

a. synthesized by the fetoplacental unit with participation from the fetal liver

b. decrease in maternal serum (MS) with increasing gestational age

c. MS-uE₃ MoM values are significantly affected by maternal weight.

d. MS-uE₃ MoM values are not affected by diabetes status.

e. MS-uE₃ is increased in trisomy 18.

39. A 20-year-old woman who is 18 weeks pregnant had the following results for her second trimester serum screening:

MS-AFP	0.75 MoM
uE3	0.75 MoM
Total hCG	3.0 MoM
Inhibin-A	2.2 MoM

a. The patient's fetus is at high risk for an ONTD.
b. The fetus is at high risk for trisomy 21.
c. The fetus' risk for trisomy 21 is 1 in 200 based on maternal age alone.
d. The fetus is at high risk for trisomy 18.
e. The fetus is at low risk for both trisomy 21 and an ONTD.

40. What is the next step in the diagnostic process for the patient in #13?
a. repeat the quadruple screen to see if the results change
b. repeat maternal serum AFP only
c. no follow-up testing necessary, the fetus is at low risk for an anomaly
d. verify the patient's gestational age by ultrasound
e. measure the AF-AFP concentration

41. Which diagnostic tests are part of the first trimester Down syndrome screening panel?
1. maternal serum pregnancy associated plasma protein-A (PAPP-A)
2. maternal serum free hCGβ
3. nuchal translucency by ultrasound
4. maternal serum AFP
   a. 1, 2, and 3
   b. 1 and 3
   c. 2 and 4
   d. 4 only
   e. all of the above

Answers

1. d. Aldosterone, a poorly protein-bound steroid hormone, is an important regulator of electrolyte balance and extracellular fluid balance. In the kidneys, aldosterone regulates blood volume and blood pressure through active sodium reabsorption, potassium and hydrogen excretion, and passive water reabsorption. *Cook J. Adrenal Disorders. In: Clarke W, Dufour DR, eds. Contemporary Practice in Clinical Chemistry. American Association for Clinical Chemistry, 2006:375–86.*

2. b. Laboratory results associated with hyperaldosteronism include hypokalemia (if not on low sodium diet), hyperkaluria (if not on low sodium diet), metabolic alkalosis, a high normal serum sodium (very rarely hypernatremia is present), elevated serum and urine aldosterone, nonsuppressable plasma and urine aldosterone, and suppressed plasma renin activity. *Cook*

J. Adrenal Disorders. In: Clarke W, Dufour DR, eds. Contemporary Practice in Clinical Chemistry. American Association for Clinical Chemistry, 2006:375–86.

3. e. Cortisol is a glucocorticoid produced by the adrenal cortex, and along with aldosterone, is essential for human life. Cortisol is steroid hormone synthesized from cholesterol.

Because of cortisol's lipid solubility, it is found in circulation bound to specific carrier proteins, including cortisol-binding globulin (CBG) (90%–97%), albumin, and sex hormone binding globulin (SHBG). The small portion of cortisol that is free in circulation is biologically active.

Cortisol also exhibits a diurnal variation pattern; cortisol levels are highest about 8 am and lowest in late evening in those individuals on a normal sleep–wake cycle. *Cook J. Adrenal Disorders. In: Clarke W, Dufour DR, eds. Contemporary Practice in Clinical Chemistry. 2006:375–86.*

4. a. The most diagnostically sensitive screening test for Cushing's syndrome is the 24 h urinary free cortisol level, which represents the free or unbound fraction and normalizes cortisol's diurnal variation. In the absence of stress, a urinary free cortisol >250 μg/d (70 μmol/d) is indicative of Cushing's syndrome. False negative results may occur due to intermittent cortisol production, a small adrenal adenoma, or renal impairment. Because cortisol is secreted in a pulsatile fashion, a single serum cortisol determination lacks the requisite diagnostic sensitivity and specificity. Because of the ease of sample collection, salivary cortisol has also been considered as a screening test for Cushing's syndrome. Unfortunately, salivary cortisol levels also respond to stress. ACTH and dexamethasone suppression testing are used to differentiate between the different causes of Cushing's syndrome. *Cook J. Adrenal Disorders. In: Clarke W, Dufour DR, eds. Contemporary Practice in Clinical Chemistry. 2006:375–86.*

5. b. The adrenal glands, two small, triangular-shaped glands, are composed of a yellow outer cortex and a pearly gray inner medulla. The cortex is comprised of three distinct zones of cells—the outermost zona glomerulosa cells which are primarily responsible for the synthesis and secretion of the major mineralocorticoid hormone aldosterone; the central cortical zona fasciculata which produces glucocorticoids, such as cortisol; and the inner zona reticularis that surrounds the medulla and is the site of production of adrenal sex steroid hormones. The medulla is composed of chromaffin cells, or pheochromocytes, that secrete catecholamines. *Cook J. Adrenal Disorders. In: Clarke W, Dufour DR, eds. Contemporary Practice in Clinical Chemistry. American Association for Clinical Chemistry, 2006:375–86.*

6. e. The adrenogenital syndromes result from the autosomal recessive inheritance of two copies of a gene coding for either the absence or deficiency of one of the synthetic enzymes involved in the synthesis of the adrenal steroid hormones. In all adrenogenital syndromes, plasma cortisol production is decreased and ACTH is increased. The most common adrenal enzyme deficiency involves the 21-hydroxylase enzyme, accounting for 90% of cases of congenital adrenal hyperplasia and has an incidence of 1/ 15,000 live births. 21-Hydroxylase is involved in the synthetic pathways for both glucocorticoids and mineralocorticoids. Thus, glucocorticoids (cortisol) and mineralocorticoids (aldosterone) are present in low concentrations, while androgenic steroids accumulate. *Cook J. Adrenal Disorders. In: Clarke W, Dufour DR, eds. Contemporary Practice in Clinical Chemistry. 2006:375–86.*

7. e. Tumors of mature adrenal chromaffin cells are known as pheochromocytomas, which are a rare (800/y in US) cause of potentially lethal secondary hypertension, occurring in <1% of all hypertensives. Pheochromocytomas produce excess catecholamines. Fractionated urine and plasma catecholamines and their metabolites are the most common laboratory diagnostic analyses available for pheochromocytoma. Generally, in a patient with a pheochromocytoma, plasma and urinary catecholamines and urinary VMA and metanephrines are increased to usually greater than twice the upper reference interval limit. The classic triad of symptoms includes palpitations, severe headaches, and sweating. Cardiovascular effects include tachycardia, postural hypotension, diaphoresis, palpitations, and chest pain while other symptoms include headache, nausea, weakness, nervousness, tremors, pallor, flushing, diarrhea, and hyperglycemia. *Cook J. Adrenal Disorders. In: Clarke W, Dufour DR, eds. Contemporary Practice in Clinical Chemistry, 2006:375–86.*

8. d. Primary Cushing's syndrome is excessive glucocorticoid secretion from an adrenal adenoma or carcinoma. Secondary Cushing's syndrome, also known as Cushing's disease, results from an ACTH-secreting pituitary tumor that stimulates the adrenal cortex to release cortisol. In Cushing's disease, the normal negative feedback mechanism fails and bilateral adrenal hyperplasia ensues. Also, an ectopic, nonpituitary, ACTH-secreting tumor, usually originating in the lungs, pancreas, or thymus, can cause Cushing's syndrome. ACTH, CRH stimulation testing, and dexamethasone suppression testing are primarily used to differentiate between the causes of Cushing's syndrome. *Cook J. Adrenal Disorders. In: Clarke W, Dufour DR, eds. Contemporary Practice in Clinical Chemistry, 2006:375–86.*

9. c. Excessive circulating adrenal hormones cause the clinical presentation of Cushing's syndrome. The characteristic clinical features are grouped by the effects caused by the classes of adrenal cortical hormones released. The glucocorticoid effects include the classical Cushingoid appearance, which is characterized by truncal obesity, moon facies, buffalo hump, purple striae, excessive bruising, and thin skin. The excessive glucocorticoids also cause poor wound healing, osteopenia, glucose intolerance, psychiatric symptoms (euphoria, mania, depression), and muscle weakness. The effects related to mineralocorticoid overproduction include hypertension, hypokalemia, metabolic alkalosis, and edema. In females, the sex steroids effects cause amenorrhea and hirsutism, while in males impotence and gynecomastia are evident. *Cook J. Adrenal Disorders. In: Clarke W, Dufour DR, eds. Contemporary Practice in Clinical Chemistry, 2006:375–86.*

10. a. Addison's disease, or chronic primary adrenal insufficiency, is a rare autoimmune-mediated destruction of the adrenal cortex, primarily through tuberculosis, metastatic carcinoma, or amyloidosis. Addison's is characterized by a slow presentation of symptoms, with the disease severity related to the amount of functioning gland remaining. Its clinical features are related to the progressive loss of adrenocortical hormones, especially cortisol, aldosterone, and the adrenal sex hormones. Laboratory findings associated with Addison's disease include a decreased random serum cortisol, decreased serum and urine aldosterone, increased plasma ACTH, decreased serum sodium, and increased serum potassium and urine sodium. *Cook J. Adrenal Disorders. In: Clarke W, Dufour DR, eds. Contemporary Practice in Clinical Chemistry. 2006:375–86.*

11. c. Catecholamines are fractionated into epinephrine, norepinephrine, and dopamine, whose release is stimulated by stressors such as fear and pain—"fight or flight." Epinephrine is the major adrenal catecholamine (80%–90%) which functions as a hormone and thus affects metabolism (mobilizes energy stores) and increases heart rate and blood pressure. Epinephrine also functions also as a neurotransmitter. Norepinephrine (10%–20%) and dopamine function solely as CNS neurotransmitters. The catecholamine metabolites are metanephrine, normetanephrine, homovanillic acid (HVA), and vanillylmandelic acid (VMA). *Cook J. Adrenal Disorders. In: Clarke W, Dufour DR, eds. Contemporary Practice in Clinical Chemistry, 2006:375–86.*

12. b. Cushing's syndrome is characterized by excessive glucocorticoids. Causes can be either endogenous or exogenous (chronic long-term glucocorticoid therapy) in origin. Primary endogenous causes are adrenal adenoma or carcinoma (15%). Secondary causes (Cushing's disease) are due to an ACTH-secreting pituitary tumor (70%), which results in bilateral adrenal hyperplasia because of the failure of the normal negative feedback mechanism. Ectopic causes include a nonpituitary ACTH-producing tumor (15%) located primarily in the lung, pancreas, or thymus. The familial forms of Cushing's syndrome are due to either Multiple Endocrine Neoplasia (MEN) Type 1 or Primary Pigmented Micronodular Adrenal disease. *Cook J. Adrenal Disorders. In: Clarke W, Dufour DR, eds. Contemporary Practice in Clinical Chemistry, 2006:375–86.*

13. a. Dynamic testing is performed to elucidate the cause of recognized Cushing's syndrome. The dexamethasone suppression tests are used to differentiate the source of the excess ACTH. These tests involve administration of dexamethasone, a potent cortisol analog, to suppress pituitary ACTH secretion according to one of several protocols.

Overnight	1 mg at 11 pm	Screen	Serum cortisol <5 µg/dL (150 nmol/L)
Low Dose	0.5 mg q 6 h × 2 d	Confirmation of Cushing's syndrome	Urine cortisol <20 µg/d (5.6 µmol/d)
High Dose	2 mg q 6 h × 2 d or 8 mg at 11 pm	Differential diagnosis- Cushing's disease: ↓ cortisol	Urine cortisol <20% of baseline or serum cortisol <50% of baseline

In the patient with an ACTH-producing pituitary adenoma, ACTH and thus cortisol are partially suppressible by dexamethasone, but this is not so with adrenal neoplasms or ectopic ACTH production. *Cook J. Adrenal Disorders. In: Clarke W, Dufour DR, eds. Contemporary Practice in Clinical Chemistry, 2006:375–86.*

14. e. Patients with pseudo Cushing's syndrome have increased serum cortisol but fail to develop the progressive effects associated with true Cushing's syndrome. This disorder is common in depressed and alcohol-abusing individuals. Cortisol resistance is a rare genetic syndrome in which the patient has increased cortisol levels but do not have Cushing's syndrome. The symptoms associated with cortisol resistance include hypertension and androgen excess. The use of the dexamethasone/CRH combo test will aid in the diagnosis of pseudo Cushing's syndrome, with an increased cortisol level in response to the challenge implying true Cushing's syndrome. *Cook J. Adrenal Disorders. In: Clarke W, Dufour DR, eds. Contemporary Practice in Clinical Chemistry, 2006:375–86.*

15. c. The clinical features of hyperandrogenism, or androgen excess, depend upon age of onset and degree of androgen excess. The female infant presents with ambiguous genitalia while the male child exhibits premature puberty. In the adult male, hyperandrogenism is associated with gynecomastia, infertility, and impotence while the adult female displays hirsutism, oligo/amenorrhea, and acne in the mild form and altered body habitus, male pattern balding, cliteromegaly, and deepened voice in the more severe case. *Cook J. Adrenal Disorders. In: Clarke W, Dufour DR, eds. Contemporary Practice in Clinical Chemistry, 2006:375–86.*

16. c. Long-term pharmacological doses of glucocorticoids can cause adrenal insufficiency with illness or trauma due to the chronic suppression of CRH and ACTH by these exogenous glucocorticoids. Only 2 weeks' exposure to pharmacological doses of glucocorticoids can cause CRH-ACTH-adrenal axis suppression. Recovery of the hypothalamic-pituitary-adrenal axis can take weeks to months depending upon the duration of glucocorticoid

exposure. Because of the induced secondary adrenal insufficiency, pituitary-adrenal function cannot be reliably assessed from the glucocorticoid dose, the duration of therapy, or the basal plasma cortisol concentration. Stimulation testing is required to assess pituitary-adrenal function. The major diagnostic test is the ACTH stimulation test because it accurately reflects adrenal reserve. *Schlaghecke R, et al. The effect of long-term glucocorticoid therapy on pituitary-adrenal responses to exogenous corticotropin-releasing hormone. N Engl J Med 1992;326:226–30.*

17. d. Adrenal insufficiency can be classified as either central or acquired. Central types of adrenal insufficiency include iatrogenic (long-term use of glucocorticoids), congenital and acquired hypopituitarism, or ACTH unresponsiveness. The acquired types include autoimmune destruction of the adrenal cortex that can occur in isolation or part of a polyglandular autoimmune disorder, infections (e.g., TB, HIV), neoplastic destruction, metabolic disorders (e.g., various forms of adrenal leukodystrophy, Wolman disease, Smith-Lemli-Opitz syndrome), and hemochromatosis. The congenital types of adrenal insufficiency include the three forms of adrenal hypoplasia congenital, congenital adrenal hyperplasia (CAH) due to the deficiency of an enzyme required for the adrenal synthesis of cortisol, lipoid adrenal hyperplasia due to mutations in the steroid acute regulatory protein or cholesterol side chain cleavage gene, and mutations or deletions of P_{450} oxidoreductase. *Bhansali A, et al. A preliminary report on basal and stimulated plasma cortisol in patients with acquired immunodeficiency syndrome. Indian J Med Res 2000;112:173–7.*

18. a. Adrenal autoantibodies occur in about two-thirds of those with autoimmune Addison disease. Autoantigens include members of the cytochrome P_{450} enzyme family involved in adrenal steroidogenesis. Addison's disease is also associated with Type 1 and 2 autoimmune polyglandular diseases. These two forms differ in genetics, age at onset, prevalence of adrenal cortex/21-hydroxylase autoantibodies, and associated autoimmune diseases. Autoimmune polyendocrine disorder 1 is an autosomal regressive disease that is directly associated with mutations in the autoimmune regulator gene AIRE. *Bøe AS, et al. Mutational analysis of the autoimmune regulator (AIRE) gene in sporadic autoimmune Addison's disease can reveal patients with unidentified autoimmune polyendocrine syndrome type I. Eur J Endocrinol. 2002;146:519–22; Betterle C, et al. Autoimmune adrenal insufficiency and autoimmune polyendocrine syndromes: autoantibodies, autoantigens, and their applicability in diagnosis and disease prediction. Endocrinol Rev 2002;23:327–64.*

19. b. Adrenal insufficiency has a prevalence of 5%–8% in HIV-infected patients and $\sim$20% in those with advanced AIDS. Primary causes include adrenal gland infection with cytomegalovirus (CMV), mycobacterium avian complex, or tuberculosis; adrenal metastases or hemorrhage; and antiadrenal cell antibodies that interfere with corticosteroid action in vitro and are thought to contribute to cortisol deficiency in vivo. Secondary causes include hypothalamic or pituitary involvement by infectious agents or infiltration by lymphoma or Kaposi's sarcoma. Because the basal serum cortisol is frequently elevated in AIDS patients, this measurement does not reliably exclude the diagnosis. The best diagnostic approach is the cosyntropin stimulation test with a normal result manifested as a peak serum cortisol concentration >20 mg/dL. *http://www.hopkins-aids.edu/educational/caserounds/ caserounds_19.html, assessed 1/2007. http://www.endocrinology.med. ucla.edu/adrenal_axis.htm, assessed 1/2007.*

20. d. Acute adrenal insufficiency associated with sepsis is common, but often undiagnosed, finding in critically ill patients (prevalence of 30%–60%). The mortality risk associated with acute adrenal insufficiency in septic patients is assessed through the administration of 250 mg corticotropin. An increased risk of death with a 28-day mortality rate up to 82% is associated with a baseline serum cortisol is <34 μg/dL and cortisol response to corticotropin change is ≤9 μg/dL. A decreased risk of death is found when the baseline cortisol is >34 μg/dL and cortisol response to corticotropin >9 μg/dL. If the baseline serum cortisol is >34 μg/dL, then adequate endogenous levels of cortisol exist and no replacement therapy is required. If the baseline serum cortisol is 15–34 μg/dL, then 250 mg corticotropin is administered. If the serum cortisol response is ≤9 μg/dL, replacement therapy is initiated. If random serum cortisol is <15 μg/dL in the severely septic patient, then adrenal insufficiency is diagnosed and glucocorticoid replacement therapy is given. *Pene F, et al. Prognostic value of relative adrenal insufficiency after out-of-hospital cardiac arrest. Int Care Med 2005;31:627–33.*

21. a Cushing's syndrome is caused by excessive cortisol production. Screening tests include a 24-h urinary free cortisol level (results exceeding 120 μg/d indicate Cushing's syndrome) and by the overnight low-dose dexamethasone suppression test (1 mg at midnight) followed by the documentation of a high morning cortisol level. *McPherson RA, et al., eds. Henry's Clinical Diagnosis and Management by Laboratory Methods, 22nd ed., 2011, pp. 393–4.*

22. b., c., and d. TSH is typically paradoxically normal or even mildly elevated in cases of pituitary failure because current TSH assays measure bioinactive TSH isoforms that tend to be released in these conditions. NTI can cause abnormal TSH in hospitalized euthyroid patients. Sensitive TSH assays are needed for discriminating true hyperthyroidism (TSH, <0.01 mIU/L) from the mild transient TSH suppression (0.01–0.1 mIU/ L) caused by NTI. Low TSH is common in pregnancy, especially during the first trimester when hCG exerts direct thyroidal stimulation. A sensitive TSH assay is needed to distinguish true hyperthyroidism (TSH, <0.01 mIU/L) from mild, transient, hCG-mediated TSH suppression

(0.01–0.1 mIU/L). After thyroidectomy, DTC patients often receive high doses of L-T4 designed to suppress the trophic influence of TSH. Sensitive TSH assays are needed to target the degree of TSH suppression relative to recurrence risk vs iatrogenic subclinical hyperthyroidism. Hashimoto's thyroiditis is characterized by high TSH, usually associated with detectable TPOAb. Highly sensitive TSH assays are not critical for the diagnosis of this condition. *Hovens GC, et al. Associations of serum thyrotropin concentrations with recurrence and death in differentiated thyroid cancer. J Clin Endocrinol Metab 2007;92:2610–15. Spencer CA. Applications of a new chemiluminometric thyrotropin assay to subnormal measurement. J Clin Endocrinol Metab 1990;70:453–60. Hollowell JG, et al. Serum TSH, T₄, and thyroid antibodies in the United States population (1988 to 1994): National Health and Nutrition Examination Survey (NHANES III) J Clin Endocrinol Metab 2000;85:3631–5.*

23. c. There is a log/linear relationship between TSH and free T4 together with the genetically determined free T4 setpoint dictates that the free T4 population reference range is an insensitive parameter to gauge the free T4 status of an individual. Because only ~20% of the serum T3 is normally derived directly from the thyroid gland, most being produced from T4 in peripheral tissues, serum T3 concentration is not a useful test of thyroid status. Most case finding detects Hashimoto's thyroiditis as subclinical hypothyroidism (high TSH/normal free T4). Free T4 will not become abnormal until the hypothyroidism becomes overt and T3 is a very insensitive test for thyroid deficiency because T4 to T3 conversion increases as hypothyroidism develops. As described in (a), free T3 has no role in thyroid case finding because most of the T3 is not derived from the thyroid. The combination of TSH and TPOAb is most sensitive for case finding especially for detecting mild thyroid deficiency in early pregnancy that can be detrimental to both mother and fetus. The log/linear TSH/free T4 relationship dictates that free T4 will remain within reference limits unless TSH becomes markedly abnormal (>10 mIU/L or <0.05 mIU/L). Tg can be abnormal with many different thyroid pathologies and is a nonspecific marker of thyroid dysfunction that has no role in thyroid case finding. The primary use of Tg and TgAb measurements is for the follow-up monitoring of patients with differentiated thyroid cancers (DTC). *Spencer CA, et al. Applications of a new chemiluminometric thyrotropin assay to subnormal measurement. J Clin Endocrinol Metab 1990;70:453–60. Spencer CA, et al. National Health and Nutrition Examination Survey III thyroid-stimulating hormone (TSH)-thyroperoxidase antibody relationships demonstrate that TSH upper reference limits may be skewed by occult thyroid dysfunction J Clin Endocrinol Metab 2007;92:4236–40. Negro R, et al. Levothyroxine treatment in euthyroid pregnant women with autoimmune thyroid disease: effects on obstetrical complications. J Clin Endocrinol Metab 2006;91:2587–91.*

24. d. This patient has increased T4 and T3, and results would suggest hyperthyroidism had the TSH been decreased. However, given that the patient is clinical euthyroid, it indicates that there is some hormone resistance resulting in slightly higher circulating thyroid levels needed to maintain the euthyroid state. *Persani L, et al. Evidence for the secretion of thyrotropin with enhanced bioactivity in syndromes of thyroid hormone resistance. J Clin Endocrinol Metab 1994;78:1034–9.*

25. a. The elevated T4 and T3 with a very low TSH establishes the diagnosis of hyperthyroidism. *McPherson RA, et al., eds. Henry's Clinical Diagnosis and Management by Laboratory Methods, 22nd ed., 2011, pp. 374–81.*

26. b. Low FT4, low T3, and low TSH suggest either a primary (thyroidal), secondary (pituitary), or tertiary (hypothalamic) hypothyroidism. An additional provocative test may be necessary to differentiate between the two. *McPherson RA, et al., eds. Henry's Clinical Diagnosis and Management by Laboratory Methods, 22nd ed., 2011, pp. 374–81.*

27. c. Proteinuria defined as excretion of 300 mg or more protein every 24 h or 300 mg/L in at least two random urine samples taken at least 4–6 h apart or −1 + on dipstick in at least 2 random urine samples taken at least 4–6 h apart. *Wagner LK. Diagnosis and Management of Preeclampsia. Am Fam Phys 2004;70(12):2317–24.*

28. c. The strength of fetal fibronectin is its high negative predictive value, >99% for predicting birth within 7 days of sampling. Therefore a symptomatic patient with a negative fFN result has <1% chance of delivering within the next week and can be sent home. *Gronowski AM. Handbook of Clinical Laboratory Testing During Pregnancy, 2004, pp. 37–41.*

29. d. fFN testing is FDA approved for measurement in symptomatic and asymptomatic women, but patients must conform to certain criteria including: specified gestational ages, intact fetal membranes, cervical dilation less than 3 cm, no placenta previa, no cervical cerclage, no moderate to gross vaginal bleeding, and no sexual intercourse within 24 h of sampling. *Gronowski AM. Handbook of Clinical Laboratory Testing During Pregnancy, 2004, pp. 37–41.*

30. b. Maternal blood volume increases ~45% during pregnancy lowering hemoglobin and hematocrit values along with erythrocyte counts. The dramatic increase in estrogen induces an increase in hepatic production of several hormone binding proteins (TBG, CBG, and SHBG). An increase in glomerular filtration rate to $\sim 170 \, mL/min/1.73 \, m^2$ leads to decreased BUN and Creatinine until just prior to delivery. Increased cortisol binding globulin combined with decreased hepatic clearance results in an increase in total and free cortisol during pregnancy. *Burtis CA, et al., eds. Tietz Textbook of Clinical Chemistry and Molecular Diagnostics, 4th ed., 2006, pp. 2156–8.*

31. e. Fetal lung maturation requires ample expression of pulmonary surfactant. Surfactant maintains alveolar stability by preventing collapse and reduces the inspirational pressure for lung inflation. Infants with RDS have a higher surface tension at the alveolar air-liquid interface because of surfactant deficiency. *Burtis CA, et al., eds. Tietz Textbook of Clinical Chemistry and Molecular Diagnostics, 4th ed., 2006, pp. 2166, 2188–9.*

32. b. The endocrine society's 2007 Recommendations for management of thyroid dysfunction during pregnancy and postpartum suggest screening only women at high risk for thyroid disease. *Abalovich M, et al. Management of thyroid dysfunction during pregnancy and postpartum: an Endocrine Society Clinical Practice Guideline. J Clin Endocrinol Metab 2007;92:S1–47.*

33. b. Calculating an MoM first involves establishing medians for each gestational week being measured. Medians are calculated from maternal serum measurements of markers using the laboratory's specific method(s) collected from at least 100 patients that are representative of the population that will be screened. Medians must be reestablished at least annually to adjust for change in patient population. MoMs are calculated by dividing the individual test result by the laboratory-specific median. *Burtis CA, et al., eds. Tietz Textbook of Clinical Chemistry and Molecular Diagnostics, 4th ed., 2006, pp. 2168–75.*

34. d. AFP is elevated in the serum of a woman pregnant with a fetus affected by a neural tube defect. Only open neural tube defects (ONTDs) allow access of fetal serum proteins to amniotic fluid, and ~90% of ONTDs can be detected by measuring maternal serum AFP. Maternal serum AFP is elevated in twin pregnancies and increases dramatically with gestational age. Women who have a moderately elevated MS-AFP result should be retested. If the second result is negative, the test is screen negative. *Burtis CA, et al., eds. Tietz Textbook of Clinical Chemistry and Molecular Diagnostics, 4th ed., 2006, pp. 2165–9.*

35. a. Amniotic fluid AFP is measured using the same assay as maternal serum AFP with results 50- to 200-fold higher in amniotic fluid. The presence of fetal blood in amniotic fluid samples can increase AFP results and laboratories should note the presence of blood on the report. The laboratory should test for the presence of fetal blood in screen positive samples (MoM above 2.0 or 2.5 depending upon the laboratory), by measuring fetal hemoglobin concentrations. Acetylcholinesterase testing of amniotic fluid is recommended for confirmation of an ONTD. *Burtis CA, et al., eds. Tietz Textbook of Clinical Chemistry and Molecular Diagnostics, 4th ed., 2006, pp. 2169, 2182–3.*

36. d. The inhibins are members of the transforming growth factor-β and are negative feedback regulators of follicle stimulating hormone. The activins are also members of this super family of proteins. *Burtis CA, et al., eds. Tietz Textbook of Clinical Chemistry and Molecular Diagnostics, 4th ed., 2006, p. 2186.*

37. d. Inhibins are members of the TGFβ superfamily of proteins. They form covalent heterodimeric molecules consisting of an α and β subunit. Dimeric inhibin A (DIA) peaks at 8–10 weeks' gestation and declines to a minimum at 17 weeks and then rises slowly until term. There is very little increase in DIA between 17 and 20 weeks as compared to other Down syndrome screening markers. *Burtis CA, et al., eds. Tietz Textbook of Clinical Chemistry and Molecular Diagnostics, 4th ed., 2006, pp. 2171–2, 2186.*

38. a. The fetal adrenal gland secretes DHEAS which is converted by the fetal liver to 16-OH-DHEAS and subsequently to unconjugated Estriol (uE$_3$) in the placenta. MS-uE$_3$ concentrations increase dramatically in the second trimester. MS-uE$_3$ concentrations are minimally affected by maternal weight, but are ∼7%–8% lower in patients with diabetes mellitus. MS-uE$_3$ MoMs are suppressed in trisomy 18. *Gronowski AM. Handbook of Clinical Laboratory Testing During Pregnancy, 2004, pp. 96–9, 119.*

39. b. MoMs are institution dependent; however, it is possible to interpret the patients' results based upon several previous studies showing that in Down syndrome, maternal serum AFP and unconjugated estriol (uE$_3$) are ∼25% lower, while concentrations of DIA and HCG are on average twice as high as in an unaffected pregnancy. MoM values for each marker are ∼1 in unaffected pregnancies. Based on these values the patient has screened positive for Down syndrome. *Burtis CA, et al., eds. Tietz Textbook of Clinical Chemistry and Molecular Diagnostics, 4th ed., 2006, p. 2167.*

40. d. Because most of these markers show an association with gestational age it is critical to verify the patient's week of gestation by ultrasound. Repeat testing is not recommended for Down syndrome screening. *Burtis CA, et al., eds. Tietz Textbook of Clinical Chemistry and Molecular Diagnostics, 4th ed., 2006, p. 2170.*

41. a. Studies have shown that a first trimester screening panel consisting of a combination of nuchal translucency measurements by ultrasound, maternal serum PAPP-A, and maternal serum free hCGβ concentrations detect ∼85% of Down syndrome cases with a 5% false positive rate. *Burtis CA, et al., eds. Tietz Textbook of Clinical Chemistry and Molecular Diagnostics, 4th ed., 2006, p. 2176.*

Therapeutic drug monitoring and toxicology

Chapter 14

Clinical toxicology

Sarah Delaney and Paul J. Jannetto
Department of Laboratory Medicine and Pathology, Mayo Clinic, Rochester, MN, United States

1. A patient with attention-deficit hyperactivity disorder (ADHD) is prescribed methylphenidate (Ritalin). As part of routine compliance monitoring,
   the physician orders a urine amphetamine/methamphetamine immunoassay
   (cutoff 500 ng/mL) test which comes back negative. What is the best explanation for the negative urine immunoassay result?
   a. The patient is noncompliant with the methylphenidate.
   b. The patient missed taking the last daily dosage of methylphenidate.
   c. The patient is a fast metabolizer of methylphenidate which results in a
      negative immunoassay test.
   d. Methylphenidate and its primary metabolite does not cross-react with the
      amphetamine/methamphetamine immunoassay.
   e. The cutoff for the amphetamine/methamphetamine immunoassay was
      too high.
2. A post liver transplant patient prescribed tacrolimus decided to begin a new
   supplementation (high-dose biotin; 300 mcg/day) routine. During the past
   6 months, the patient's tacrolimus concentrations monitored using a competitive immunoassay at the local hospital ranged from 6.0 to 7.0 ng/mL.
   However, the latest concentration obtained showed a concentration of
   11.6 ng/mL. The physician requested that the sample be analyzed at an outside reference laboratory using LC-MS/MS and they obtained a value of
   6.8 ng/mL. The most likely explanation for the discordant results is:
   a. The competitive immunoassay reacts to tacrolimus metabolites leading
      to falsely elevated concentrations.
   b. The patients sample was hemolyzed causing false elevations on the
      tacrolimus immunoassay result.
   c. High-dose biotin has been shown to cause up to 71% false elevation in
      tacrolimus competitive immunoassays.
   d. The patient took twice his/her usual dosage of tacrolimus.
   e. The sample collected was a peak instead of a trough draw for tacrolimus.

Self-assessment Q&A in Clinical Laboratory Science, III. https://doi.org/10.1016/B978-0-12-822093-1.00014-4
© 2021 Elsevier Inc. All rights reserved.

3. A comatose patient is brought to the Emergency Room by his friends who stated he was using fentanyl (not prescribed). The patient responded after several doses of naloxone (Narcan). Urine toxicology testing was performed and immunoassay screen and reflex confirmatory results are shown as follows:

Immunoassay Screens	Result	Cutoff (ng/mL)	Reference range
Amphetamine/ Methamphetamine	Negative	500	Negative
Benzodiazepines	Negative	100	Negative
Opiates	Negative	300	Negative
Methadone	Negative	300	Negative
Oxycodone	Negative	300	Negative
Fentanyl	**Presumptive Positive**	2	Negative
THC	Negative	50	Negative
PCP	Negative	25	Negative

LC-MS/MS confirmation	Result	Limit of quantification (ng/mL)	Reference range
Fentanyl	Negative	0.2	Negative
Norfentanyl	Negative	1.0	Negative

What is the most likely explanation for the discordant fentanyl immunoassay and confirmation results?

a. The patient really did use fentanyl, but the drug is not stable in urine.

b. While the patient might have thought he was abusing fentanyl, it was actually a fentanyl analog.

c. The LC-MS/MS fentanyl result was a false-negative result.

d. The patient did not use an opioid at all, but instead used methamphetamine.

e. Naloxone commonly causes false positives on fentanyl immunoassays.

4. Which of the following stat serum tests is not recommended to support an Emergency Department from a clinical toxicology perspective:

a. acetaminophen

b. salicylate

c. ethyl alcohol

d. ethylene glycol

e. cocaine

5. Toxic syndromes or toxidromes are clinical syndromes commonly used to recognize and suggest a specific class of poisoning. A distinguishing feature to differentiate anticholinergic findings from sympathomimetic findings is:

a. Anticholinergic agents produce warm, flushed, dry skin, but sympathomimetic agents produce diaphoresis.

b. Anticholinergic agents produce agitation, but sympathomimetic agents produce lethargy.

c. Anticholinergic agents produce hallucinations, but sympathomimetic agents do not.

d. Anticholinergic agents produce bradycardia, but sympathomimetic agents produce tachycardia.

e. none of the above

6. Respiratory depression, slurred speech, diplopia, nystagmus, hypotension, and ataxia are characteristic of what toxidrome?

a. anticholinergic

b. cholinergic

c. opioid

d. sedative-hypnotic

e. sympathomimetic

7. Specimen validity testing is required for regulated workplace drug testing. Which of the following is the correct interpretation of one of the tests required by Health and Human Services (HHS):

a. pH $\geq$4.0 and <4.5 means the urine sample is adulterated

b. Creatinine $\geq$2.0 mg/dL and specific gravity $\leq$1.0010 means the urine sample is dilute

c. Oxidant present/positive and nitrite positive ($\geq$500 mcg/mL) means the urine sample is adulterated

d. a. and c.

e. none of the above

8. A chronic pain management patient prescribed oxycodone came in for routine urine drug testing for compliance. While the opioid confirmation test identified oxycodone and its metabolite (oxymorphone), the patient also tested positive for the carboxy metabolite of tetrahydrocannabinol (THC-COOH) using a LC-MS/MS confirmation test. When questioned about the "positive/confirmed" THC, the patient denied using any marijuana, but did say he was frequently taking hempseed oil he ordered online. What is the most likely interpretation in this scenario?

a. The THC confirmation assay (LC-MS/MS) cross-reacts with the cannabidiol present in hempseed oil causing a false-positive result.

b. THC content of hempseed oil is 5 mcg/mL or lower in the United States, so it is impossible to cause a positive THC confirmation test.

c. Studies have previously shown that hempseed oil from different countries may have THC concentrations as high as 1500 mcg/mL. As a result, the hempseed oil could result in a positive THC-COOH confirmation test.

d. The patient is definitely using marijuana since hempseed oil and other over-the-counter products like cannabidiol do not contain any THC.

e. none of the above

9. A previously healthy 5-year-old male was brought to the Emergency Department because of nausea, vomiting, and increased difficulty breathing. An empty container of roach poison was found near the child. The child had tearing bilaterally, rhinorrhea, gurgling sonorous respirations when supine, and was incontinent (urine/stool). Based on his clinical presentation, the most likely explanation for the toxicity and treatment would be:
   a. organophosphate poisoning; treatment is atropine/pralidoxime
   b. mercury poisoning; treatment is fresh frozen plasma
   c. arsenic poisoning; treatment is ascorbic acid
   d. thallium poisoning; treatment is EDTA chelation therapy
   e. iron poisoning, treatment is deferoxamine

10. A 78-year-old man was found unresponsive in his trailer after the first cold day of winter. There is a kerosene heater still on from the previous night. On attempts to arouse him, officers note the cherry hue of his lips, cheek, and mucous membranes. The most likely mechanism of his death was:
    a. accidental ingestion of kerosene
    b. hepatic necrosis with fatty changes due to excessive alcohol consumption
    c. inhibition of hemoglobin synthesis causing a shift in the hemoglobin-oxygen dissociation curve to the right
    d. decrease oxygen content of blood and oxygen availability to tissues
    e. inhibition of iron incorporation into hemoglobin

11. A 6-year-old male is brought to the Emergency Room by his mother. The child has been nauseous and vomiting for 2 days and complains his stomach hurts. His physical exam showed the following: Temperature = 98.6°F, Blood pressure = 110/70 mmHg, heart rate = 103 bpm. He was jaundiced and diaphoretic. Laboratory testing showed the following:

Analyte	Patient result	Reference range
Sodium	128 mmol/L	135–145 mmol/L
Potassium	4.2 mmol/L	3.4–4.7 mmol/L
Chloride	103 mmol/L	98–109 mmol/L
Glucose	90 mg/dL	75–99 mg/dL
Alanine aminotransferase (ALT)	1345 U/L	7–55 U/L
Aspartate aminotransferase (AST)	1250 U/L	8–60 U/L
Prothrombin time (PT-INR)	2.9	0.9–1.1
Urine drug of abuse panel	Negative	Negative
Salicylate	Negative	Negative
Acetaminophen	35 mg/L	10–30 mg/L

The most likely diagnosis/treatment for the child is:
a. salicylate overdose/activated charcoal
b. acetaminophen overdose/N-acetyl-L-cysteine (NAC)
c. alcohol overdose/Dialysis

d. digoxin overdose/Digibind

e. warfarin overdose/Vitamin K

12. A patient presents to the Emergency Department with significant metabolic acidosis with a large anion gap not accounted for by lactate and without ketosis. The patient also has a positive osmolal gap and his urine contains envelope-shaped crystals. The most likely ingestant was:

   a. ethyl alcohol

   b. ethylene glycol

   c. acetone

   d. salicylate

   e. propylene glycol

13. Methotrexate is used to treat certain types of cancer of the breast, skin, head, neck, or lung. In patients with delayed clearance due to impaired renal function, glucarpidase (Voraxaze) can be administered for the treatment of toxic plasma methotrexate concentrations ($>1\,\mu M/L$). However, what special precautions need to be taken to monitor methotrexate concentrations after the administration of the glucarpidase?

   a. Use a regular serum collection tube, not a serum separator

   b. Monitor methotrexate concentrations hourly using an immunoassay

   c. Use a chromatographic method to measure methotrexate concentrations

   d. Serum samples should be protected from light and kept at room temperature

   e. none of the above

14. A teenager is brought to the Emergency Department with hyperthermia, hypertension, tachycardia, hyponatremia, nausea, vomiting, and chest pain after attending an all-night dance party (rave). The patient is having auditory and tactile hallucinations. Routine urine toxicology testing is done:

Immunoassay	Result	Cutoff (ng/mL)	Reference range
Amphetamine/ Methamphetamine	Negative	500	Negative
Benzodiazepines	Negative	100	Negative
Cocaine Metabolite	Negative	150	Negative
Opiates	Negative	300	Negative
THC	Negative	50	Negative
PCP	Negative	25	Negative

The most likely agent ingested was:

   a. methamphetamine

   b. phencyclidine

   c. cocaine

   d. cathinones

   e. morphine

15. A recently divorced 28-year-old woman was admitted with altered mental status, acute abdominal pain, and weakness in the lower extremities. In the hospital, she developed alopecia, numbness of the hands and feet with asymmetric reflexes with the upper extremities being preserved over the lower extremities, tachycardia, and hypertension.

Initial chemistry test results are shown as follows:

Analyte	Result	Reference range
Sodium	139 mmol/L	135–145 mmol/L
Potassium	4.0 mmol/L	3.6–5.2 mmol/L
Glucose	130 mg/dL	70–140 mg/dL
Creatinine	0.4 mg/dL	0.6–1.1 mg/dL
Blood urea nitrogen	15 mg/dL	6–21 mg/dL
Chloride	102 mmol/L	98–107 mmol/L
Bicarbonate	25 mmol/L	22–29 mmol/L
Alanine aminotransferase	421 U/L	7–45 U/L
Aspartate aminotransferase	199 U/L	8–43 U/L

Based on the peripheral neuropathy, high transaminases, and alopecia which heavy metal testing should be considered?
a. thallium
b. zinc
c. iron
d. manganese
e. copper

16. Which of the following is true regarding the monitoring of the tricyclic antidepressant imipramine:
a. monitor the active metabolite only (desipramine)
b. monitor the inactive metabolite (trimipramine)
c. monitor parent drug (imipramine) only
d. monitor both imipramine and desipramine
e. none of the above

17. A baby boy is born that displays neonatal abstinence syndrome. The mother, a known heroin addict, admits to using heroin throughout her pregnancy. Meconium testing is performed and reveals the following test results:

ELISA test	Result	Reference range
Amphetamine/Methamphetamine	Negative	Negative
Opiates	Negative	Negative
Benzoylecgonine	Negative	Negative
Tetrahydrocannabinol carboxylic acid	Negative	Negative

Based on the clinical scenario and test results, the most likely explanation is:
a. The baby was not exposed to heroin or other illicit drugs in utero.
b. The mother is lying about using heroin.
c. A negative meconium does not rule out intrauterine drug exposure.

  d. Heroin and its metabolites do not get detected in any of the testing ordered.

  e. none of the above

18. An 85-year-old male presented to the Emergency Department with complaints of nausea, vomiting, abdominal pain, bradycardia, and dysrhythmias. He was prescribed digoxin (0.4 mg/day) for atrial fibrillation. Initial laboratory testing showed an elevated potassium (6.5 mmol/L; reference range 3.6–5.2 mmol/L) and a critical digoxin concentration (5 ng/mL; reference interval 0.5–2.0 ng/mL). As a result, the patient was administered Digibind and the patient's potassium and digoxin concentrations were monitored hourly.

Time	Potassium (mmol/L)	Digoxin (ng/mL)
1500 (Admission)	6.5 H	5.0 H
1600	6.0 H	55.0 H
1700	4.5	65.0 H
1800	3.6	45.0 H

  What is the most likely explanation for the digoxin results?

  a. It is an acute ingestion and the patient is still absorbing the digoxin causing the concentrations to increase.

  b. The wrong collection tube was used to collect the patient's digoxin concentration.

  c. An immunoassay was being used to monitor digoxin concentrations.

  d. An LC-MS/MS assay was being used to monitor digoxin concentrations.

  e. none of the above

19. A patient in a pain management program has a random urine sample collected as part of routine controlled substance monitoring. While the provider finds the prescribed opioid (hydrocodone), they also confirmed the presence of amphetamine by LC-MS/MS. A list of the patients medications is shown as follows:

  Hydrocodone (20 mg; bid), Ibuprofen (200 mg; prn), and Lisdexamfetamine (50 mg/day)

  What is the most likely cause of the positive amphetamine confirmatory test?

  a. hydrocodone

  b. ibuprofen

  c. lisdexamfetamine

  d. patient is using illicit amphetamine

  e. none of the above

20. A 2-year-old boy had eaten some rodenticide (warfarin). The best test to monitor the effect of the rodenticide would be:

  a. prothrombin time

  b. serum vitamin K1 concentrations

    c. plasma warfarin concentrations

    d. activated partial thromboplastin time

    e. thrombin Time

21. Which of the following drugs may cause a false positive on an amphetamine immunoassay drug screen, but will give a negative confirmatory result?

    a. methylphenidate (Ritalin)

    b. bupropion (Wellbutrin, Zyban)

    c. MDMA (Ecstasy)

    d. lisdexamfetamine (Vyvanse)

    e. all of the above

22. A 22-year-old female presented to the Emergency Department 4h after ingesting of half a bottle of extra-strength Tylenol with the intent of self-harm. Upon arrival her serum toxicology testing revealed an acetaminophen concentration of 178mcg/mL (therapeutic: 10–30mcg/mL; toxic: >150mcg/mL). The patient received N-acetylcysteine (NAC) an initial dose of 150mg/kg over 15min, 50mg/kg over 4h, followed by 100mg/kg over 16h.

    If the patient were to have her blood drawn within this timeframe, which of the following tests would yield a falsely decreased results?

    a. prothrombin Time

    b. triglycerides

    c. lactate dehydrogenase

    d. enzymatic creatinine

    e. b. and d.

    f. a. and b.

23. Which of the following drugs do not typically require therapeutic drug monitoring?

    a. alprazolam

    b. carbamazepine

    c. voriconazole

    d. vancomycin

    e. valproic acid

24. Which of the following is true about methanol?

    a. Causes an increase in blood pH

    b. The CNS effects of methanol are substantially less severe than those of ethanol.

    c. Will lead to the formation of the toxic metabolite formic acid

    d. Lowers osmolar gap with the formation of ethanol

    e. all of the above

    f. b. and c.

    g. a. and d.

25. Which of the following signs and symptoms is not consistent with morphine overdose?
  a. lowered blood pressure
  b. mydriasis (dilated pupils)
  c. hypothermia
  d. respiratory depression
  e. bradycardia

26. A 42-year-old male with a history of opioid use disorder presents for his biweekly Suboxone (buprenorphine/naloxone) compliance testing. His prescribed daily dose is dose 24/6 mg buprenorphine/naloxone tablets. Following are his results:

LC-MS/MS confirmation, urine	Result	Limit of quantification (ng/mL)	Reference range
Buprenorphine	440	5	Positive
Naloxone	135	25	Positive

Which of the following is the best answer?
  a. Based on the results provided, the patient is definitively compliant.
  b. Based on the results provided, the patient is only partially compliant; he should have higher urine concentrations based on his high dose.
  c. Based on the results provided, the patient is definitively noncompliant.
  d. There is not enough information in the results provided to tell if the patient is compliant or not.
  e. The patient is probably a rapid metabolizer and should switch medications.

27. Which of the following cannabinoids is not a psychotropic compound?
  a. cannabinol
  b. cannabidiol
  c. Δ^8- tetrahydrocannabinol
  d. Δ^9- tetrahydrocannabinol
  e. 11-nor-9-carboxy-Δ^9- tetrahydrocannabinol

28. A patient with systemic lupus erythematosus prescribed 450 mg hydroxychloroquine per day presents for his monthly serum monitoring. Last month, his hydroxychloroquine serum concentration was 307 ng/mL (therapeutic target: 1000 ng/mL). His most recent serum result was 238 ng/mL. What is the most likely explanation for this patient's subtherapeutic hydroxychloroquine serum concentration?
  a. The specimen was collected at trough instead of peak
  b. Tube underfilled which underestimated the result
  c. The specimen was past acceptable stability
  d. The patient was noncompliant
  e. The patient's liver enzymes were induced causing increased metabolism of hydroxychloroquine

29. Which of the following is true regarding the action of buprenorphine on opioid receptors?
    a. buprenorphine is a mu partial agonist and a kappa antagonist
    b. buprenorphine is a mu full agonist and a kappa antagonist
    c. buprenorphine is a mu full antagonist and a kappa full agonist
    d. buprenorphine is a mu partial antagonist and a kappa full agonist
    e. buprenorphine is a mu partial agonist and a kappa partial agonist

30. Which of the following is the therapeutic target for mycophenolic acid (MPA)?
    a. MPA activates inosine monophosphate dehydrogenase
    b. MPA inhibits inosine monophosphate dehydrogenase
    c. MPA activates mammalian target of rapamycin (mTOR)
    d. MPA inhibits mammalian target of rapamycin (mTOR)
    e. MPA activates calcineurin

31. A 34-year-old male with end-stage renal disease presents for evaluation of kidney transplant eligibility. The patient was a self-reported "light smoker" of tobacco products and e-cigarettes for the last 10 years but endorsed to have been abstinent for over two weeks and is currently using a nicotine patch to help him quit.
    What is the most likely interpretation from the following results?

LC-MS/MS confirmation, urine	Result (ng/mL)	Reference range (nontobacco user with no passive exposure) (ng/mL)
Nicotine	1201	<5.0
Cotinine	2486	<2.0
Anabasine	18	<2.0
Nornicotine	34	<2.0

    a. The patient is only using a high-dose nicotine patch therapy
    b. The patient is using a nicotine patch and actively smoking e-cigarettes
    c. The patient is using a nicotine patch and actively using a tobacco product
    d. The patient is only smoking e-cigarettes
    e. The patient has abstained from using all nicotine products

32. An increased percentage of free phenytoin is most likely in which of the following clinical conditions:
    a. uremia
    b. hyperalbuminemia
    c. hypoglycemia
    d. hypernatremia
    e. all of the above

33. Which of the following is not true regarding codeine?
    a. Poor CYP2D6 metabolizers will receive inadequate analgesia from codeine

b. It is considered a prodrug
c. Codeine has a higher affinity for the mu opioid receptor than morphine
d. Long-term, high-dose administration will lead to codeine metabolism to hydrocodone
e. Approximately 10% of codeine is metabolized to norcodeine

34. A patient with spinal stenosis is prescribed 40 mg oxycodone every 4–6 h as needed. As part of the patient's monthly compliance testing, his pain management physician orders an opioid urine confirmatory test. What is the most likely interpretation from the following results?

LC-MS/MS confirmation	Result (ng/mL)	Cutoff concentration (ng/mL)
Codeine	Negative	<25
Dihydrocodeine	Negative	<25
Hydrocodone	26	<25
Norhydrocodone	Negative	<25
Hydromorphone	Negative	<25
Oxycodone	2900	<25
Noroxycodone	2330	<25
Oxymorphone	830	<25
Noroxymorphone	1250	<25
Morphine	Negative	<25

a. The patient is abusing hydrocodone
b. The patient is compliant with his oxycodone prescription but is also abusing hydrocodone
c. The patient is compliant with his oxycodone
d. The patient is using hydrocodone and oxymorphone
e. The patient took a recent dose of oxymorphone

Answers

1. d. Methylphenidate is not detected by amphetamine/methamphetamine-based immunoassays due to its poor cross-reactivity with the antibody. In addition, the parent drug is in low concentrations and the predominant form found in urine is ritalinic acid. *Rifai N, et al. Textbook of Clinical Chemistry and Molecular Diagnostics. 6th ed., 2018. p. 858.*

2. c. Immunoassays commonly use the interaction between streptavidin and biotin which can be susceptible to interference from high-dose biotin intake in patients. *Mrosewski I, et al. Interference from high dose biotin intake in immunoassays for potentially time-critical analytes by Roche. Arch. Pathol. Lab Med 2020.*

3. b. While patients or the friends of patients may report the victim used one specific illicit drug, it may have been another drug or mixed with another drug. Most fentanyl immunoassay kits are designed to detect fentanyl, but have fairly good cross-reactivity to many fentanyl analogs (e.g., furanylfentanyl, B-hydroxyfentanyl, and despropionylfentanyl). Based on the response to naloxone, presumptive positive fentanyl immunoassay, negative confirmatory test for fentanyl/norfentanyl, and report of friends, the most likely explanation is the patient actually used a fentanyl analog. However, additional testing could be performed to confirm this interpretation. *Jannetto PJ, et al. The fentanyl epidemic and evolution of fentanyl analogs in the United States and European Union. Clin Chem 2020;65(2):242–53.*

4. e. Quantitative measurement of serum drug concentrations is meaningful if it correlates to clinical toxicity or influences patient management. *Magnani BJ, et al, ed., Clinical toxicology testing: a guide for laboratory professionals. 2nd ed., 2020. p. 69.*

5. a. Anticholinergic agents produce warm, flushed, dry skin, but sympathomimetic agents produce diaphoresis. *Rifai N, et al., eds. Tietz Textbook of Clinical Chemistry and Molecular Diagnostics. 6th ed. 2018. p. 834.*

6. d. Characteristics of the sedative-hypnotic toxidrome include: respiratory depression, slurred speech, diplopia, nystagmus, hypotension, blurred vision, confusion, dysesthesia, lethargy/coma, sedation, and ataxia. *Rifai N, et al., eds. Tietz Textbook of Clinical Chemistry and Molecular Diagnostics. 6th ed. 2018. p. 835.*

7. c. When an oxidant is present at or above specific levels (e.g., 200 mcg/dL) and is verified using a confirmatory test (e.g., nitrite positive ≥500 mcg/mL) it means the urine sample is adulterated. *Magnani BJ, et al, ed., Clinical toxicology testing: a guide for laboratory professionals. 2nd ed., 2020. p. 69.*

8. c. Studies have previously shown that hempseed oil from different countries may have THC concentrations as high as 1500 mcg/mL while in America the hempseed oil typically was less than 118 mcg/mL. Based on the source and frequency of usage, it is possible that the hempseed oil could result in a positive THC-COOH confirmation test. *Jang E, et al. Concentrations of THC, CBD, and CBN in commercial hemp seeds and hempseed oil sold in Korea. Forensic Sci. Int. 2020.*

9. a. Organophosphates exert their toxicity by inhibiting the action of acetylcholinesterase which causes a pronounced cholinergic response. Specific therapy includes administration of atropine to block the muscarinic actions of acetylcholine. In addition, pralidoxime is given to reactivate cholinesterase. *Rifai N, et al., eds. Tietz Textbook of Clinical Chemistry and Molecular Diagnostics. 6th ed. 2018. p. 853.*

10. d. Carbon monoxide is a colorless, odorless, tasteless gas which is the product of incomplete combustion. The binding affinity of hemoglobin for carbon monoxide is much greater than oxygen. Once bound, it also increases the oxygen affinity for the remaining subunits so less oxygen dissociates causing the hemoglobin-oxygen dissociation curve to shift to the left. As a result, carbon monoxide decreases the oxygen content of the blood and oxygen availability to the tissues. *Rifai N, et al., eds. Tietz*

Textbook of Clinical Chemistry and Molecular Diagnostics. 6th ed. 2018. p. 839.

11. b. Initial clinical findings in acetaminophen toxicity can be relatively mild and nonspecific symptoms not predictive of impending hepatic necrosis which begins 24–36 h after toxic ingestion and becomes more severe by 72 h. Specific therapy for acetaminophen overdose is administration of N-acetyl-L-cysteine which acts as a glutathione substitute and provides substrate to replenish hepatic glutathione. *Rifai N, et al., eds. Tietz Textbook of Clinical Chemistry and Molecular Diagnostics. 6th ed. 2018. p. 846.*

12. b. Both methanol and ethylene glycol can result in a metabolic acidosis with large anion gap and positive osmolal gap. However, the calcium oxalate crystals are consistent with ethylene glycol ingestion. *Magnani B, et al. Clinical toxicology testing: a guide for laboratory professionals. 2nd ed. Northfield, IL, College of American Pathologists 2020. p. 203.*

13. c. Methotrexate concentrations within 48 h following administration of glucarpidase can only reliably be measured using a chromatographic method. The inactive metabolite of methotrexate (DAMPA; 4-deoxy-4-amino-N10-methylpteroic acid) formed as a result of the glucarpidase treatment interferes with the measurement of methotrexate if immunoassays are used. As a result, the immunoassay provides an overestimation of the methotrexate concentration. *https://www.accessdata.fda.gov/drugsatf-da_docs/label/2012/125327lbl.pdf.*

14. d. Designer stimulants and "club drugs" are commonly used in raves because they produce feelings of euphoria and energy and a desire to socialize. Cathinones have structural similarities to methamphetamine and is a naturally occurring stimulant found in the leaves of the Khat plant. Cathinones are abused because of their psychostimulant and hallucinatory effects. *Rifai N, et al., eds. Tietz Textbook of Clinical Chemistry and Molecular Diagnostics. 6th ed. 2018. p. 857.*

15. a. Thallium is a colorless, odorless, nearly tasteless, and is rapidly absorbed by dermal contact, inhalation, or ingestion, making it an ideal poison. Fatal exposures occur from criminal or suicidal attempts and environmental exposure (used in photoelectric cells, lamps, semiconductors). Prussian blue, hemodialysis, and hemoperfusion have been suggested as chelators to remove thallium. *Jannetto PJ, et al. Till death do us part. Clin Chem 2018;64(10):1548–9.*

16. d. Both the parent drug (imipramine) and its active metabolite (desipramine) should be monitored. *Rifai N, et al., eds. Tietz Textbook of Clinical Chemistry and Molecular Diagnostics. 6th ed. 2018. p. 850.*

17. c. Illicit drug use during pregnancy is a major medical issue. Meconium is the first intestinal discharge from the newborn and is a viscous, dark green substance. The presence of drugs in meconium is indicative of in utero

drug exposure up to 5 months before birth. While a positive result indicates intrauterine drug exposure, a negative result does not rule it out. In this case the clinical presentation of the infant and confirmed usage of heroin by the mother fits the baby's presentation. *Rifai N, et al., eds. Tietz Textbook of Clinical Chemistry and Molecular Diagnostics. 6th ed. 2018. p. 884.*

18. c. Immunoassays are the most commonly used method for the measurement of digoxin in serum. Use of digibind, digoxin-specific Fab fragments, which neutralize the drug and used in settings of acute toxicity can complicate digoxin measurements by immunoassays. As a result, it is recommended that free digoxin measurements be monitored using ultrafiltration which eliminates this interference. *Rifai N, et al., eds. Tietz Textbook of Clinical Chemistry and Molecular Diagnostics. 6th ed. 2018. p. 838.*

19. c. Lisdexamfetamine (Vyvanse) is a prescription drug commonly used for attention deficit hyperactivity disorder. It will be detected as amphetamine in the patients' urine sample. As a result, the amphetamine is expected based on the list of patients prescribed medications. *Rifai N, et al., eds. Tietz Textbook of Clinical Chemistry and Molecular Diagnostics. 6th ed. 2018. p. 858.*

20. a. While it is possible to measure warfarin directly, the concentrations do not correlate well with clinical effects. The most useful test to follow the anticoagulant status of someone exposed to warfarin is the prothrombin time which is usually reported out as the international normalized ratio (INR). *Magnani BJ, et al, ed., Clinical toxicology testing: a guide for laboratory professionals. 2nd ed., 2020. p. 327.*

21. b. Bupropion is a commonly prescribed drug used as a smoking cessation agent and antidepressant drug. Therapeutic use of bupropion has been known to cause false positives on some amphetamine immunoassay screens; however, will result in a negative confirmation. Methylphenidate does not cross-react well with amphetamine immunoassays and MDMA and Lisdexamfetamine will both screen and confirm positive. *Casey ER, et al. Frequency of false positive amphetamine screens due to bupropion using the Syva Emit II Immunoassay, J Med Toxicol (2011);7:105–8.*

22. e. N-acetylcysteine is a commonly used treatment to help prevent liver toxicity in the setting of acetaminophen overdose. It has been reported to cause interference with Trinder-based assays such as enzymatic creatinine, cholesterol, high-density lipoprotein cholesterol, triglycerides, and uric acid. *Genzen JR, et al. N-acetylcysteine interference of Trinder-based assays Clin Biochem. 2016;49(1–2):100–4.*

23. a. Therapeutic drug monitoring (TDM) is valuable for drugs that have variable pharmacokinetics, a narrow therapeutic range, and are used chronically. Alprazolam is a benzodiazepine drug that is generally considered to be safe, effective, and have minimal side effects; therefore, TDM is not warranted. However, benzodiazepine drugs are commonly abused and require compliance testing if a patient is at risk of developing dependence. *Rifai N, et al., eds. Tietz Textbook of Clinical Chemistry and Molecular Diagnostics. 6th ed. 2018. Chapters 40 and 41.*

24. f. Methanol itself is significantly less toxic than ethanol. The toxic effects result from the production of formaldehyde and formic acid. Metabolism to formic acid leads to an elevated anion gap, metabolic acidosis, and potentially blindness or death. *Rifai N, et al., eds. Tietz Textbook of Clinical Chemistry and Molecular Diagnostics. 6th ed. 2018. Chapter 41.*

25. b. The two cardinal signs of morphine toxicity are CNS depression and miosis. Mydriasis can result from meperidine or propoxyphene toxicity. *Rifai N, et al., eds. Tietz Textbook of Clinical Chemistry and Molecular Diagnostics. 6th ed. 2018. Chapter 41.*

26. d. Interpretation of suboxone compliance testing is complex. It is recommended that laboratories offer buprenorphine and/or naloxone metabolites (norbuprenorphine and nornaloxone/noroxymorphone, respectively) in their test menu in order to help correctly interpret suboxone compliance testing. The absence of buprenorphine and/or naloxone metabolites could indicate specimen adulteration (noncompliance). Based on the results provided and the absence of any metabolite information, it is challenging to determine if the patient is compliant or not. *Donroe JH, et al. Interpreting quantitative urine buprenorphine and norbuprenorphine levels in office-based clinical practice. Drug and Alcohol Dependence, 2017;180:46–51.*

27. b. Δ^9-Tetrahydrocannabinol (Δ^9-THC) is the main psychoactive constituent of cannabis, with Δ^8-tetrahydrocannabinol, cannabinol, and 11-nor-9-carboxy-Δ^9-tetrahydrocannabinol being less potent psychoactive agents than Δ^9-THC. Cannabidiol is a cannabinoid lacking psychoactive properties. *Sharma P, et al. Chemistry, Metabolism, and Toxicology of Cannabis: Clinical Implications. Iran J Psychiatry 2012;7(4):149–56.*

28. d. Approximately 40%–50% of patients with chronic illness do not take their medications as prescribed. When assessing the reason for subtherapeutic drug concentrations in TDM, compliance should be considered. *Kleinsinger F. The unmet challenge of medication nonadherence. Perm J 2018; 22:18–033.*

29. a. Buprenorphine is a mu opioid receptor (MOR) partial agonist and kappa opioid receptor (KOR) antagonist. Pain relief is provided through partial MOR activation, but in contrast to full MOR agonists, analgesia will plateau and prevent toxic effects such as respiratory depression and sedation as seen with full agonists. *Rifai N, et al., eds. Tietz Textbook of Clinical Chemistry and Molecular Diagnostics. 6th ed. 2018. Chapter 41.*

30. b. Mycophenolic acid (MPA)—the active metabolite of the prodrug mycophenolate mofetil (MMF)—inhibits inosine monophosphate dehydrogenase, which is a key enzyme involved in purine synthesis in T lymphocytes. *Rifai N, et al., eds. Tietz Textbook of Clinical Chemistry and Molecular Diagnostics. 6th ed. 2018. Chapter 40.*

31. c. Patients actively using tobacco products or receiving high-dose nicotine patch therapy can have urine nicotine concentrations in the range of 1000–5000 ng/mL and cotinine concentrations from 1000 to 8000 ng/mL. However, tobacco products contain unique biomarkers such as anabasine, which can help differentiate tobacco use from nicotine replacement therapies. The presence of anabasine in this patient indicates that in addition to the nicotine patch, the patient is likely using tobacco products as well. *Moyer TP, et al. Simultaneous analysis of nicotine, nicotine metabolites, and tobacco alkaloids in serum or urine by tandem mass spectrometry, with clinically relevant metabolic profiles. Clin Chem 2002;48:1460–71.*

32. a. Phenytoin is approximately 90% bound to albumin in healthy patients. An increase in nonprotein nitrogen compounds as seen in uremia can lead to a decrease in albumin causing free phenytoin concentrations to increase. *Rifai N, et al., eds. Tietz Textbook of Clinical Chemistry and Molecular Diagnostics. 6th ed. 2018. Chapter 40.*

33. c. Codeine has poor affinity for the mu opioid receptor and approximately one-tenth the analgesic potency of morphine. *Rifai N, et al., eds. Tietz Textbook of Clinical Chemistry and Molecular Diagnostics. 6th ed. 2018. Chapter 41.*

34. c. Oxycodone is metabolized to oxymorphone, noroxycodone, and noroxymorphone, which are all detectable in the patient's urine, suggesting the patient was compliant with his oxycodone prescription. Oxycodone preparations can contain up to 1% of hydrocodone as a pharmaceutical impurity. *Pesce A, et al. Interpretation of urine drug testing in pain patients. Pain Med 2012;13:868–85.*

Chapter 15

Forensic toxicology

Luke Rodda[a] and Alan H.B. Wu[b]
[a]Office of the Chief Medical Examiners, San Francisco, CA, United States, [b]University of
California, San Francisco, CA, United States

1. Select the recommended testing protocol following a positive alcohol result in a postmortem blood specimen.
   a. Retest the blood specimen to determine if it is a false positive.
   b. Test another blood specimen for alcohol, if available, to confirm the original alcohol result.
   c. Test the vitreous humor, or urine if vitreous humor is unavailable.
   d. No need for any further testing, unless it was blood collected from a central region of the body, which therefore requires retesting of the central blood to determine if it is a false positive.
   e. a., b., and c.
2. A 50-year-old female was found dead after being last seen 2 weeks prior, during the summer months. Significant putrefaction and insect infestation were present. In close proximity, investigators found an empty bottle of wine in addition to empty bottles of Lunesta and Klonopin which both appeared to be abused in consideration of the date filed being only days before the female was last seen. At autopsy, the lining of the gastrointestinal tract appeared as shaded blue. Due to the advanced stages of decomposition there was minimal gastric fluid. Combined alcohol and CNS depressants polysubstance toxicity was expected. However, toxicology on the central blood specimen returned a BAC of 0.16, with a vitreous alcohol of 0.19, and only 25 ng/mL of eszopiclone/zopiclone and 35 ng/mL of clonazepam. Although the drugs' results appeared to not be in fatal concentration ranges and are relatively well tolerated when compared to clinical ranges, the cause of death was determined as polydrug and alcohol toxicity. Select how was this conclusion was justified.
   a. The difference in BAC and vitreous humor alcohol indicates that the BAC was likely much higher at death.
   b. Significant drug degradation likely occurred during the postmortem interval.

Self-assessment Q&A in Clinical Laboratory Science, III. https://doi.org/10.1016/B978-0-12-822093-1.00015-6
177

   c. It is not justified as the toxicology testing could not discern between the zopiclone enantiomers.

   d. Any concentrations of clonazepam and eszopiclone/zopiclone are fatal when combined with even small amounts of alcohol consumption.

   e. a and d

3. Select which of the following does not alter postmortem redistribution (PMR).

   a. specimen collection container preservative

   b. specimen collection site

   c. postmortem interval

   d. lipophilicity of drug

   e. pK_a and V_d of drug

4. In LC-MS/MS and GC-MS/MS (triple quadrupole) instruments, select what typically takes place in the second quadrupole.

   a. an ion's mass is calculated.

   b. ions are sorted from lowest to highest mass ranges.

   c. ions are sorted from highest to lowest mass ranges.

   d. fragmentation occurs.

   e. a. and b.

5. A sexual assault survivor arrived to her small local hospital 4 hours after a suspected drug spiking. A clinical, nonpreserved, blood specimen was taken during her regular medical care. She was then transferred to the city hospital 3 hours later where forensic services were available. Here, a gray-top blood specimen and a urine specimen were collected using tubes provided from a forensic laboratory evidence kit. All three (3) specimens were obtained by the investigating officer with chain of custody maintained throughout all specimens. Select what specimen(s) should be delivered and analyzed by the forensic toxicology laboratory.

   a. All specimens should be delivered, and the gray-top blood (due to it being preserved) and urine should be analyzed.

   b. All specimens should be delivered, and the gray-top blood (due to it being provided from an official forensic kit) and urine should be analyzed.

   c. All specimens should be delivered, but only the urine should be analyzed as it is the most useful DFC specimen.

   d. All specimens should be delivered, and the unpreserved blood and urine should be analyzed.

   e. Just the urine should be delivered and analyzed because it is the most useful DFC specimen.

6. Select appropriate characteristics that a quality assurance program should incorporate.

   a. Clear guidelines on management of quality control acceptance.

   b. A training program for laboratory personnel.

   c. Traceability of reference materials.

d. Application and successful performance in the proficiency testing programs.

e. all of the above

7. As a minimum for quantitation techniques, select the validation studies that must be performed before performing forensic casework.

a. Selectivity/interference, matrix effects, accuracy and precision, limits of detection and quantitation, and processed sample stability.

b. Selectivity/interference, matrix effects, accuracy and precision, limits of detection and quantitation, calibration model, and carry over studies.

c. Selectivity/interference, matrix effects, accuracy and precision, limits of detection and quantitation, processed sample stability, calibration model, carryover, and dilution integrity.

d. Selectivity/interference, matrix effects, accuracy and precision, limits of detection and quantitation, processed sample stability, calibration model, carryover, dilution integrity, and applicability.

e. all studies mentioned

8. Select which would likely not be in a routine forensic toxicology scope of testing.

a. drugs of abuse

b. cardiac medications

c. carboxyhemoglobin

d. alcohol

e. b. and c.

9. Select the most useful postmortem specimen for the biochemical analysis of potassium, sodium, chloride, calcium, magnesium, glucose, and urea.

a. whole blood

b. plasma

c. vitreous humor

d. urine

e. serum

10. At 10 pm, a long-term abuser of benzodiazepines ingests what was thought to be an alprazolam bar/pill that was purchased illicitly. However, unbeknownst to the opioid naive male, it contained only fentanyl. Over the next 10–20 min, he was seen slumping over and died in front of friends. An autopsy was performed the following morning with pulmonary edema evident. His toxicology report showed fentanyl (5 ng/mL) and norfentanyl (3 ng/mL) in the femoral blood. The cause of death was given as fentanyl toxicity. In another case, a long-term heroin user injects heroin that has also been laced with fentanyl which was also unknown to the user. Although he experiences the intended euphoric effects, he does not succumb to the heroin and/or fentanyl and attempts to drive. He is found asleep in the driver's seat with the engine running at an intersection and is arrested for DUID. His blood toxicology reports showed fentanyl (25 ng/mL), norfentanyl (10 ng/mL), 6-monoaceytlmorphine (5 ng/mL),

and morphine (30 ng/mL). No opioid antagonists such as naloxone or naltrexone were detected in either case and there was no indication that they were given to either of the two males. Both the decedent's and driver's bloods were shown to contain fentanyl. However, the decedent's blood contained five times less fentanyl. Select the best explanation to describe how fentanyl was a cause of death at a lower concentration, and yet was higher in a living driver.

a. Their opioid tolerances were different.

b. Their routes of administration were different.

c. The heroin provides a protective mechanism when used in combination with fentanyl.

d. The fentanyl concentration in the decedent was likely to be higher than 20 ng/mL; however, due to degradation and instability in postmortem blood, decreased significantly.

e. a., b., and c.

11. A driver was stopped on the roadside by law enforcement following signs of erratic driving. Upon a drug recognition evaluation from a Drug Recognition Expert (DRE), the following observations were made: dilated pupils, increased pulse, increased temperature, increased blood pressure, no horizontal gaze nystagmus (HGN), no vertical gaze nystagmus (VGN), and no lack of convergence (LOC). Select the drug the driver is likely to be under the influence from.

a. fentanyl.

b. methamphetamine.

c. edibles (cannabis).

d. alprazolam.

e. alcohol (no drugs suspected).

12. At 9:00 pm, a car is observed driving erratically and subsequently a traffic stop was performed. After roadside assessment showed signs of intoxication, a blood sample for the purposes of a toxicology screen was authorized to be collected. A BAC of 0.07% was given for the specimen collected at 11:00 pm. Select what would be the best retrograde estimate of the BAC at the time of driving. The driver admitted to drinking hours prior to driving, therefore full absorption should be assumed.

a. ~0.08% (w/v)

b. ~0.01% (w/v)

c. ~0.10% (w/v)

d. ~0.07% (w/v) or ~0.14% (w/v)

e. After 1 h, retrograde blood alcohol analysis is unreliable.

13. An infant was found dead in the early hours of the morning. Following ~6 weeks of forensic investigations, subsequent toxicology testing helped to determine the cause of death as methadone toxicity. As there were 2- and 3-year-old siblings in the same household, law enforcement and child protective services were concerned of the potential risk to the siblings'

safety. A warrant was obtained for collection of samples from the siblings for toxicology testing. Choose which biological specimen would be the most useful and appropriate to collect from the children.
a. blood
b. saliva
c. urine
d. hair
e. none of the above

14. Cocaine (35 ng/mL) and benzoylecgonine (720 ng/mL) were detected following the analysis of a driver's blood who was suspected to be under the influence of a stimulant. Prior to a scheduled court hearing, the defense requested that the blood be reanalyzed by a second credible laboratory of their choice in order to verify the blood toxicology results. This second test occurred approximately a year after the original analysis. Over this time, the gray-top blood tube had since been stored in appropriate refrigerated conditions. Upon the second reanalysis, there was no cocaine detected (cocaine's limit of detection = 20 ng/mL); however, benzoylecgonine was detected at 800 ng/mL. An argument was made that the original analysis was incorrect and that there was no psychoactive drug in the driver at the time of driving. Select what could be the possible reason(s) for the difference in these two results.
a. The initial result was likely a false positive.
b. The second result was likely incorrect and that the laboratory, although credible, was hired by the defense and so the testing results are inadmissible in court.
c. There was breakdown of cocaine during storage.
d. The two cocaine results ("35 ng/mL" and "not detected") are within the normal acceptable variance of typical analytical forensic drug testing and uncertainty measurements.
e. The samples were inadvertently switched.

15. Select a description that is not a suitable feature when choosing an appropriate internal standard for an analytical method.
a. It is does not degrade during the analytical process.
b. It is not expected to be within samples of casework.
c. The response is similar to the target analyte(s) response at similar concentrations.
d. It interferes with a target analyte(s).
e. It is a stable deuterated version of a target analyte.

16. Select what opinions and testimony would be inappropriate by a forensic toxicologist.
a. Use of the phrases "scientific certainty" or "reasonable degree of scientific certainty."
b. Calculation of the dose of a drug based on a postmortem drug concentration in blood.

  c. Performing of extrapolation calculations for drugs other than ethanol.

  d. All are acceptable opinions or testimonies.

  e. All are unacceptable opinions or testimonies.

17. Determine which of the following are not considered a requirement of good traceability.

  a. Calibration of analytical equipment including balances, reference standards, volumetric glassware, pipettes, diluters, and syringes.

  b. Linearity studies to determine either linear or quadratic fit during method validation.

  c. Elements such as documentation of measurement uncertainty and technical competence.

  d. The use of certified reference material.

  e. All are considered relevant.

18. Select when analysis of stomach (gastric) contents is most useful

  a. In every postmortem case in order to finalize a complete medicolegal evaluation.

  b. To confirm a determination of natural deaths as the manner.

  c. In manner of death determination following relatively high drug concentrations in blood.

  d. It should be analyzed in all homicide casework.

  e. To determine the extent of postmortem redistribution.

19. Select what information from a death investigation can help demonstrate likely heroin use over morphine use when morphine is detected and in the absence of 6-monoacetylmorphine (6-MAM) detection.

  a. There was no evidence of an alternate source of morphine identified (i.e., morphine sulfate medication) and the cause of death is consistent with acute or secondary complication of heroin overdose, or an intoxication-related traumatic death.

  b. Toxicological reports detected morphine:codeine less than or equal to ∼10:1 in blood.

  c. Toxicological reports detected morphine:codeine less than or equal to ∼1:10 in blood.

  d. a. and b.

  e. a. and c.

20. A 67-year-old relatively recluse male was found unresponsive in his home in the evening by EMS after a welfare check was made following no answer to a phone call from a friend throughout the day. Upon admission to hospital, the subject did not regain consciousness and continued to deteriorate. The male was pronounced dead on the 17th day of care. Only a benzodiazepine positive drug result was reported following urine toxicology using a typical immunoassay testing kit used clinically. A subsequent plasma analysis for benzodiazepines returned a lorazepam concentration of 50 ng/mL. The treating physician was uncomfortable with a natural medical cause of death. Although the decedent was diagnosed with cancer 8 months prior, he was in remission following 6 months of chemotherapy. The potentially supratherapeutic lorazepam result did not appear toxic in isolation. Subsequently, the case came under jurisdiction of the Medical Examiner. Medical Examiner Investigators observed an empty bottle of 2 mg Ativan tablets, prescribed 4 months prior to death, to treat chemotherapy-induced nausea. Investigations also learned that the decedent had a caretaker who started approximately 6 months prior to the death to assist with the chemotherapy side effects. The friend reported concern over the heir to his estate as the decedent had no children or other known relatives. Additionally, the decedent's oncologist reported no recent medical or health issues with his energy improving. It was discovered that the caretaker had continued to visit the male following the chemotherapy beyond the decedent needing assistance and appeared to befriend the decedent. Further investigations revealed that the caretaker had a history of Kratom abuse. At autopsy, signs of CNS depression were observed and as a result, drug intoxication was suspected. However, all antemortem blood specimens collected during hospital care, besides blood collected on the last day of care, were destroyed. Subsequently, select which specimen(s) are most suitable for forensic postmortem toxicological analysis.

   a. Antemortem blood specimens collected on day 17 following hospital admission.

   b. Postmortem blood and urine specimens collected at autopsy.

   c. Postmortem gastric contents and liver specimens collected at autopsy.

   d. all of the above

   e. none of the above

21. What is the major advantage of high-resolution mass spectrometry for toxicological analysis?

   a. differentiation between stereoisomers

   b. the determination of compound identity by mass fragmentation

   c. the determination of compound identity by molecular formula

   d. enables the combination of mass spectrometry and infrared scanning

   e. resolves liquid chromatography peaks to separate coeluting compounds

22. Which of the ionization methods produces significant fragmentation, and therefore is not useful for detection of the molecular ion?

   a. electron impact

   b. chemical ionization

   c. atmospheric pressure ionization

   d. matrix-assisted laser desorption ionization

   e. electrospray ionization

23. A driver is involved with a motor vehicle accident. Using the alcohol dehydrogenase method, the serum ethanol from the individual is 80 mg/dL from a sample collected in the emergency room. The result was used in a civil case against the patient. Which of the following is true?

    a. The driver may be negligent because results exceed the legal driving limit of 0.08% at the time of the accident.

    b. Results are not admissible because most states require whole blood analysis.

    c. Results are not admissible because chain-of-custody procedures were not employed.

    d. Results are not admissible because the head-space gas chromatography method was not used.

    e. none of the above

24. A urine sample from a diabetic patient is sent to a drug testing laboratory for ethanol testing. The sample was sent through the mail and arrived 5 days after collection. The patient had a urinary tract infection. The urine ethanol result was 20 mg/dL using the head-space gas chromatographic method. Which of the following may be true?

    a. ethanol was produced through in vitro fermentation

    b. the assay used was not specific for ethanol

    c. the patient drank methanol

    d. the sample was contaminated with an alcohol wipe

    e. value represents endogenous ethanol production

25. Which of the following regarding poppy seed consumption is false?

    a. can produce a false positive result for codeine

    b. can produce a false positive result for hydromorphone

    c. consumption can mask a morphine use/abuse

    d. poppy seed use can be detected by measuring thebaine

    e. opiate screening cutoffs for workplace drug testing are set high to minimize likelihood of opiate positivity due to consumption

26. According to the Substance Abuse and Mental Health Services Administration, which of the following is considered a substituted urine when submitted for workplace drug testing?

    a. creatinine <5 mg/dL

    b. creatinine <2 mg/dL and specific gravity ≤1.0010

    c. creatinine <2 mg/dL, or specific gravity ≤1.001

    d. creatinine <2 mg/dL and specific gravity ≤1.001 or specific gravity ≥1.0200

    e. creatinine <5 mg/dL and specific gravity ≤1.001 or specific gravity ≥1.0200

27. If the urine sample is not considered substituted, what is the upper creatinine cutoff concentration that is considered diluted?

    a. 5 mg/dL

    b. 10 mg/dL

c. 20 mg/dL

d. 25 mg/dL

e. not specified

28. Which of the following drugs were changed to the SAMHSA workplace drug testing panel effective October 1, 2017?

a. added 6-acetylmorphine at 10 ng/mL for screening and confirmation

b. added MDA/MDMA at 500 ng/mL screening, 250 ng/mL confirmation

c. removed THC as it is now legalized in many states now

d. added hydrocodone/hydromorphone at 300 and oxycodone/oxymorphone at 100 ng/mL for screening

e. changed screening amphetamine/methamphetamine cutoff from 1000 to 500 ng/mL

29. Which opiate/opioid is responsible for the most deaths in the United States in 2019?

a. heroin

b. morphine

c. fentanyl

d. tramadol

e. oxycodone

30. Which of the following is best explanation for a low result for serum cholinesterase?

a. reversible inhibition by organophosphate pesticide exposure

b. presence of a genetic deficiency

c. analytic error, low results are not meaningful

d. irreversible inhibition by carbamates

e. a. and b.

31. Which of the following is correct regarding the difference between red cell and pseudocholinesterase measurements?

a. results are interchangeable

b. red cell cholinesterase is analytically more reliable to measure

c. red cell cholinesterase is a more reliable indicator of pesticide exposure

d. red cell cholinesterase is influenced by liver disease and malnutrition

e. results of both tests can be used to select optimum therapy

Answers

1. c. Postmortem microbial (bacteria, yeast, and molds) contamination and fermentation of available sugars into ethanol can occur. The synthesis of postmortem ethanol increases as storage temperature increases and the interval between death and sample collection at autopsy increases. Additionally, in cases where there is a relatively high concentration of alcohol in the gastric contents, or in a traumatized body where the gastric lining may have ruptured, diffusion and/or contamination of ethanol into surrounding tissues, such as blood, may occur. For these reasons, it is recommended that vitreous humor be analyzed following a positive BAC. Vitreous humor is a relatively distant tissue from the central region

and, importantly, is within a relatively closed environment that is less susceptible to microbial growth. Subsequently, it is common to see a positive BAC while detecting no ethanol in the vitreous humor in a decomposed case, suggesting postmortem production of alcohol. *Garriott's Medicolegal Aspects of Alcohol, 6th edition. Y.H. Caplan and B.A. Goldberger, Eds; Lawyers and Judges Publishing Company, Tucson, AZ, 2015, p. 702. Kugelberg FC et al. Interpreting results of ethanol analysis in postmortem specimens: a review of the literature. Forc Sci Int 2007; 165:10–29.*

2. b. The drugs eszopiclone and clonazepam can undergo significant degradation in unpreserved blood. As the body was undergoing significant putrefaction with increased microbial action, with the postmortem interval likely up to 2 weeks, and was exposed to warmer summer days, it could be expected that the eszopiclone and clonazepam concentrations were much higher at the time of death. Considering the investigative information indicating overuse of the drugs, the blue residue in the gastric contents as a likely result from Lunesta and Klonopin, and in conjunction with alcohol, the combination of the three CNS depressants taken concurrently and in high amounts is sufficient for the cause of death to be determined as polydrug and alcohol toxicity. The difference between the BAC and vitreous humor alcohol concentration is typical considering the increased water content in the vitreous humor and together, these concentrations indicate antemortem consumption of alcohol. The lack of selectivity in reporting between the zopiclone enantiomers is inconsequential as their pharmacological properties are very similar and in this complex postmortem context, any such differences are minimal when interpreting the results in totality. *Michael D. et al. Stability of nitrobenzodiazepines in postmortem blood. J Forensic Sci. 1998;43(1):5–8. Gunnel H. et al. Stability tests of zopiclone in whole blood. For Sci Int 2010;200:130–5. Peters FT, et al. Antemortem and postmortem influences on drug concentrations and metabolite patterns in postmortem specimens. WIREs For Sci. 2019;1:e1297. https:// doi.org/10.1002/wfs2.1297. Butzbach DM. The influence of putrefaction and sample storage on post-mortem toxicology results. Forensic Sci Med Pathol 2010;6:35–45.*

3. a. PMR is a phenomenon in postmortem toxicology casework in which the concentration of an analyte changes in the tissue after death. The mechanism involved is the movement of drugs between adjacent tissues. There are several factors that influence the extent to which PMR of a drug may occur. Properties such as an increased lipophilicity or volume of distribution (V_d) tend to predict that drugs will exhibit greater PMR. Additionally, blood collected from a central site are typically exposed to higher concentrations of drug(s) in adjacent tissue than peripheral blood (e.g., femoral), and thus, often shows increased drug concentrations to that of the peripheral blood. An increased period of time between time of death and autopsy

sample collection (e.g., postmortem interval) can also exacerbate PMR. As the effects of PMR are a phenomenon that occurs within the decedent, the preservative used during specimen collection is not a factor that alters the degree of PMR. *Jones GR, et al. Site dependence of drug concentrations in postmortem blood—a case study. J Anal Toxicol 1987;11:186–90. Tolliver S. An investigation of the relationship between antemortem and postmortem drug concentrations in blood. FIU Electronic Theses and Dissertations 2010, 321. https://digitalcommons.fiu.edu/etd/321. Gerostamoulos D, et al. Forensic Sci Med Pathol 2012;8:373. https://doi.org/10.1007/s12024-012-9341-2. Brockbals L, et al. Time-dependent postmortem redistribution of opioids in blood and alternative matrices. J Anal Toxcol 2018;42:365–74, https://doi.org/10.1093/jat/bky017. Pélissier-Alicot AL, et al. Mechanisms underlying postmortem redistribution of drugs: a review. J Anal Toxicol 2003;27:533–44. Yarema MC, et al. Key concepts in postmortem drug redistribution. Clin Toxicol 2005;43:235–41. Luckenbill K, et al. Fentanyl postmortem redistribution: preliminary findings regarding the relationship among femoral blood and liver and heart tissue concentrations. J Anal Toxicol 2008;32:639–43.*

4. d. In the second of three quadrupoles, collision gas can be allowed to fill the space causing incoming selected ions from the first quadrupole to collide into the inert gas. This leads to fragmentation of the precursor ion(s) into product ions that can be allowed to pass selectively through the third quadrupole and subsequently to the detector. A range of different MS/MS modes exist by applying different configurations of the three quadruples. The most common mode used in forensic toxicology is multiple reaction monitoring (MRM). *Pitt JJ. Principles and applications of liquid chromatography-mass spectrometry in clinical biochemistry. Clin Biochem Rev 2009;30:19–34. Clarke's Analytical Forensic Toxicology, 2nd edition. Adam Negrusz et al, Eds. UK, 634 pp., ISBN 978-0857110541. Mass Spectrometry for the Clinical Laboratory, 1st edition. Hari Nair, William Clarke, eds.; 304 pp., ISBN: 978-0128008713.*

5. d. Forensic laboratories typically provide evidence envelopes and specimen tubes for use in drug-facilitated crime (DFC) casework (which includes drug-facilitated sexual assault, DFSA, casework). Following a suspected drug-facilitated crime, it is imperative to collect blood and urine as soon as possible and within 24h and 120h (5 days) of the incident, respectively (note, these time points may be increased if the analytical methodologies used are superior in sensitivity). Although preserved blood is preferential, the earliest collected specimens should be analyzed if chain of custody is maintained. In this example, the second and only preserved blood was collected 7h later. Relatively fast eliminating drugs such as GHB may have been removed from the blood by this time and therefore a blood collected 4h after the incident is preferred. Although urine is typically most useful in DFC cases, if blood is collected within these stated

times, the detection of drugs in blood allows for inferences as to when a drug was likely consumed, compared to urine interpretation. Therefore, the earliest blood and urine should be analyzed in this case. *Society of Forensic Toxicologists (SOFT): Drug-Facilitated Crimes Committee (DFC) formerly the DFSA Committee. Recommended minimum performance limits for common DFC drugs and metabolites in urine samples. (2017) https://www.soft-tox.org/files/MinPerfLimits_DFC2017.pdf. Laboratory and Scientific Section-United Nations Office on Drugs and Crime, Vienna. Guidelines for the forensic analysis of drugs facilitating sexual assault and other criminal acts. United Nations, December 2011. New York, U.S. https://www.unodc.org/documents/scientific/forensic_analys_of_drugs_facilitating_sexual_assault_and_other_criminal_acts.pdf].*

6. e. ISO/IEC 17025 is an international standard for laboratories involved in testing, such as a forensic toxicology laboratory, and specifies the general requirements for the competence, impartiality, and consistency of results obtained from laboratories. Another standard frequently achieved by laboratories is the American Board of Forensic Toxicology, a specialized standard aimed to enhance the practice for the detection, identification, and quantitation of alcohol, drugs, and other toxins in human biological specimens. The earlier selections are all key characteristics of an effective quality assurance program. *ISO/IEC 17025:2017. General requirements for the competence of testing and calibration laboratories. American Board of Forensic Toxicology Accreditation Program. https://www.abft. org/files/ABFT_LAP_Outline_September_2014.pdf.*

7. b. Although encouraged and required in certain uses of the developed method, there are only seven studies which would be characterized as mandatory for quantitative analysis. However, if the proposed method is to be used in certain scenarios, then further studies should be performed. Further studies may include dilution accuracy and precision studies to ensure the method can allow for the dilution of samples when an original result falls above the upper limit of the curve. Fewer studies are required for qualitative methods. *Scientific Working Group for Forensic Toxicology (SWGTOX) Standard Practices for Method Validation in Forensic Toxicology. J Anal Toxicol 2013;37:452–74. Peters FT, et al. Validation of new methods. For Sci Int 2007;165:216–24.*

8. e. Recommendations of toxicology testing for typical forensic toxicology casework such as medicolegal death investigation (MDI), driving under the influence of drugs (DUID), and drug-facilitated crimes (DFC) casework always include alcohol and drugs of abuse screening at a minimum. Tests to examine specific or expanded investigations such as carbon monoxide poisoning, or the compliance or toxicity of cardiac medications, typically require special request. *ASB 119, Standard for the Analytical Scope and Sensitivity of Forensic Toxicology Testing for Medicolegal Death Investigations (OSAC draft, under review). Logan et al.,*

Recommendations for toxicological investigation of drug-impaired driving and motor vehicle ratalities-2017 update. J Anal Toxicol 2018;42:63–8. SOFT DFC Committee, Drug-Facilitated Crimes Cutoffs, 07/2017, https://www.soft-tox.org/files/MinPerfLimits_DFC2017.pdf.

9. c. Vitreous humor is the fluid within the globe of the eye and is ~99% water, viscous, colorless, normally clear, and acellular. Forensic biochemical analysis of constituents within vitreous humor is useful in death investigation. Disease states such as death from hypothermia, alcoholic acidoketosis, dehydration, water or salt intoxication, beer potomania, glycemia, and intoxication from sodium-based bleaches, among other disease states, can be assessed from vitreous results. Furthermore, potassium has been examined extensively to estimate the postmortem interval (PMI) as its concentration increases over time. Additionally, as postmortem changes can be problematic for the analysis of the constituents in postmortem blood specimens, the relatively isolated vitreous humor can be a suitable alternative in postmortem analyses. Therefore, vitreous humor is also an alternative, or complementary, specimen for the analysis of drugs. *B. Zilg, et al. Interpretation of postmortem vitreous concentrations of sodium and chloride. Foren Sci Int 2016;263:107–13. Pérez-Martínez C, et al. Influence of the nature of death in biochemical analysis of the vitreous humour for the estimation of post-mortem interval. Aust J Foren Sci 2019, doi:10.1080/00450618.2019.1593503. Muñoz J, et al. A new perspective in the estimation of postmortem interval (PMI) based on vitreous [K+]. J Foren Sci 2001;46:209–14. Bévalot F et al. Vitreous humor analysis for the detection of xenobiotics in forensic toxicology: a review. Foren Toxicol 2016;34:12–40.*

10. a. Drug use history must be considered to determine such factors as tolerance. The μ-opioid receptor develops particularly significant tolerance following repeated exposure to opioid agonists. This tolerance is recognized by a reduced responsiveness to the opioid agonist, requiring increased doses to achieve the desired effect. As such, opioid concentrations overlap considerably between therapeutic, toxic, and fatal. Although postmortem redistribution may have changed the blood concentration, this would likely have been minimal due to the peripheral site of collection and expedient autopsy following death. It is justified to include the relatively low fentanyl concentration as a factor in the cause of death considering the pulmonary edema pathology findings indicating respiratory depression and the opioid naivety of the male. Conversely, the survived driver who was impaired for the purposes of driving used fentanyl potentially for the first time and did not succumb to the full extent of the fentanyl and active heroin metabolites in his system due to the tolerance developed over prolong heroin use. *Pelletier DE et al. Common findings and predictive measures of opioid overdoses. Acad Foren Pathol 2017;7:91–8. Lee D, et al. Illicit fentanyl-related fatalities in Florida: Toxicological findings. J Anal*

Toxicol 2016;40:588–94. Pearson J, et al. Postmortem toxicology findings of acetyl fentanyl, fentanyl, and morphine in heroin fatalities in Tampa, Florida. Acad Foren Pathol 2015;5:676–89. Erratum in: Acad Foren Pathol 2017;7:667–704.

11. b. Following suspicion of an impairing substance while operating a vehicle, a Standardized Field Sobriety Test (SFST) is typically performed. If the individual shows signs of impairment and no alcohol is detected on the roadside breath testing device, then further assessments may be necessary. The Drug Evaluation and Classification (DEC) program is a systematic and standardized procedure to determine impairment by a drug category (s) and is performed by a DRE. The program is based on a variety of observable signs and symptoms such as appearance, behavior, performance of psychophysical tests, eyes in different lighting conditions, and vital signs. All information observed during the assessment can be used in conjunction with the toxicology report, driving behavior, and drug use history. The observations in this example best describe the use of central nervous system stimulants, such as methamphetamine. Body tremors, talkativeness, exaggerated reflexes, excitement, nasal area redness, anxiety, and bruxism are other effects observed following use of CNS stimulants. *Hartman RL, et al. Drug recognition expert (DRE) examination characteristics of cannabis impairment. Accid Analy Prev 2016;92:219–29. Ramaekers JG, et al. Marijuana, alcohol and actual driving performance. Hum. Psychopharmacol. Clin. Exp 2000;15:551–8. Logan BK. Methamphetamine and driving impairment. J Foren Sci 1996;41:457–64. Yeakel JF, et al. Butalbital and driving impairment. J Forensic Sci 2013;58:941–5. Dubois S, et al. The impact of benzodiazepines on safe driving. Traf Injury Prev 2008;9:404–13.*

12. c. Alcohol follows zero order kinetics down to 0.01% at an average elimination rate of 0.015% per hour. Full absorption of alcohol from the gastrointestinal tract into the body is rapid and typically occurs with 15–60 min after consumption. Two hours elapsed from the time of driving to the blood collection. Therefore, 0.07% plus 0.03% (2h × 0.015%) equals ∼0.10%. Elimination rates can range from 0.01% to 0.02% per hour, with alcoholics typically eliminating alcohol at a higher rate. Even if considering that the defendant eliminates alcohol at the slower rate of 0.01% per hour, this would result in a ∼0.09% retrograde calculation. The defendant would have been above the 0.08% Per Se legal limit for regular drivers in all U.S. states (besides Utah, which has a Per Se legal limit of 0.05%). *Garriott's Medicolegal Aspects of Alcohol, 6th edition. Y.H. Caplan and B.A. Goldberger, Eds, 2015, 702 pp., ISBN 978-1936360888. Jones AW. Alcohol, its absorption, distribution, metabolism, and excretion in the body and pharmacokinetic calculations. WIREs Forensic Sci. 2019;1:e1340.* https://doi.org/10.1002/wfs2.1340. *Jones*

AW. Evidence-based survey of the elimination rates of ethanol from blood with applications in forensic casework. For Sci Int 2010;200:1–20.

13. d. Following suspicion of drugs being allegedly given over a period of time of more than 1–2 days, blood and saliva typically are unable to detect most drugs. After 5–7 days, urine typically also becomes unusable for toxicology testing. Hair testing is able to potentially provide an indication of drug exposure over longer periods of time. Scalp hair typically grows approximately one (1) cm per month however, does not always reach that growth until at least a year old. Both whole strand analysis and segmental hair analysis are possible. Segmental hair analysis was performed and could show that over a period of several months; methadone was continuously detected in the siblings' hair. These results were used to determine custody of the other children. *Kintz P. Hair analysis in forensic toxicology. WIREs Forensic Sci. 2019;1:e1196.* https://doi.org/10.1002/wfs2.1196. *Alvarez JC, et al. Hair analysis does not allow to discriminate between acute and chronic administrations of a drug in young children. Int J Legal Med 2018;132:165–72.*

14. c. Even during appropriate refrigerated storage conditions and preservation of blood, cocaine is highly susceptible to chemical transformation over an extended period of time due to hydrolysis into a degradation product (and expected metabolite) benzoylecgonine. *Isenschmid DS, et al. A comprehensive study of the stability of cocaine and its metabolites. J Anal Toxicol 1989;13:250–6.*

15. d. An internal standard is an essential component of any forensic analytical method. When selecting an appropriate internal standard, A, B, C, and E are essential characteristics. Labeling of a deuterium in replacement of hydrogen has allowed for the advent of deuterated forms of commonly analyzed drugs and poisons. Deuterated standards are used primarily as internal standards. *Hearn WML, et al. Common methods in postmortem toxicology, in Drugs Abuse Handbook, 2nd edition. Steven B. Karch; Ed. CRC Press, CRC, 2007. Levine B. (ed.) (2013). Principles of forensic toxicology, 4th edition. American Association for Clinical Chemistry Press, Washington, DC.*

16. e. The earlier opinions and/or testimony are generally considered to be inappropriate for a forensic toxicologist to offer in a medicolegal setting due to the fact that they currently lack consensus within the scientific community or are generally beyond the scope of a forensic toxicologist's expertise. Unless required by jurisdictional regulations, phrases such as "scientific certainty" or "reasonable degree of scientific certainty" should not be used. Further reading on what opinion and testimony statements are generally considered acceptable are available as a recommended national guideline following its creation through the Organization of Scientific Area Committees (OSAC) toxicology subcommittee. *ANSI/ASB Best*

Practice Recommendation 037, First Edition 2019: Guidelines for Opinions and Testimony in Forensic Toxicology http://www.asbstandards board.org/wp-content/uploads/2019/01/037_BPR_e1.pdf.

17. b. Traceability can be established by producing measurements using equipment that has been calibrated through the use of certified reference materials in the test or calibration method and/or with established metrological traceability. Additionally, proper handling and storage procedures must be undertaken for equipment and certified reference materials used to establish and maintain measurement traceability. *ANSI/ASB Standard 017, First Edition 2018. Standard Practices for Measurement Traceability in Forensic Toxicology https://asb.aafs.org/wp-content/uploads/2018/06/017_Std_e1.pdf.*

18. c. In a case where blood and autopsy/pathology results indicate drug toxicity as the cause of the death, analysis of stomach contents may assist in determining either accident or suicide in manner where there is a lack of suicidal intention, investigative evidence, or otherwise. For example, after analysis of a decedent's blood, 80 ng/mL of hydromorphone was detected and subsequently attributed as the cause of death. However, although the manner suggested suicide, it was not obvious. As such, examination of the stomach contents revealed at least 15 partially consumed yellow hydromorphone 4 mg tablets which provided sufficient information of intent to determine the manner as suicide. *Levine B. (ed.) (2013). Principles of forensic toxicology, 4th edition. American Association for Clinical Chemistry Press, Washington, DC. Politi L, et al. A rapid screening procedure for drugs and poisons in gastric contents by direct injection-HPLC analysis. Foren Sci Int 2004;141:115–20.*

19. d. Heroin (diacetylmorphine) metabolizes within minutes to 6-monoacetylmorphine which is a key indicator of heroin use. However, 6-MAM is metabolized and eliminated from blood relatively quickly and as such, is often not detected either. Codeine is typically found in trace amounts as a by-product during the manufacturing process of illicit heroin from the opium poppy and consequently, is often found at blood concentrations of a tenth to that of morphine. When codeine administration occurs and metabolizes into morphine, an inverse is typically seen where ~10% of the morphine concentration in blood is seen compared to codeine. *Stam NC, et al. The attribution of a death to heroin: a model to help improve the consistent and transparent classification and reporting of heroin-related deaths. For Sci Int 2017;281:18–28.*

20. e. Typical clinical immunoassays, for example UTOX, do not screen for Kratom (that contains the psychoactive components mitragynine and metabolite 7-hydroxymitragynine). Therefore, a more comprehensive toxicological analysis is required. However, due to the extended period of time from hospital admission to the only available antemortem specimen (that was collected on day 17), useful toxicology testing is unachievable due to elimination of drugs and metabolites. This is therefore also consistent for all blood, urine, and other internal biological tissues postmortem specimens. Hair was not available due to chemotherapy treatment, and even so, hair is typically inadequate to reveal singular exposure in an acute fatal overdose. Antemortem specimens are often pivotal for successful and complete medicolegal death investigations. Hospitals are encouraged to retain antemortem specimens when a potential homicide, accident, suicide, or suspicious death is possible. Although significant central nervous depression due to Kratom and Ativan toxicity was suspected, with the given lorazepam antemortem result and insufficient further toxicology analysis possible, the cause of death for this decedent was subsequently given as undetermined. *Walsh EE, et al. To test or not to test?: the value of toxicology in a delayed overdose death. J Forensic Sci, 2019;64:314–7. Organization of Scientific Area Committees, Medicolegal Death Investigation Subcommittee. Recommendations for medical examiner/coroner drug-related death investigations. 2018. https://www.nist.gov/sites/default/files/documents/2018/02/14/osac_mdi_drug_related_ investigation_recom mendations_final_2-14-18.pdf.*

21. c. High-resolution mass spectrometry enables mass identifications to several positions past the decimal point. This enables the determination of the exact molecular mass, not just the nominal mass. The molecular formula can be deduced from the mass. *Wu AHB et al. Role of liquid chromatography-high resolution mass spectrometry (LC-HR/MS) in clinical toxicology. Clin Toxicol (Phil) 2012;50:733–42.*

22. a. Electron impact produces a reproducible mass spectrum. When compared to library spectra, the mass spect pattern is used to identify unknowns. *Mass spectrometry tutorial. Iowa State University chemical instrumentation facility. https://www.cif.iastate.edu/mass-spec/ms-tutorial.*

23. a. Driving while intoxicated cutoffs are defined in whole blood, which are lower than the corresponding serum by 10%–15%, and this value is below the DUI cutoff of 0.08%, However, considering that there was some delay from the accident to presentation, thus this value may be higher than the DUI cutoff. For civil cases, the forensic standards do not necessarily need to be met. *Barnhill MT.Comparison of hospital laboratory serum alcohol levels obtained by an enzymatic method with whole blood levels forensically determined by gas chromatography. J Anal Toxicol. 2007;31:23–30.*

24. a. Trace amounts of ethanol can be produced through in vitro fermentation. The requirements are glucose in the urine, as can be seen from a diabetic; fermentation enzymes such as those found in infections; and sufficient incubation time, such as the time it took for the lab to receive the sample. *Sulkowski HA, et al. In-vitro production of ethanol in urine by fermentation. J Foren Sci 1995;40:990–3.*

25. b. Poppy seeds contain codeine, morphine, thebaine, and other naturally occurring opiates but not hydromorphone. *Cassella G, et al. The analysis*

of thebaine in urine for the detetion of poppy seed consumption. J Anal Toxicol 1997;21:376–83.

26. d. Substitution with water or saline will produce values outside these limits. These tests will not detect substitution of one urine sample for another. *Urine Drug Testing for Specific Substances. In: Drug Testing in Primary Care. SAMHSA Technical Assistance Publication Series TAP 32.2012, Chapter 5.*

27. c. Creatinine values between 2 and <20 mg/dL are considered dilute by SAMHSA regulations. Dilution can be the result of excess fluid intake or the addition of liquid after collection. *Urine Drug Testing for Specific Substances. In: Drug Testing in Primary Care. SAMHSA Technical Assistance Publication Series TAP 32.2012, Chapter 5.*

28. d. New to 2017 was the addition of the semisynthetic opioids. Each of the other changes was made earlier, except that THC is still being tested. *Fed Reg 2017;82:7920–70. https://www.samhsa.gov/sites/default/files/work place/frn_vol_82_7920_.pdf.*

29. c. Synthetic opioids such as fentanyl and fentanyl analogs are the leading cause of the opioid crisis in the United States. Most comprehensive analysis of fentanyl crisis urges innovative action. *Science News, 2019. https://www.sciencedaily.com/releases/2019/08/190829081407.htm.*

30. b.Organophosphates are irreversible inhibitors and carbamates are reversible inhibitors of acetylcholinesterase. Low levels can be seen in genetic variances. These individuals are at risks for prolonged apnea due to succinylcholine anesthesia. *Colović MB, et al. Acetylcholinesterase inhibitors: pharmacology and toxicology. Curr Neuropharmacol. 2013;11 (3):315–35. doi:10.2174/1570159X11311030006. Geyer BC. Reversal of succinylcholine induced apnea with an organophosphate scavenging recombinant butyrylcholinesterase. PLos One 2013;8:e59159.*

31. c. Although psuedocholinesterase is easier to measure, red cell cholinesterase tracks organophosphate exposures better. *Brahmi MD, et al. Prognostic value of human erythrocyte acetylcholinesterase in acute organophosphate poisoning. Am J Emerg Med 2006;24:822–7.*

Chapter 16

Mass spectrometry

Y. Victoria Zhang[a], Brandy Young[a], Putuma P. Gqamana[a],
Wayne B. Anderson[a], and Alan H.B. Wu[b]
[a]University of Rochester, Rochester, NY, United States, [b]University of California, San Francisco,
CA, United States

1. Which mobile phase system would work best for an LC-MS/MS assay for
   fentanyl using a C18 reversed-phase column with electrospray ionization in
   positive ion mode?

Fentanyl

a. Mobile Phase A: 95:5 (5 mM Sodium Chloride/Methanol) with 0.1% formic acid; Mobile Phase B: 5:95 (5 mM Sodium Chloride/Methanol) with
   0.1% formic acid
b. Mobile Phase A: 95:5 (5 mM Ammonium Formate/Methanol) with 0.1%
   formic acid; Mobile Phase B: 5:95 (5 mM Ammonium Formate/Metha-
   nol) with 0.1% formic acid
c. Mobile Phase A: 95:5 (5 mM Ammonium Bicarbonate/Methanol) with
   0.1% ammonium hydroxide; Mobile Phase B: 5:95 (5 mM Ammonium
   Bicarbonate/Methanol) with 0.1% ammonium hydroxide
d. Mobile Phase A: 95:5 (5 mM Ammonium Bicarbonate/ Methanol) with
   0.1% potassium hydroxide; Mobile Phase B: 5:95 (5 mM Ammonium
   Bicarbonate/Methanol) with 0.1% potassium hydroxide

Self-assessment Q&A in Clinical Laboratory Science, III. https://doi.org/10.1016/B978-0-12-822093-1.00016-8
195

2. You have developed a quantitative MRM LC-MS/MS method on a triple quadrupole mass spectrometer for a proprietary drug in urine. The method incorporates a stable isotope-labeled internal standard. The method has a limit of detection of 50 pg/mL and an LLOQ of 0.25 ng/mL. The compound has a retention time of 2.1 min and a peak width of 8 seconds. Which cycle time should you use in your method?
   a. 0.1 s
   b. 0.5 s
   c. 2.0 s
   d. 4.0 s
   e. 8.0 s

3. You have obtained a high-resolution accurate mass spectrum (shown as follows) of a doubly protonated compound by electrospray ionization in positive ion mode. What is the monoisotopic mass of the compound?

Isotope number	m/z	Percent total	Percent maximum
0	346.69757	64.11	100.00
1	347.19903	27.86	43.46
2	347.70040	6.70	10.44
3	348.20171	1.16	1.80
4	348.70299	0.16	0.25
5	349.20426	0.02	0.03
6	349.70553	0.00	0.00
7	350.20698	0.00	0.00

   a. 173.349 g/mol
   b. 346.698 g/mol
   c. 347.199 g/mol
   d. 691.380 g/mol
   e. 1043.951 g/mol

4. You have obtained a high-resolution accurate mass spectrum of the peptide FAWRIRM (average mass = 979.197, monoisotopic mass = 978.511) by electrospray ionization in positive ion mode. Which of the following ions are you likely to see in the mass spectrum?
   a. 327.415
   b. 490.268
   c. 978.511
   d. 979.197
   e. 1957.022

5. You have developed and validated an LC-MS method on a triple quadrupole mass spectrometer for the quantification of fentanyl in urine. Your method employs SRM (selected reaction monitoring) analysis to monitor the following MS transition: m/z 337.2 → m/z 188.1. A patient presents with clinical symptoms of fentanyl exposure; however, his urine samples have repeatedly tested negative (<1 ng/mL) for fentanyl by your assay. Using the existing assay on the triple quadrupole mass spectrometer as a starting point, which of the following experimental approaches would have the best chance of detecting the presence of fentanyl or a fentanyl analog in the patient's urine?

m/z 188.1434

Fentanyl
$[M+H]$+: 337.2274 Da

CH$_3$

a. Concentrate a large volume (>10 mL) of the patient's urine on a SPE cartridge, elute with organic solvent, evaporate the eluate, and reconstitute in a small volume (~500 µL) of mobile phase. Analyze the concentrated sample by the validated method.

b. Reanalyze the processed urine on the validated LC/MS system. Instead of performing the validated SRM scan, perform a SIM (selected ion monitoring) Q1 scan for m/z 337.2.

c. Reanalyze the processed urine on the validated LC/MS system. Instead of performing the validated SRM scan, perform a precursor ion scan for precursors of m/z 188.1.

d. Reanalyze the processed urine on the validated LC/MS system. Instead of performing the validated SRM scan, perform a full Q1 scan across the range m/z 100–1500.

6. Which of the following statements is FALSE, regarding ACS/Sigma-Aldrich reagents and grading for LC-MS/MS use?

a. The purity of analytical reagents and solvents is very critical in the performance of the LCMS instrument.

b. Only LC-MS grade reagents are suitable for use in LC-MS/MS analysis.

    c. Analytical grade reagents are 99.5% pure, so they automatically qualify for use in LC-MS/MS analysis.

    d. Technical grade reagents should never be used in any LC-MS/MS application.

    e. Life science grade often refers to high chemical purity as well as biological purity, but it does not necessarily imply suitability for LC-MS/MS applications unless this is explicitly stated in writing.

7. During the flow of solvent through the LC-MS instrument, a distribution of sharp evenly spaced peaks of high intensity shows up on the readout. What could be the cause?

    a. using detergent on solvent glassware

    b. reusing old mobile phase

    c. using Millipore water with $9\,M\Omega\,cm$ (megohm) reading

    d. using biodegradable plastic tubing in the LC connections

    e. any of the above

8. Which one of the following aqueous buffers has the highest pH? (pKa-formate $= 3.75$; pKa-acetate $= 4.76$; Both $\sim100\%$ acid stocks have a pH around 2.4)

    a. 10 mM ammonium formate in 0.1% (v/v) formic acid

    b. 10 mM ammonium acetate in 0.1% (v/v) acetic acid

    c. 5 mM ammonium formate in 0.1% (v/v) formic acid

    d. 5 mM ammonium acetate in 0.1% (v/v) acetic acid

    e. they all have the same pH

9. Which one of the following statements is true regarding the analytical performance of immunoassays (IA) when compared to LC-MS/MS assays?

    a. Immunoassays are better for the high-throughput analysis of a lot of analytes.

    b. LC-MS/MS assays are better for the high-specific analysis of all analytes.

    c. Immunoassays are better for the high-specific analysis of most analytes.

    d. LC-MS/MS assays are better for the high-throughput analysis of a lot of analytes.

    e. Both assays can be used equally for the high-throughput and high specific analysis of most analytes.

10. During tandem mass spectrometry using a triple quadrupole instrument (QqQ), the analyte precursor ions are selected in Q1, fragmented in the collision cell (q), and the fragment ions detected in Q3. The detection of a filtered precursor ion-fragment ion pair is called single reaction monitoring (SRM). For multiple reaction monitoring (MRM)...

    a. multiple fragment ions need to be detected alone in Q3.

    b. multiple precursor ions need to be fragmented in Q1.

  c. multiple combinations of precursor ion-fragment ion pairs can be scanned and used conveniently.

  d. multiple precursor ion-fragment ion pairs can be filtered from multiple or single analytes.

  e. one precursor ion-fragment ion pair has to be filtered from the fragmentation of multiple analytes.

11. When developing a quantitative LC-MS/MS assay what measure can be taken to help control for matrix impacts on the sample analysis?

  a. the use of calibrants and QCs.

  b. the use of a stable isotope labeled internal standard.

  c. a qualitative assessment of the potential interferences.

  d. a longer LC column.

12. Discuss why it is necessary to calibrate a quadrupole mass spectrometer?

  a. to set the mass accuracy along the measurement range.

  b. to set the relative abundance of the signal in limited regions along the mass axis.

  c. to set the width of the mass peaks in the upper mass range.

  d. to improve the liquid chromatography.

13. There are many MS scan modes—full scan, precursor ion scan, product ion scan, multiple reaction monitoring (MRM), etc. However, MRM is the most commonly employed in a toxicology setting. Why would one choose a MRM scan over a full MS scan?

  a. Because it offers a survey scan, which will allow the detection of all molecules with similar functional groups.

  b. Because it offers all product ion information—in a near simultaneous manner—that is not selectively filtered.

  c. Because it offers information specific to the analyte of interest. It offers the analysis of the fragment ion that is specific to the analyte of interest. All other ions are filtered out and this offers improved sensitivity for the analyte of interest.

  d. Because it offers a survey of all the molecules in the biological sample.

14. There are many MS scan modes—full scan, precursor ion scan, product ion scan, multiple reaction monitoring (MRM)—to name a few—however, MRM is the most commonly employed in a toxicology setting. Identify the MRM scan in the following schematic?

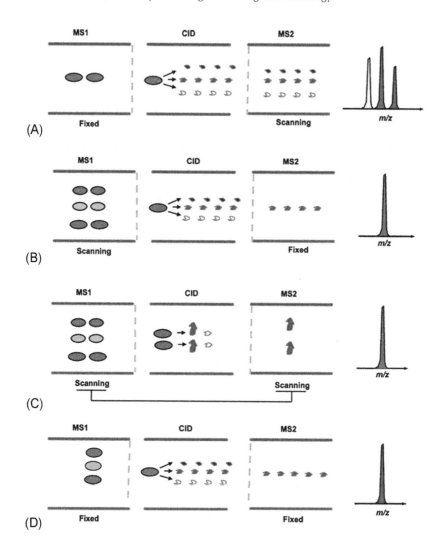

15. The molecules below (THC and THC-metabolites) were separated on a C18 stationary phase (i.e., hydrophobic stationary phase) using a methanol mobile phase. Match the correct molecule (A–B) with the correct elution peaks (I–III).
   a. A:I, B:II, C:III
   b. A:III, B:I, C:II
   c. A:II, B:III, C:I
   d. A:III, B:II, C:I

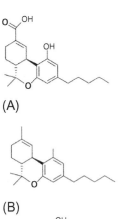

(A)

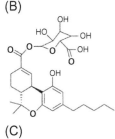

(B)

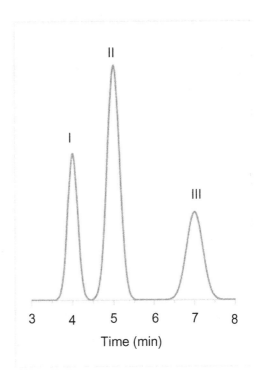

(C)

16. When the analyte is *ionized* what will form?
    a. anions or cations
    b. neutral molecules
    c. clusters (i.e., dimers or trimers)
    d. nominal mass species
17. What are the three main components of the mass spectrometer?
    a. LC, mass analyzer, and source
    b. mass analyzer, LC, and detector
    c. source, mass analyzer, and detector
    d. source, detector, and LC
18. Describe why the LC-MS/MS assay might offer some advantages over the traditional immunoassay testing platform?
    a. more analyte cross reactivity
    b. more specific and selective analyte detection
    c. better antibody affinity
    d. limited analyte resolution

19. Which of the following is not considered a high-resolution mass spectrometer, capable of part-per-million resolution of molecular mass?
    a. ion trap
    b. Fourier transform ion cyclotron
    c. time of flight
    d. orbitrap
    e. all of the above are high resolution
20. Which of the following is used as the detector in most mass spectrometers?
    a. electron capture
    b. photomultiplier
    c. electronmultiplier
    d. ion sensing electrode
    e. flame ionization
21. Which application has not been described using mass spectrometry?
    a. serum proteins for monoclonal gammopathy
    b. whole blood analysis for immunosuppressants
    c. newborn screening using dried blood spots
    d. ambient ionization for point-of-care testing
    e. CD4 and CD8 counts for monitoring HIV infections
22. Which of the following mechanisms best describes how an ion-pairing reagent is used to improve liquid chromatography?
    a. Ionic molecules of opposite charge to the target analyte for chromatographic retention.
    b. Ionic molecules of the same charge to the target analyte for chromatographic retention.
    c. Nonionic molecules of the opposite charge to the target analyte for chromatographic elution.
    d. Nonionic molecules of the same charge to the target analyte for chromatographic elution.
    e. Nonionic molecules of the same charge to the target analyte for chromatographic retention.

Answers

1. b. First, under acidic conditions, the oxygen of the amide group can accept a proton, leading to the formation of a positively charged oxonium ion. (The stability of the oxonium ion is enhanced through resonance stabilization with the nitrogen of the amide group, forming a quaternary ammonium ion.) Second, nonvolatile components should be avoided in LC/MS mobile phases. Therefore, a. and d. should be avoided due to the presence of nonvolatile components. c. and d. should be avoided because they are basic. B is the best choice, volatile components and acidic. *A guide to HPLC and LC-MS buffer selection, John Dolan, www.ace-hplc.com, Advanced Chromatography Technologies, 1 Berry Street, Aberdeen, Scotland. Making LC*

methods MS friendly, 08Oct2013, Mark Powell, www.agilent.com, Agilent Technologies.

2. b. For accurate and reproducible integration, you should collect 15–20 data points across a chromatographic peak; more data points are usually unnecessary and result in long processing times and wasted storage space. Since you are collecting data for at least two SRM transitions (analyte and IS), the mass spectrometer must switch between them. The cycle time is the amount of time that the mass spectrometer spends acquiring data for all transitions in the method before repeating the process. Hence, if you use a cycle time of 0.5 s, you will collect 16 data points across a peak with a width of 8 s, which is optimal. Parameters that influence MS quantitation. https://www.ionsource.com/tutorial/msquan/tips.htm. *Lange V, et al. Selected reaction monitoring for quantitative proteomics: a tutorial. Mol Syst Biol 2008;4:222.*

3. d. $m/z = [M+2H]/+2$, $346.69757*(2) = M+2 * (1.007825)$, $693.39514 = M +2.01565$, $M = 693.39514 - 2.01565$, $M = 691.380$ g/mol. http://prospector. ucsf.edu/prospector/cgi-bin/msform.cgi?form = msisotope, for peptide FAWRI, charge state +2. *De Vijlder T, et al. Mass Spec Rev 2018;37:607–29.*

4. b. Ions corresponding to the average mass (i.e., 327.415, 979.197) will not be observed in the HRAM spectrum. Ions with m/z ratios directly proportional to the monoisotopic mass of the peptide (i.e., 978.511, 1957.022) will not be observed because they fail to account for the mass of the adducted proton. The doubly charged $[M+2H]^{2+}$ ($m/z = 490.268$) species will be abundant, especially since the peptide contains two arginine residues, which are easily protonated. *http://prospector.ucsf.edu/prospector/cgi-bin/msform. cgi?form=msisotope, or http://prospector.ucsf.edu/prospector/cgi-bin/ msform.cgi?form = msproduct for peptide FAWRIRM, charge state +2.*

5. c. Concentrating the sample will not help, the patient is exhibiting symptoms of fentanyl exposure and the validated assay has adequate sensitivity. Performing a SIM scan is less selective and sensitive than the validated SRM scan. Scanning for precursors of m/z 188.1 may detect fentanyl analogs that are chemically modified on the "left" or amide side of the molecule. (Additionally, scanning for a neutral loss of 149.1 may detect fentanyl analogs that are chemically modified on the "right" side of the molecule.) d. Performing a full Q1 scan will create a mountain of less sensitive, unselective data that would have to be searched manually for fentanyl analogs. The analog would only be detected if it had a known molecular mass and was present at relatively high concentration. *Hoffmann E. Tandem mass spectrometry: a primer. J. Mass Spectrom 1996;31:129–37. Klingberg J, et al. Collision-induced dissociation studies of synthetic opioids for non-targeted analysis. Front Chem 2019;7:331. doi:10.3389/fchem.2019.00331.*

6. c. AR grade is no guarantee for LCMS compatibility. The solvents need to be explicitly declared and guaranteed as LC-MS/MS grade. *MacNeill R, Solvent and additive purity selections in bioanalytical LC-MS, in https://www.bioanalysis-zone.com/2018/10/02/solvent-additive-purity-selections-bioanalytical-lc-ms/. https://www.labicom.cz/cogwpspogd/ uploads/2018/06/Purity-of-lab-reagents-K.-Pokajewicz-Honeywell.pdf.*

7. e. The evenly distributed peaks typically originate from a polymer impurity, e.g., PEG. Any one of the above could be the source. *Guo X, et al. Characterization of typical chemical background interferences in atmospheric pressure ionization liquid chromatography mass spectrometry. Rapid Commun Mass Spectrom 2006;20:3145–50. Bartlett MG. Preface. J Chromatogr B 2005;825:97. Dolan JW. LCGC North Am 2005;23:1256. https://www.acros.com/myBrochure/LCMS_wht% 20paper_FINAL.pdf.*

8. b. pH = pKa + log ratio (B/A). Hence, pH is proportional to pKa if ratio is equal, or proportional to ratio if pKa is equal. Acetate pKa > Formate pKa. 10mM > 5mM, etc. 1. *http://www.molbiotools.com/buffercalculator. html.*

9. a. IAs have been dominant for a while as automated clinical analyzers; used in single analyte assays, as well as in some multiplex assays. The recent clinical usage of LC-MS/MS continues to increase rapidly, owing to several significant advantages of LC-MS/MS over the traditional IA methods, such as specificity and multiplex advantage, both which have yet to be harvested for high-throughput clinical analysis in the future. *Hoofnagle AN, et al. The fundamental flaws of immunoassays and potential solutions using tandem mass spectrometry. J Immunol Methods 2009;347(1–2):3–1.*

10. d. SRMs pertain to the filtering of a single precursor-ion/fragment-ion pair, whereas MRM pertains to multiple SRMs. *Kondrat RW et al. Direct analysis of mixtures by mass spectrometry. Analy Chem 1978;50:81A–92A. Johnson JV, et al. Tandem-in-Space and Tandem-in-time mass spectrometry: triple quadrupoles and quadrupole ion traps. Anal Chem 1990;62:2162–72. Hoffmann E. Tandem mass spectrometry: a primer. J Mass Spectrom 1996;31:129–37.*

11. b. An internal standard mimics the analyte and is added to the sample to compensate for the variability encountered during the sample handling/ processing, due to the matrix, and instrument analysis. *Panuwet P et al. Biological matrix effects in quantitative tandem mass spectrometry-based analytical methods: advancing biomonitoring. Crit Rev Analyl Chem 2016;46:93–105.*

12. a. The mass spectrometer is essentially an "atomic scale" that can measure charged particles within a specified mass range and calibrating is the process of ensuring the mass accuracy. The importance of tuning and calibration in liquid chromatography-mass spectrometry (LC-MS). CHROMacademy, LCGC's *http://www.chromatographyonline.com/importance-tuning-and-calibration-liquid-chromatography-mass-spectrometry-lc-ms.*

13. c. A—describes a neutral loss scan, where ions with a particular functional group will be detected. B—describes a product ion scan where all fragment ions are detected. C—is the answer. D—describes a full ion scan, where

every ionized species is detected. *Hoffmann E. Tandem mass spectrometry: a primer. J Mass Spectrom 1996;31:129–37.*

14. d. The MRM experiment is accomplished by fixing the parent/daughter ion pairs to be monitored at the detector. Hence, MRM uses Q1 and Q3 filters. *Johnson JV, et al. Tandem-in-Space and Tandem-in-time mass spectrometry: triple quadrupoles and quadrupole ion traps. Anal Chem 1990;62:2162–72. Hoffmann E. Tandem mass spectrometry: a primer. J. Mass Spectrom 1996;31:129–37.*

15. c. With these conditions the most hydrophilic molecule elutes 1st and the most hydrophobic molecule will elute last. The 11-methyl functional group dictates the polarity of the molecule. *McDonald PD, et al. Understanding stationary phases for reversed-phase separations: new notions for a new century. HPLC 2001 Maastricht; 18–19 June 2001. https://www.waters.com/waters/library.htm?locale=en_US&lid=1539734.*

16. a. Charged particles are ions. In order for a molecule to be detected in the mass spectrometer it must be ionized. These particles can be negatively charged (anions) or positively charged (cations). *Ionization: an overview | ScienceDirect Topics. https://www.sciencedirect.com/topics/agricultural-and-biological-sciences/ionization.*

17. c. For an instrument to be considered a mass spectrometer it must have three components. First, a source to ionize the analyte(s) of interest. Second, a mass analyzer to "weight" the analyte(s) of interest. Third, a detector to record the mass/charge of the analyte(s). *Squires G, et al. Francis Aston and the mass spectrograph. J Chem Soc Dalt Trans 1998;38:93–9.*

18. b. The LC-MS/MS assay offers a direct assessment of the analyte and three levels of analytical selectivity. This makes the assay highly specific and selective. The immunoassay is an indirect assessment of the analyte of interest. *Armbruster DA, et al. Clinical chemistry laboratory automation in the 21st Century - Amat Victoria curam (Victory loves careful preparation). Clin Biochem Rev 2014;35:143–53. Bertol E, et al. Comparison of immunoassay screening tests and LC-MS-MS for urine detection of benzodiazepines and their metabolites: results of a national proficiency test. J Anal Toxicol 2013;37: 659–64.*

19. a. High resolution mass spectrometers enable mass resolutions of 5 ppm or less and are used for molecular formula assignments for low molecular weight compounds and for proteomic discovery. An ion trap measures nominal molecular mass. *May M. Pros and cons of three high-resolution mass spec approaches. https://www.biocompare.com/Editorial-Articles/338099-Pros-and-Cons-of-Three-High-Resolution-Mass-Spec-Approaches/.*

20. c. Under vacuum, when an ion strikes the cathode of the electron multiplier, it stimulates release of other electrons, which in turn are focused to a series of dynodes thereby multiplying the signal. The last stage is the anode where an electrical signal is produced proportional to the

number of ions entering the detector, thereby multiplying the signal. *Wikipedia. Electron multiplier. https://en.wikipedia.org/wiki/Electron_multiplier.*

21. e. Answers a.–c. are in routine clinical use in some specialty hospitals. Mass spectrometry for point-of-care is not currently being conducted, but ambient temperature ionization has demonstrated its feasibility. Currently, there are no preliminary data for T-cell counts. *Ferreira CR, et al. Ambient ionization mass spectrometry for point-of-care diagnostics and other clinical measurements. Clin Chem 2016;62:99–110.*

22. a. Ion paring reagents form a complex with the target analyte of opposite charge enabling better separation of other compounds through chromatographic retention. It works by increasing the hydrophobicity of the target for reversed-phase liquid chromatography. For example, the sulfonic acids and quaternary ammonium salts are used to interact with negatively and positively charged analytes, respectively. *Yerneni CK. Ion pair chromatography: a critical perspective. J Anal Pharm Res 2017;4(6). http://medcraveonline.com/JAPLR/JAPLR-04-00121.pdf.*

Chapter 17

Metal testing

Wieslaw Furmaga[a] and He Sarina Yang[b]
[a]University of California, San Francisco, CA, United States, [b]Weill Cornell Medicine, New York, NY, United States

1. Which of the following metals is considered to be an essential trace element in humans?
   1. manganese
   2. copper
   3. selenium
   4. magnesium
   5. iodine
      a. 2, 3, 5
      b. 1, 2, 3, 5
      c. 2, 5
      d. 2, 3, 4, 5
      e. all of the above
2. What are the common symptoms of heavy metal poisoning?
   a. peripheral neuropathy
   b. impaired renal function
   c. vomiting
   d. nausea
   e. all of the above
3. Which of the following hereditary disease involves the metabolism of copper?
   a. Menkes disease
   b. Wilson disease
   c. pantothenate kinase-associated neurodegeneration (PKAN)
   d. acrodermatitis enteropathica
   e. a and b
4. Cobalt is an essential component of which of the following vitamins?
   a. vitamin C
   b. vitamin B12
   c. vitamin E
   d. vitamin D
   e. vitamin K

Self-assessment Q&A in Clinical Laboratory Science, III. https://doi.org/10.1016/B978-0-12-822093-1.00017-X

5. Based on the 97.5th percentile of the National Health and Nutrition Examination Survey from 2007 to 2010, what are the CDC reference values of blood lead level for adult and children, respectively?
   a. 10 μg/dL, 5 μg/dL
   b. 10 μg/dL, 10 μg/dL
   c. 5 μg/dL, 5 μg/dL
   d. 20 μg/dL, 10 μg/dL
   e. 40 μg/dL, 10 μg/dL

6. Which of the following clinical measurement are related to iron status?
   a. hemoglobin
   b. erythrocyte indices
   c. serum ferritin
   d. zinc protoporphyrin
   e. all of the above

7. Which of the following are symptoms of lead poisoning?
   a. anemia
   b. loss of appetite
   c. abdominal pain
   d. constipation
   e. all of the above

8. Lead inhibits _____, an enzyme that catalyzes synthesis of heme from porphyrin.
   a. tyrosinase
   b. aminolevulinic acid dehydratase
   c. sulfhydryl oxidase
   d. glutamine synthetase
   e. pyruvate carboxylase

9. Erythrocyte protoporphyrin concentration is a sensitive indicator of low-level lead exposure.
   a. true
   b. false

10. Selenium is a cofactor required to maintain activity of _____, an enzyme that catalyzes the degradation of organic hydroperoxides.
    a. phosphoenolpyruvate carboxykinase
    b. sulfhydryl oxidase
    c. glutathione peroxidase
    d. glycosyl transferase
    e. prolidase

11. What disease is caused by selenium deficiency:
    a. aceruloplasminemia
    b. hemochromatosis
    c. Plummer-Vinson syndrome
    d. Minamata disease
    e. Keshan disease

12. What metals may be detected in patients who are surgically inserted metal-on-metal (MoM) implant?
 1. chromium
 2. cobalt
 3. lead
 4. nickel
 5. magnesium
   a. 1, 2
   b. 1, 2, 4
   c. 1, 2, 3, 4
   d. 1, 2, 4, 5
   e. all of the above

13. Which of the following is not true concerning lead?
   a. Lead forms covalent bonds with the sulfhydryl group of cysteines in protein and causes the tertiary structure of the protein to change.
   b. A significant fraction of the absorbed lead is rapidly incorporated into bone.
   c. Lipid-dense tissues are particularly sensitive to lead.
   d. Lead inhibits ferrochelatase causing an increase of protoporphyrin in red blood cells.
   e. An estimated 90% of the lead in whole blood is found in the plasma.

14. Which test has the strongest correlation with lead toxicity?
   a. blood lead
   b. urine lead
   c. hair lead
   d. zinc protoporphyrin
   e. red cell distribution width

15. What is the specimen of choice for blood lead testing?
   a. whole blood containing no anticoagulant or preservative
   b. whole blood containing EDTA as an anticoagulant
   c. serum containing no anticoagulant
   d. plasma containing sodium heparin as an anticoagulant
   e. red blood cells

16. Which form of arsenic is the most toxic?
   a. As (III)
   b. As (V)
   c. monomethylarsine (MMA)
   d. dimethylarsine (DMA)
   e. arsenobetaine

17. Which of the following is not true concerning arsenic (As) toxicity?
   a. As inhibits pyruvate dehydrogenase which converts pyruvate to acetyl coenzyme A.
   b. As competes with phosphate for the reaction converting adenosine diphosphate (ADP) to adenosine triphosphate (ATP).

    c. As binds with hydrated sulfhydryl group on protein, distorting the three-dimensional configuration of the protein and causing it to lose activity.

    d. Arsenobetaine and arsenocholine, two most common forms of organic arsenic that are found in food, are toxic.

    e. As is a carcinogen and increases risk for bladder, skin, and lung cancers.

18. Organic mercury, e.g., methylmercury that is commonly found in shellfish and seafood, is harmless to human health.

    a. true

    b. false

19. Which of the following is not true about cadmium?

    a. Cadmium binds to protein and forms protein-Cd adduct that changes the conformational structure of the protein.

    b. Smokers have higher levels of cadmium in blood.

    c. In the liver and kidney, cadmium is found bound to a small protein called metallothionein.

    d. β_2-Microglobulin is a specific measurement of cadmium exposure.

    e. The half-life of cadmium is approximately 16 years.

20. Which of the following analytical method is not used in elemental testing?

    a. flame atomic absorption spectrometry

    b. inductively coupled plasma emission spectrometry

    c. graphite furnace atomic absorption spectrometry

    d. nephelometry

    e. inductively coupled plasma mass spectrometry

21. Which of the following statement is true about inductively coupled plasma mass spectrometry (ICP-MS)?

    a. The plasma is formed in the ICP torch and is made up of argon ions and free electrons.

    b. Singly charged positive ions are formed when collision-induced ionization occurs in the plasma.

    c. ICP-MS is capable of simultaneously detecting multiple elements in a single run.

    d. In the standard mode of data acquisition, detection of arsenic is interfered by argon chloride.

    e. all of the above

Answers

  1. b. Trace minerals that are essential in humans and animals include chromium, cobalt, copper, iodine, iron, manganese, molybdenum, selenium, and zinc. *Clarke W, et al., eds. Contemporary Practice in Clinical Chemistry, Second Edition, Chapter 38, p. 535.*

  2. e. Symptoms of heavy metal poisoning vary a lot and early symptoms are often nonspecific. Symptoms may include, but not limited to, diarrhea, nausea, abdominal pain, vomiting, weakness, fever, peripheral neuropathy, and impairment of renal and hepatic functions. *National Organization for Rare Disorders, Heavy metal poisoning.*

3. e. (A) Menkes disease is an X-linked recessive disorder of copper metabolism. It is characterized by a defect in the transport and incorporation of copper into copper-containing enzyme, leading to deficiency of different copper enzymes. (B) Wilson disease is an inherited disorder resulting from an excessive accumulation of copper, first in the liver and then in the brain. This disease has an autosomal recessive mode of inheritance. (C) PKAN is an inherited neurodegenerative disorder characterized by the excessive accumulation of iron in brain causing progressive gait disturbance and cognitive impairment. (D) Acrodermatitis enteropathica is inherited in an autosomal recessive mode, and the responsible gene is SLC39A4, encoding a zinc transport protein. The disease is a disturbance of zinc homeostasis resulting from a partial block in the intestinal absorption of zinc. *Ferreira CR, et al. Disorder of metal metabolism, Translational Science Rare Disease, 2017, 2(3–4): 101–39.*

4. b. Cobalt is the essential cofactor in vitamin B12. Vitamin B12 is a collective term for a group of cobalt-containing compounds. *Burtis CA, et al, eds. Tietz Fundamentals of Clinical Chemistry and Molecular Diagnosis, 7th ed., Chapter 32, p. 599.*

5. c. CDC now uses a blood lead reference value of 5 μg per deciliter (μg/dL) to identify both adult and children with blood lead levels. The blood lead reference range is based on the 97.5th percentile of the National Health and Nutrition Examination Survey blood lead distribution from 2007 to 2008 and 2009 to 2010. *Centers for Disease Control and Prevention website.*

6. e. Clinically important measures related to iron status include hemoglobin, hematocrit, erythrocyte indices (mean cell volume, mean cell hemoglobin, mean cell hemoglobin concentration), red cell distribution width, serum iron, total iron binding capacity, transferrin saturation, serum ferritin, zinc protoporphyrin, free zinc protoporphyrin, and soluble transferrin receptor (sTfR). *Clarke W, et al., eds. Contemporary Practice in Clinical Chemistry, Second Edition, Chapter 38, p. 536.*

7. e. Initial symptoms of lead poisoning may be nonspecific and hard to detect, which may include vomiting, loss of appetite, weight loss, abdominal pain, and fatigue. The development of lead toxicity follows a progressive pattern. Lead inhibits the synthesis of heme from porphyrin, causing anemia. High levels of lead exposure can cause damage to the kidneys and central nervous system, eventually leading to seizures, unconsciousness, coma, and even death. *Burtis CA, et al., eds. Tietz Fundamentals of Clinical Chemistry and Molecular Diagnosis, 7th ed., Chapter 32, p. 601.*

8. b. Lead inhibits aminolevulinic acid dehydratase (ALAD), which is an enzyme that catalyzes the synthesis of heme from porphyrin. Inhibition of ALAD causes accumulation of protoporphyrin in erythrocytes, which is a significant marker for lead exposure. *Burtis CA, et al., eds., Tietz Fundamentals of Clinical Chemistry and Molecular Diagnosis, 7th ed., Chapter 32, p. 601.*

9. b. Erythrocyte protoporphyrin concentrations are not a sensitive indicator of low-concentration lead exposure, but they are definitive markers for significant lead exposure. *Burtis CA, et al, eds., Tietz Fundamentals of Clinical Chemistry and Molecular Diagnosis, 7th ed., Chapter 32, p. 602.*

10. c. Selenium is a cofactor required to maintain activity of glutathione peroxidase. Absence of Se correlates with loss of glutathione peroxidase activity and is associated with damage to cell membranes caused by the accumulation of free radicals. *Burtis CA, et al., eds., Tietz Fundamentals of Clinical Chemistry and Molecular Diagnosis, 7th ed., Chapter 32, p. 604.*

11. e. Keshan disease is a congestive cardiomyopathy caused by a combination of dietary deficiency of selenium and the presence of a mutate strain of Coxsackievirus, named after Keshan region of China. Children living in the Keshan region who receive no Se supplement develop cardiomyopathy. *Burtis CA, et al., eds., Tietz Fundamentals of Clinical Chemistry and Molecular Diagnosis, 7th ed., Chapter 32, p. 604.*

12. d. Metal-on-metal implants are made up of various combinations of aluminum, chromium, cobalt, iron, magnesium, molybdenum, nickel, and/or vanadium. Continuous motion at the MoM surfaces may result in release of microparticles into the surrounding tissues. These microparticles can corrode, resulting in the release of metal ions into the systemic circulation. Lead is not used in the MoM implant. *Burtis CA, et al., eds., Tietz Fundamentals of Clinical Chemistry and Molecular Diagnosis, 7th ed., Chapter 32, p. 599.*

13. e. Approximately 99% of the lead in blood is associated with red blood cells; the remaining 1% resides in blood plasma. *CDC Agency for Toxic Substances & Disease Registry: Environmental Health and Medicine Education website.*

14. a. The definitive test for lead toxicity is measurement of blood lead. Blood lead has the strongest correlation with lead toxicity. *Burtis CA, et al., eds., Tietz Fundamentals of Clinical Chemistry and Molecular Diagnosis, 7th ed., Chapter 32, p. 601.*

15. b. Blood containing EDTA as an anticoagulant is the specimen of choice for lead analysis. Approximately 99% of the lead in blood is associated with red blood cells; the remaining 1% resides in blood plasma. Serum and plasma are not optimal specimens for lead testing. *Burtis CA, et al., eds., Tietz Fundamentals of Clinical Chemistry and Molecular Diagnosis, 7th ed., Chapter 32, p. 602.*

16. b. As (V) is more toxic than As (III). MMA and DMA are partially detoxified metabolites. Arsenobetaine is a nontoxic form of As that is present in many foods. *Burtis CA, et al. eds., Tietz Fundamentals of Clinical Chemistry and Molecular Diagnosis, 7th ed., Chapter 32, p. 597.*

17. d. Arsenobetaine and arsenocholine are nontoxic forms of arsenic that are present in many foods. Inorganic species of Arsenic, such as As^{3+} and As^{5+} are highly toxic. *Burtis CA, et al., eds., Tietz Fundamentals of Clinical Chemistry and Molecular Diagnosis, 7th ed., Chapter 32, p. 597.*

18. b. Organic mercury is highly toxic to the central nervous system. People who are exposed to methylmercury exhibit symptoms of organic mercury poisoning, which include ataxia, impaired speech, visual field constriction, hearing loss, and somatosensory change. *Burtis CA, et al., eds., Tietz Fundamentals of Clinical Chemistry and Molecular Diagnosis, 7th ed., Chapter 32, p. 603.*

19. d. Cadmium exposure may cause renal damage. β_2-Microglobulin is a marker of renal function, but it is not a specific marker of cadmium exposure. Urinary concentration of cadmium is a more specific measure of cadmium exposure. *Burtis CA, et al., eds., Tietz Fundamentals of Clinical Chemistry and Molecular Diagnosis, 7th ed., Chapter 32, p. 598.*

20. d. Analytical techniques used to measure metals in biological fluid include atomic absorption spectrometry with flame, graphite furnace atomic absorption spectrometry, inductively coupled plasma emission spectrometry, and inductively coupled plasma mass spectrometry. Nephelometry is a technique used in immunology to determine the level of protein in blood, and it is not used in elemental testing. *Burtis CA, et al, eds., Tietz Fundamentals of Clinical Chemistry and Molecular Diagnosis, 7th ed., Chapter 32, p. 594.*

21. e. For a. the argon gas flows through the ICP torch where some electrons are stripped from their argon atoms caused by a high-voltage spark applying to the gas. These electrons, which are caught up and accelerated in the magnetic field, then collide with other argon atoms, stripping off more electrons. Argon gas is broken down into argon atoms, argon ions, and free electrons, which are collectively called inductively coupled plasma. b. When tiny droplets of the samples are injected at a high speed into the ICP plasma torch, the hot plasma dries the aerosol, dissociates the molecules, and forms singly charged positive ions by stripping off an electron from the analyte atoms. c. ICP-MS is capable of detecting multiple elements simultaneously whereas atomic absorption spectrometry can only detect one element at a time. d. In the standard mode, detection of arsenic is interfered by argon chloride because they have the same mass of 75. *Practical Guide to ICP-MS, a tutorial for beginners. 3rd ed.*

Chapter 18

Pharmacogenomics and therapeutic drug monitoring

Alan H.B. Wu

University of California, San Francisco, CA, United States

1. The nomenclature for identifying pharmacogenomic variants has been standardized. Which of the following is incorrectly listed for *CYP2C19*2*?
   a. CYP is the name of the metabolizing enzyme
   b. The *1 variant identifies the variant as wild type
   c. C identifies the subfamily
   d. 19 identifies the chromosome location of the isoenzyme
   e. 2 refers to the family of CYP enzymes
2. What is the difference between the terms "pharmacogenomic" and "pharmacogenetic"?
   a. no difference
   b. Pharmacogenetics refers to the effect of a single gene while pharmacogenomics refers to the effect of all relevant genes and nongenetic factors.
   c. Pharmacogenomics refers to somatic variants while pharmacogenetics refers to germ line variants.
   d. Pharmacogenetics refers to pharmacokinetic interactions while pharmacogenomics refers to pharmacodynamic interactions.
   e. Pharmacogenetics refers to pharmacokinetic interactions while pharmacogenomics refers to the effect of drug transporters.
3. What is not part of the definition of a "companion diagnostic" laboratory test?
   a. The genetic variant is usually somatic.
   b. Helps identify patients who are most likely to benefit from the therapeutic product.
   c. Often the test and the drug are coapproved by the FDA.
   d. Most widely used for cardiovascular disease medications.
   e. Does not include pharmacogenetic tests.

Self-assessment Q&A in Clinical Laboratory Science, III. https://doi.org/10.1016/B978-0-12-822093-1.00018-1

4. Which antiplatelet drug is affected by CYP2C19 variants?

   a. clopidogrel
   b. prasugrel
   c. ticagrelor
   d. rivaroxaban
   e. aspirin

5. A mother on codeine for pain control is breast feeding her infant child. The child dies of an opiate overdose. Which of the following is most likely the explanation?

   a. The mother is a poor CY2D6 metabolizer.
   b. The mother is an ultra-rapid CYP2D6 metabolizer.
   c. The child is an ultra-rapid CYP2C19 metabolizer.
   d. The child is a poor CYP2C19 metabolizer.
   e. Both the mother and child are poor CYP2D6 and CYP2C19 metabolizer.

6. Which of the following is not a barrier for implementing pharmacogenomic tests into medical practice today?

   a. translating pharmacogenomic information into medical decisions
   b. availability of FDA-approved tests that can produce results quickly
   c. education of primary users
   d. reimbursement of tests
   e. no trials that PGx demonstrate that reduces adverse outcomes

7. Which of the following is false regarding the differences in measuring thiopurine methyltransferase activity vs genotyping?

   a. Many genotyping assays only detect the *TPMT*2, *3A,* and *3C,* which may miss 10% of the variants.
   b. The presence of drug inhibitors can affect the enzyme assay.
   c. Blood transfusions can affect the genotyping assay.
   d. Genotyping and phenotyping produce equivalent results most of the time.
   e. Sample processing is a bigger issue for the enzyme assay.

8. Which of the following is true regarding the pharmacogenomic testing for antiplatelet medications?

   a. CYP2C19 testing is important for both clopidogrel and ticagrelor.
   b. Enzyme immunoassays are readily available for the clopidogrel active metabolite.
   c. Platelet aggregometry is useful to determine the pharmacodynamics of antiplatelet medications.
   d. CYP2C19 testing can determine the optimum dose required.
   e. Individuals with CYP2C19 *2 or *3 variants must control their diet of leafy vegetables.

9. Your hospital has implemented a point-of-care (POC) molecular assay for detecting CYP2C19 polymorphisms for patients undergoing percutaneous intervention (PCI) and receiving clopidogrel therapy. Patients genotyped to be CYP2C19 a slow metabolizer should be treated with:

a. higher doses of clopidogrel to account for CYP2C19 function.

b. lower doses of clopidogrel to account for CYP2C19 function.

c. treat with direct acting P2Y12 inhibitors (e.g., ticagrelor and prasugrel)

d. no change in clopidogrel therapy CYP2C19 is responsible for inactivating the drug.

e. follow-up with CYP3A4/A5 and CYP2D6 genotyping.

10. Which medical complications are accompanied by variants in HLAB*5701 and HLAB*1502?

1. Stevens-Johnson syndrome

2. anaphylactic shock

3. toxic epidermal necrolysis

4. toxic shock syndrome

   a. 1,3

   b. 1,2,3

   c. 2,4

   d. 4 only

   e. All of the above

11. Which CYP P450 enzyme is responsible for the metabolism of 25% of the drugs used in clinical practice today?

a. CYP 2D6

b. CYP 2C9

c. CYP 2C19

d. CYP 3A4

e. CYP 1B1

12. Which of the following is not a prodrug for which therapeutic efficacy is limited in poor metabolizers?

a. codeine

b. clopidogrel

c. warfarin

d. tamoxifen

e. all of the above

13. Which drug is incorrectly matched with the pharmacogenomics marker?

a. HLA-b*5701: abacavir

b. CYP2D6: tamoxifen

c. CYP2C9: Plavix

d. VKORC1: warfarin

e. TPMT: azathioprine

14. What is the correct order of metabolism rate among the various ages?

a. premature neonate > toddler > teenager > adult > geriatric

b. premature neonate < toddler < teenager < adult < geriatric

c. toddler > neonate > teenager > adult > geriatric > premature neonate

   d. geriatric > adult > premature neonate > toddler > teenager

   e. all about the same

15. Which form of a drug is best suited for passive gastrointestinal absorption?

   a. basic environment of an acidic drug

   b. pinocytosis

   c. bound to albumin or α1-acid

   d. nonionized form

   e. hydrophilic form

16. Which of the following is not a rationale for therapeutic drug monitoring?

   a. genetic variances

   b. wide therapeutic index

   c. drug-drug interactions

   d. presence of diseases

   e. none of the above

17. Which class of drugs requires testing in whole blood?

   a. anticonvulsants

   b. antiarrhythmics

   c. antibiotics

   d. immunosuppressants

   e. antidepressants

18. Therapeutic drug monitoring is important for busulfan for which medical indication?

   a. multiple sclerosis

   b. hematopoietic stem cell transplantation

   c. breast cancer

   d. hepatic cirrhosis

   e. congestive heart failure

19. Which drug has declined in its usage in the United States, greatly reducing the need for therapeutic drug monitoring?

   a. vancomycin

   b. digoxin

   c. theophylline

   d. phenytoin

   e. methotrexate

20. Which serum drug test is used for overdose detection and not therapeutic monitoring?

   1. salicylate

   2. ibuprofen

   3. acetaminophen

   4. morphine

      a. 1, 3

      b. 1, 2, 3

      c. 2,4

d. 4 only

e. all of the above

21. Which of the following is the major clinical utility of measuring pentobarbital levels?

    a. drug used for treating epilepsy

    b. determining if an individual on pentobarbital can be declared brain dead

    c. determine the proper dose due to pharmacogenomic interactions

    d. determine the proper dose due to pharmacokinetic interactions

    e. determine the proper dose due to pharmacodynamic interactions

22. Which is the correct equation for the volume of distribution (Vd)?

    a. drug dose/blood concentrations

    b. blood concentration/drug dose

    c. $\text{dose} \times e^{-kd \times t}$

    d. $0.693/k_d$

    e. fraction of the dose absorbed × dose/area under the serum concentration vs time curve

23. From the following figure, which curve represents drug concentrations vs time for intravenous bolus?

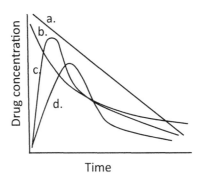

Answers

1. b. By convention *1 is designated as the wild-type variant for the study population when first identified, but is not necessarily the most common variant. The 2C19 refers to the isoenzyme but not the chromosomal location. *Kalman LV et al. Pharmacogenetic allele nomenclature: international workgroup recommendations for test result reporting. Clin Pharmacol Ther 2016;99:172–85.*

2. b. An example of a pharmacogenomic interaction is CYP2C9 and VKORC1 for warfarin, part of a complex mechanism of factors, each with a minor effect. HLAB variant is an example for pharmacogenetics where a single variant has a major impact. *Pirmohamed M. Pharmacogenetics and phar-macogenomics. Br J Clin Pharmacol 2001;52:345–7.*

3. d. Currently, companion tests are most widely used for cancer, e.g., her2neu testing with trastuzumab for breast cancer therapy. *In vitro companion diag-nostic devices guidance for industry and Food and Drug Administration staff. https://www.fda.gov/media/81309/download.*

4. a. Clopidogrel is affected by loss-of-function variants such as CYP2C19*2 and CYP2C19*3. Prasugrel, ticagrelor, and aspirin are not affected by these variants while rivaroxaban is an oral anticoagulant. *Mega JL, et al. Cyto-chrome P-450 polymorphisms and response to clopidogrel. N Engl J Med 2009;360:354–62.*

5. b. Codeine is a prodrug and is metabolized through CYP2D6 to morphine, the active metabolite. If the mother is an ultra-rapid metabolizer, she may overdose her child by transferring morphine via the breast milk. *Madadi P, et al. Safety of codeine during breastfeeding. Fatal morphine poisoning in the breastfed neonate* of a mother prescribed codeine. Can Fam Phys 2007;53:33–5.

6. e. There are data available for some pharmacogenomic tests where out-comes are improved (e.g., HLAB-5901 for abacavir). The others listed are significant barriers. *Klein ME, et al. Clinical implementation of pharma-cogenomics for personalized precision medicine: barriers and solutions. J Pharmacut Sci 2017;1–6:2368–79.*

7. c. The enzyme assay affects the enzyme activity assay as it is conducted in red cells. *Weitzel KW, et al. Implementation of standardized clinical pro-cesses for TPMT testing in a diverse multidisciplinary population: chal-lenges and lessons learned. Am Soc Clin Pharmacol Ther 2018;11:175–81.*

8. c. Antiplatelet medications block the P2Y12 receptor onto platelets inhibit-ing binding to fibrinogen and the cross-linkage that occurs during aggrega-tion. The active metabolite is very unstable and not readily measured. Testing is more for therapeutic selection and not dosing. *Brown SA. Phar-macogenomic impact of CYP2C19 variation on clopidogrel therapy in pre-cision cardiovascular medicine. J Personal Med 2018;8.* https://doi.org/10.3390/jpm8010008.

9. c. The CYP2C19 enzyme is required to convert clopidogrel into an active metabolite. Given the presence of CYP2C19 polymorphisms indicating the patient is a slow metabolizer, treatment with ticagrelor or prasugrel may be more appropriate since they are direct acting P2Y12 inhibitors. Patients who are classified as intermediate metabolizers can be treated with double dose of clopidogrel, or alternately, given prasugrel or ticagrelor. *Gurbel PA, et al. Point-of-care CYP2C19 genotype guided antiplatelet therapy in cardiac catheterization laboratory and clinical outcomes after PCI. J Am Coll Card 2019;73:S1. Damman P, et al. P2Y12 platelet inhibition in clinical practice. J Thromb Thrombolysis 2012;33:143–53.*

10. a. Severe skin rashes that lead to detachments are caused by exposure to certain medications such as abacavir and carbamazepine in individuals with the HLA*5701 and HLAB*1502 variants, respectively. They produce Stevens-Johnson syndrome (defined as <10% of body surface area) and toxic epidermal necrolysis (defined as 30% of body surface area). *Fan WL, et al. J Immunol Res 2017:3186328.* https://doi.org/10.1155/2017/3186328.

11. a. Among the substrates (drugs) include opioids, antidepressants, antiarrhythmics, and anticancer drugs. https://www.pharmgkb.org/vip/PA166170264.

12. c. Warfarin directly blocks the action of vitamin K epoxide reductase (VKORC1) enzyme complex, thereby reducing the regeneration of vitamin K. Codeine metabolizes to morphine, clopidogrel to an active metabolite (no name given), and tamoxifen to endoxifen in order to generate pharmacologic action. https://study.com/academy/lesson/warfarin-mechanism-of-action.html.

13. c. Clopidogrel is metabolized mainly through CYP 2C19, although CYP 2C9 is also involved. *Brown SA, et al. Pharmacogenomic impact of CYP2C19 variation on clopidogrel therapy in precision cardiovascular medicine. J Personal Med 2018;8:* https://doi.org/10.3390/jpm8010008.

14. c. Young children have the highest rate of hepatic metabolism that declines with age. Premature neonates have the slowest due to a delay in the maturation of hepatic enzymes. *Poggesi I, et al. Pharmacokinetics in special populations. Drug Metab Rev 2009;41:422–54.*

15. d. Passive GI absorption requires that the drug be present in the nonionic form. *Kiela PR, et al. Physiology of intestinal absorption and secretion. Best Pract Res Clin Gastroenterol 2016;30:145–59.*

16. b. In order to avoid toxicity, TDM is useful for drugs that have a narrow therapeutic index. *Junaid T, et al. Therapeutic drug monitoring. Clin Pharm Educ Pract Res 2019,* https://www.sciencedirect.com/topics/pharmacology-toxicology-and-pharmaceutical-science/therapeutic-drug-monitoring.

17. d. Immunosuppressant drugs such as cyclosporine and tacrolimus are highly lipophilic and reside in red cells and not serum or plasma. *Winkler M, et al. Plasma vs whole blood for therapeutic drug monitoring of patients receiving FK506 for immunosuppression. Clin Chem 1994;40: 2247–53.*

18. b. Busulfan is a cell cycle nonspecific alkylating antineoplastic agent. Specific diseases include acute leukemia, myelodysplasia, chronic myeloid leukemia, thalassemia major, and sickle cell disease. *Salma B, et al. Therapeutic drug monitoring-guided dosing of busulfan differs from weight-based dosing in hematopoietic stem cell transplant patients. Hematol Oncol Stem Cell Ther. 2017;10:70–8.*

19. c. Theophylline has been replaced by beta2-adrenergic agonist inhalers for the treatment of chronic obstructive pulmonary disease. Its use is reduced to those who are unable to use inhalers. *Chronic obstructive pulmonary disease (COPD) medication. Medscape* https://emedicine.medscape.com/article/297664-medication#5.

20. a. Salicylate and acetaminophen are used for suicide attempts and nonograms are available for therapeutic intervention. Serum ibuprofen and morphine levels are not typically available. *Waseem M. What is the role of the Done nomogram in the workup of salicylate toxicity?* https://www.medscape.com/answers/1009987-177769/what-is-the-role-of-the-done-nomogram-in-the-workup-of-salicylate-toxicity. *Ferrell SE. Medscape. How is the Rumack-Matthew nomogram used in the workup of acetaminophen toxicity/poisoning and what information does it provide?* https://www.medscape.com/answers/820200-27243/how-is-the-rumack-matthew-nomogram-used-in-the-workup-of-acetaminophen-toxicitypoisoning-and-what-information-does-it-provide.

21. b. Pentobarbital is used to reduce intracranial pressures by reducing brain activity and induce coma in patients with closed head injury. If a patient needs to be declared brain dead, pharmacologic sedation must be ruled out with low pentobarbital levels. *Evans SJ. Brain death declaration.* https://www.lifesharing.org/wp-content/uploads/2016/02/Evans.pdf.

22. a. The volume of distribution is a measure of the amount of drug in the body to that found in the circulation. Highly lipid soluble drugs will have a high V_d, those that are more polar or bound to serum proteins will have a low V_d. Volume of distribution. https://sepia.unil.ch/pharmacology/?id=61.

23. b. Curve a. saturation kinetics, b. intravenous injection, c. is intramuscular injection and d. is seen after oral administration. Fundamentals of Antimicrobial Chemotherapy. https://courses.lumenlearning.com/microbiology/chapter/fundamentals-of-antimicrobial-chemotherapy/.

Chapter 19

Transplant therapeutic drug monitoring

Christine L.H. Snozek
Mayo Clinic Arizona, Scottsdale, AZ, United States

1. Match each immunosuppressive drug with its molecular target:

Drug	Target
Cyclosporine A	Calcineurin
Everolimus	Inosine monophosphate dehydrogenase (IMPDH)
Mycophenolic acid	Mammalian target of rapamycin (mTOR)
Sirolimus	
Tacrolimus	

   a. Calcineurin: tacrolimus and mycophenolic acid; IMPDH: cyclosporine; mTOR: sirolimus and everolimus.

   b. Calcineurin: cyclosporine and tacrolimus; IMPDH: mycophenolic acid; mTOR: sirolimus and everolimus.

   c. Calcineurin: cyclosporine; IMPDH: mycophenolic acid; mTOR: sirolimus, everolimus, and tacrolimus.

   d. Calcineurin: sirolimus and everolimus; IMPDH: mycophenolic acid; mTOR: cyclosporine and tacrolimus.

   e. None of the above correctly match immunosuppressive drugs with their targets.

2. Many analytical methods for cyclosporine, everolimus, sirolimus, and tacrolimus include preparatory steps such as addition of zinc sulfate, methanol, and/or water. What is the purpose of these steps?

   a. to ensure complete clotting of serum samples

   b. to precipitate proteins before analysis

   c. to lyse red blood cells and release drug

   d. both a and b

   e. both b and c

Self-assessment Q&A in Clinical Laboratory Science, III. https://doi.org/10.1016/B978-0-12-822093-1.00019-3
223

3. Comparing results between the immunoassays and liquid chromatography tandem mass spectrometry (LC-MS/MS) tests used in therapeutic drug monitoring of everolimus can be particularly challenging. Which of the following is NOT a known issue with everolimus TDM methods?

   a. Therapeutic targets were set using immunoassay rather than LC-MS/MS.

   b. Calibrator concentrations defined using both value-based and gravimetric methods.

   c. Potential for cross-reactivity to several inactive metabolites.

   d. Cross-reactivity or contamination with other immunosuppressant drugs.

   e. All of these are known issues with everolimus TDM.

4. Although each patient's situation is unique, what general trends can be seen in target trough concentrations for immunosuppressants in solid organ transplantation?

   a. Target concentrations increase with time since transplant due to prolonged exposure to the graft.

   b. Target concentrations are similar when given as monotherapy vs combination therapy.

   c. Patients deemed low immunological risk can be maintained with lower target concentrations.

   d. The type of organ transplanted (kidney, liver, etc.) has minimal effect on target concentrations.

   e. both a and c

5. Tacrolimus TDM is done in part because of the risk for drug-drug interactions. Which comedications are the most significant concerns for management of transplant patients on tacrolimus?

   a. comedication with nephrotoxic agents

   b. comedication with CYP3A4/5 inducers and inhibitors

   c. comedication with P-glycoprotein inducers and inhibitors

   d. comedication with CYP2D6 inhibitors

   e. all of the above

   f. a, b, and c only

6. Monitoring the variability (standard deviation or coefficient of variation) of serial trough measurements has gained interest in several areas of transplant. Which of the following accurately describes this approach?

   a. It requires significant bioinformatics capability and pharmacokinetics modeling software.

   b. It has not been shown to identify patients at risk for poor outcomes such as rejection.

   c. It can reflect changes in physiological conditions as well as patient adherence.

   d. It should be performed using at least five measurements drawn within 1 month.

   e. It must include measurements collection during inpatient admissions and outpatient clinic visits.

7. What is the relationship between CYP3A5 genotype and tacrolimus dosing in transplant?

    a. Nonfunctional alleles such as CYP3A5*3 are rare, but the standard tacrolimus dose should be decreased when they are present.

    b. Nonfunctional alleles such as CYP3A5*3 are rare, but the standard tacrolimus dose should be increased when they are present.

    c. Nonfunctional alleles such as CYP3A5*3 are common in several ethnicities, and the standard tacrolimus dose should be decreased in *3/*3 homozygotes.

    d. Nonfunctional alleles such as CYP3A5*3 are common in several ethnicities, and the standard dose should be increased in *1/*3 or *1/*1 expressers.

    e. Nonfunctional alleles such as CYP3A5*3 are common in several ethnicities, but the standard dose rarely needs to be adjusted.

8. Which of the following statement accurately describes pharmacokinetic features of mycophenolic acid (MPA)?

    a. The MPA 7-*O*-glucuronide (MPAG) metabolite is inactive but contributes to active drug exposure through enterohepatic recirculation.

    b. The major metabolite, MPAG, is active but contributes only slightly to overall immunosuppressive activity.

    c. MPA is an inactive prodrug, and both MPAG and the acyl glucuronide metabolite are active.

    d. MPA's glucuronide metabolites are inactive but several CYP-mediated metabolites are active.

    e. MPA is subject to enterohepatic recirculation but its metabolites are not.

9. Mycophenolic acid (MPA) is frequently used in combination with immunosuppressive regimens, often alongside a calcineurin inhibitor. What is the current state of therapeutic drug monitoring for MPA?

    a. There is strong evidence for its utility in nearly all solid organ transplant recipients.

    b. Studies suggest monitoring MPA may be most useful in select transplant patients.

    c. Trough MPA concentrations correlate well to area under the curve (AUC) calculations.

    d. Bayesian models have not yet been developed for MPA monitoring.

    e. AUC_{0-12} is the preferred TDM strategy because it is the most convenient for patients.

10. How well standardized are TDM assays for immunosuppressants?

    a. Very well standardized, with internationally recognized reference materials and methods.

    b. Moderately standardized: preanalytical variables are an issue but assays match well between labs.

c. Moderately standardized: preanalytical variables are minimal, but different assays match poorly.

d. Marginally standardized, with significant variability in both analytical and preanalytical factors.

11. Which statement most accurately reflects the use of biomarkers (e.g., cytokines, cell-free DNA) in management of patients before and after transplant?

a. Biomarkers are still in development but are expected to supplant traditional TDM in transplant.

b. Emerging biomarkers complement traditional TDM, each method providing unique information.

c. Biomarkers have been extensively studied but none have successfully been used in clinical practice.

d. No biomarkers predicting rejection risk or drug toxicity have been described to date.

e. The intent of biomarkers is to increase standardization and uniformity of drug dosing.

12. For cyclosporine in liver and kidney transplant patients, which single-timepoint sample has shown the best correlation with area under the curve (AUC)?

a. trough (C_0)

b. 1-h postdose (C_1)

c. 2-h postdose (C_2)

d. 3-h postdose (C_3)

e. 4-h postdose (C_4)

13. What is true about Bayesian pharmacokinetic models for transplant immunosuppression?

a. They incorporate both population pharmacokinetics and patient-specific blood concentrations.

b. They cannot be used for dose adjustments if a patient is on combination therapy.

c. They are less reliable than standard monitoring for complex patients (e.g., comorbidities).

d. both a and c

e. all of the above

14. Tacrolimus is preferred as first-line therapy over cyclosporine for many patients, but does have some negative effects. Which adverse outcome is more common with tacrolimus than cyclosporine?

a. graft loss

b. hyperlipidemia

c. acute rejection

d. gingival hyperplasia

e. new-onset diabetes

15. What is a key difference in immunosuppressive therapy in hematopoietic stem cell transplant (HSCT) as compared to solid organ transplant (SOT)?

    a. Calcineurin inhibitors are rarely used for immunosuppression in HSCT.

    b. Early posttransplant target concentrations are 10-fold higher in HSCT than SOT.

    c. HSCT patients can often be weaned off long-term immunosuppression.

    d. HSCT target concentrations are lower initially then raised over the first year posttransplant.

    e. None of the above are correct

16. Why have some studies recommended correction for hematocrit when performing therapeutic drug monitoring for tacrolimus?

    a. Patients with low hematocrit appear to have decreased clearance of tacrolimus.

    b. Patients with low hematocrit appear to have increased clearance of tacrolimus.

    c. Target concentrations were established using a standardized matrix of 45% hematocrit.

    d. Posttransplant patients generally have elevated hematocrit, especially in pediatrics.

    e. Variations in hematocrit affect TDM interpretation in adults, but only rarely in pediatrics.

17. How well established are therapeutic targets for immunosuppression in pediatric kidney transplant patients?

    a. Target trough concentrations have been well defined for children and adolescents, but not infants.

    b. AUCs have been well defined for pediatric patients and are the standard practice for TDM.

    c. Therapeutic targets have been well defined in pediatric patients for tacrolimus only.

    d. Adult therapeutic targets are used because no age-specific targets are defined for pediatrics.

    e. Therapeutic targets are not relevant because TDM is rarely performed in pediatric patients.

18. Generally, mTOR inhibitors are not introduced until 6 months or more after solid organ transplant. Which effect of these drugs is the reason for this delay?

    a. risk of infection

    b. impaired wound healing

    c. antineoplastic effects

    d. hyperlipidemia

    e. inter-patient variability

19. Modern transplant practice has expanded to include lung and other organs/ tissues that were not previously able to be transplanted. What factors are relevant to therapeutic drug monitoring for practices such as lung transplant, compared to more traditional organs such as kidney and liver?
    a. Immunosuppression is typically achieved using different drugs.
    b. TDM parameters such as optimal sampling times can be different.
    c. Targets are often guided by extrapolation from other transplant types.
    d. all of the above
    e. b and c only

20. Most immunosuppressant TDM assays include a preliminary step of rocking, inverting, or vortexing sample tubes. What is the predicted effect of omitting this step?
    a. No significant effect on measured drug concentration.
    b. Sampling from the top of the tube results in an artificially low drug concentration.
    c. Sampling from the bottom of the tube results in an artificially low drug concentration.
    d. Sampling from the bottom of the tube results in an artificially high drug concentration.
    e. b and c only
    f. b and d only

Answers

1. b. There are three major immunoregulatory pathways targeted by the immunosuppressive drugs for which posttransplant therapeutic drug monitoring is commonly performed. Calcineurin inhibitors (cyclosporine A and tacrolimus) prevent activation of the NFAT pathway that is required for lymphocyte proliferation and function. The mTOR inhibitors (sirolimus and everolimus) block cell cycle progression of T lymphocytes in response to cytokines. IMP dehydrogenase is required for T cell purine synthesis and is inhibited by mycophenolic acid. *Milone LM, et al. Therapeutic Drugs and their Management. Rifai N, et al. eds., Tietz Textbook of Clinical Chemistry and Molecular Diagnostics, 6th ed., 2018, Chap 40, 800-31.e8, 2018.*

2. e. These immunosuppressive drugs distribute extensively into red blood cells and are highly protein bound. Whole blood is therefore the preferred sample type for their quantitation, and sample preparation should include release of the drugs from cells and protein. Typically, zinc sulfate in a water-miscible organic solvent (e.g., methanol) is used to precipitate cellular and plasma proteins. Many protocols also include addition of water or ammonium bicarbonate, or a freeze-thaw cycle, to ensure complete lysis of red blood cells. *Milone LM, et al. Therapeutic Drugs and their Management. Rifai N, et al. eds., Tietz Textbook of Clinical Chemistry and Molecular Diagnostics, 6th ed., 2018, Chap 40, 800-831.e8,*

2018. Shipkova M, et al. Therapeutic drug monitoring of everolimus: a consensus report. Ther Drug Monit 2016;38(2):143–69.

3. a. Unlike other transplant immunosuppressants, many of the early clinical trials using everolimus monitored patients using LC-MS/MS rather than immunoassay. Immunoassay manufacturers use different strategies for assigning calibrator concentrations, including both comparison to LC-MS/MS values in patient samples, and gravimetric traceability. Immunoassays have variable cross-reactivity to everolimus metabolites and to sirolimus. Sirolimus has also been found as a contaminant in internal standards for everolimus, confounding LC-MS/MS analysis. *Shipkova M, et al. Therapeutic drug monitoring of everolimus: a consensus report. Ther Drug Monit 2016;38(2):143–6. Shipkova M et al. Therapeutic drug monitoring of everolimus: comparability of concentrations determined by 2 immunoassays and a liquid chromatography tandem mass spectrometry method. Ther Drug Monit 2017;39(2):102–8.*

4. c. Low immunological risk patients (e.g., those with well cross-matched grafts) generally can be maintained at lower target concentrations without significantly increasing the risk of rejection. Target concentrations tend to decrease over time, being highest in the early posttransplant phase then tapering to lower levels to minimize toxicity. Combination therapy is frequently used to permit lower concentrations of each drug, again reducing the risk of toxicity. Target concentrations are different for each type of organ transplant. *Milone M, et al. Therapeutic Drugs and their Management. Rifai N, et al. eds., Tietz Textbook of Clinical Chemistry and Molecular Diagnostics, 6th ed., 2018, Chap 40, 800-831.e8, 2018. Brunet et al. Therapeutic drug monitoring of tacrolimus-personalized therapy: second consensus report. Ther Drug Monit 2019;41(3):261–307.*

5. f. Tacrolimus is transported by P-glycoprotein and metabolized by CYP3A4 and CYP3A5. CYP2D6 does not contribute significantly to its metabolism. Tacrolimus has a significant risk for nephrotoxicity; comedication with other potentially nephrotoxic drugs can increase this risk in additive or synergistic fashions. *Brunet M, et al., Therapeutic drug monitoring of tacrolimus-personalized therapy: second consensus report. Ther Drug Monit 2019;41(3):261–307.*

6. c. Within-individual variability of trough measurements is a simple way to evaluate immunosuppressant concentrations in the same patient over time, without requiring detailed pharmacokinetic modeling. Trough concentrations are affected by patient factors such as suboptimal adherence to the dosing regimen, and physiological states such as changes in gastrointestinal mobility. Studies have shown that high within-individual variability is a strong predictor of risk for rejection and other complications, using as few as three measurements spanning 3–12 months. Inpatient values are likely less informative than outpatient concentrations, due to the controlled nature of drug administration during hospital admissions. *Brunet M, et al.*

Therapeutic drug monitoring of tacrolimus-personalized therapy: second consensus report. Ther Drug Monit 2019;41(3):261-307. Ivulich et al. Clinical challenges of tacrolimus for maintenance immunosuppression post-lung transplantation. Transpl Proc 2017;49:2153–60.

7. d. The CYP3A5*3 allele is very common in several ethnicities including whites and Asians, and standard tacrolimus doses were developed in largely nonexpressing populations. Several studies have shown that target concentrations are reached more rapidly in CYP3A5 expressers (e.g., *1/*1 or *1/*3) when the standard dose is increased prospectively. Interestingly, outcome studies to date show similar results for both expressers and nonexpressers, possibly due to use of therapeutic drug monitoring for dose adjustments even when CYP3A5 status is not known. *Brunet M, et al. Therapeutic drug monitoring of tacrolimus-personalized therapy: second consensus report. Ther Drug Monit 2019;41(3):261–307.*

8. a. MPA is pharmacologically active, but it is often administered as a prodrug, mycophenolate mofetil (MMF). Active MPA is transformed by glucuronyltransferases to its inactive metabolite MPAG and to the active acyl glucuronide metabolite. MPA and MPAG are both subject to enterohepatic recirculation; MPAG that is reabsorbed can be reactivated by removal of the glucuronide. Overall exposure to active drug is therefore dependent on recirculation of both MPA and MPAG: *Kuypers DRJ, et al. Consensus report on therapeutic drug monitoring of mycophenolic acid in solid organ transplantation. Clin J Am Soc Nephrol 2010;5:341–58.*

9. b. Studies have demonstrated conflicting results about the utility of TDM for MPA, particularly outside of renal transplant populations. Although universal monitoring remains controversial, select populations are more likely to benefit from TDM including patients with altered renal and gastrointestinal function, drug interactions, and noncompliance. Trough MPA concentrations show relatively weak correlation with AUC, whereas strategies such as Bayesian models or multi-timepoint sampling (e.g., limited sampling strategies) better reflect AUC_{0-12}. One major drawback to multi-timepoint strategies including full AUC calculations is that they require the patient to be available for longer than a single sample collection. *Kuypers DRJ, et al. Consensus report on therapeutic drug monitoring of mycophenolic acid in solid organ transplantation. Clin J Am Soc Nephrol 2010;5:341–58.*

10. d. As of the end of 2019, no internationally recognized reference materials or methods exist for TDM of immunosuppressants, although candidate materials and assays have been proposed. A large degree of variability has been demonstrated in the assays themselves (including methodology, accuracy, precision, specificity, and sensitivity) and preanalytical factors (including collection, stabilization, and storage). *Christians U, et al. Impact of laboratory practices on interlaboratory variability in therapeutic drug monitoring of immunosuppressive drugs. Ther Drug Monit 2015;37(6):718–24.*

11. b. Promising biomarkers have been described for predicting risk of rejection and individual responses to immunosuppression, as well as identifying graft dysfunction and injury. The goals of biomarker and TDM testing are to personalize transplant therapy and improve outcomes. Each methodology brings unique information; for example biomarkers might predict adverse response to a particular drug before it is given (TDM cannot), whereas TDM likely will remain standard of care to address factors such as noncompliance. *Brunet M, et al. Barcelona consensus on biomarker-based immunosuppressive drugs management in solid organ transplantation. Ther Drug Monit 38(2S):S1–20, 2016.*

12. c. Cyclosporine reaches an early peak around 2 h postdose, then concentrations decline sharply. Several studies demonstrated better correlation of C_2 peak concentrations with AUC and outcomes, compared to other postdose sampling times or trough levels. *Pollard SG. Pharmacologic monitoring and outcomes of cyclosporine. Transplant Proc 36(2 Suppl):404S–407S, 2004.*

13. a. Bayesian modeling uses both population pharmacokinetics and samples from an individual patient to predict individual drug exposure. They can be used in combination therapy and are particularly helpful in patients with atypical pharmacokinetics, comorbidities such as cystic fibrosis, and other complex situations. *Marquet P, et al. Pharmacokinetic Therapeutic Drug Monitoring of advagraf in more than 500 adult renal transplant patients, using an expert system online. Ther Drug Monit 2018;40(3):285–91.*

14. e. Several large meta-reviews have demonstrated better outcomes with tacrolimus compared to cyclosporine, including less risk of graft loss, acute rejection, hyperlipidemia, and cosmetic side effects. Tacrolimus does have higher risk for new-onset diabetes mellitus, often requiring insulin. *Penninga L, et al. Tacrolimus versus cyclosporine as primary immunosuppression after heart transplantation: systematic review with meta-analyses and trial sequential analyses of randomised trials. Eur J Clin Pharmacol 2010;66(12):1177–87. Webster AC, et al. Tacrolimus versus cyclosporin as primary immunosuppression for kidney transplant recipients. Cochrane Database Syst Rev 2005;19(4):CD003961.*

15. c. While SOT patients require lifelong immunosuppression, many HSCT patients develop sufficient tolerance to eventually wean off immunosuppressive therapy. Many of the same drugs including calcineurin inhibitors are used in both settings. Target concentrations are unique to each type of transplant but overlap considerably; initial concentrations are higher in both HSCT and SOT then gradually lowered. *Allison TL. Immunosuppressive therapy in transplantation. Nurs Clin North Am 2016;51(1):107–20.*

16. b. Because tacrolimus is measured in whole blood, low hematocrit decreases the measured total concentration and gives the appearance of increased clearance despite similar active (plasma) concentrations. Adult and pediatric patients' posttransplant frequently has low hematocrit, and variability in hematocrit has been shown to confound interpretation of tacrolimus concentrations in these populations. Target concentrations were not developed using a standard level of hematocrit. *Schijvens AM, et al. The potential impact of hematocrit correction on evaluation of tacrolimus target exposure in pediatric kidney transplant patients. Pediatr Nephrol 2019;34(3):507–15.*

17. d. Even the frequently used drug tacrolimus lacks well-defined therapeutic targets for pediatric patients, so both trough concentrations and AUCs are interpreted using adult target ranges. *Schijvens AM, et al. The potential impact of hematocrit correction on evaluation of tacrolimus target exposure in pediatric kidney transplant patients. Pediatr Nephrol 2019;34 (3):507–15.*

18. b. Although all of these are features of mTOR inhibitors, the impairment of wound healing is the rationale for delaying use to ensure successful healing of the graft. *Shipkova M, et al. Therapeutic drug monitoring of everolimus: a consensus report. Ther Drug Monit 2016;38(2):143–69.*

19. e. There are not many effective drugs for long-term immunosuppression, so newer transplant practices such as lung typically use the same immunosuppressive agents. However, pharmacokinetic parameters such as absorption and time to peak concentration (for postdose TDM sampling) can be very different. Targets are unique to each transplant type, but newer practices have few well-powered clinical trials for guidance, leading to some reliance on extrapolation from other organ sites. *Ivulich S, et al. Clinical challenges of tacrolimus for maintenance immunosuppression post-lung transplantation. Transpl Proc 2017;49:2153–60.*

20. f. When measuring drug concentrations in whole blood, premixing samples thoroughly is essential to evenly distribute the cells (where the majority of the drug is found) with the plasma. Sampling from an unmixed tube gives artificially low concentrations from the top (predominantly plasma) and artificially high concentrations from the bottom (predominantly cellular components). *Christians U, et al. Impact of laboratory practices on interlaboratory variability in therapeutic drug monitoring of immunosuppressive drugs. Ther Drug Monit 2015;37(6):718–24.*

Section C

Molecular diagnostics and infectious diseases

Chapter 20

Molecular diagnostics—Basics

Greg Tsongalis[a], Alan H.B. Wu[b], and Nam Tran[c]
[a]Dartmouth Hitchock Hospital, Hanover, NH, United States, [b]University of California, San Francisco, CA, United States, [c]University of California, Davis, CA, United States

1. Coding regions of the genome are known as:
   a. transposons
   b. cNvs
   c. exons
   d. SNPS
2. A method of measuring gains and losses throughout the genome is known as:
   a. Sanger sequencing
   b. chromosomal microarrays (CMA)
   c. clinical exome sequencing (CES)
   d. whole exome sequencing (WES)
3. What type of sequencing reaction analyzes the entire genomic including noncoding and coding regions?
   a. targeted sequencing
   b. chromosomal microarrays (CMA)
   c. whole exome sequencing (WES)
   d. whole genome sequencing (WGS)
4. Tumors that have no identifiable hereditary component are said to be:
   a. metachronous
   b. synchronous
   c. sporadic
   d. multifactorial
5. Prior to next-generation sequencing of patient samples, the DNA is:
   a. barcoded
   b. demethylated
   c. probed
   d. ligated
6. The sequencing performed in comparative genome analysis tool (CGAT) for various cancer types is known as:
   a. clinical exome sequencing (CES)

Self-assessment Q&A in Clinical Laboratory Science, III. https://doi.org/10.1016/B978-0-12-822093-1.00020-X

    b. targeted analysis

    c. whole exome sequencing (WES)

    d. whole genomic sequencing (WGS)

7. DNA-based response pathways include all of the following except:

    a. base excision

    b. phosphate excision

    c. mismatch repair

    d. homologous recombinant repair

8. The enzyme most involved with repairing single- and double-strand DNA breaks is:

    a. Polymerase α

    b. *E. coli* RI

    c. MSH6

    d. Poly (ADP-ribose) polymerase

9. Primary analysis for massively parallel sequencing reactions refers to:

    a. annotation

    b. alignment

    c. conversion of raw data to interpretable sequence

    d. interpretation

10. Which is the earliest data file type generated during massively parallel sequencing?

    a. Fastq

    b. Bam

    c. Vcf

    d. Pdf

11. Which quality sequencing metric refers to coverage at a specific nucleotide position?

    a. mapping quality

    b. quality score

    c. read depth

    d. variant call quality

12. Mutations that affect only one copy of a gene are known as:

    a. monoallelic

    b. homozygous

    c. heterozygous

    d. single nucleotide variants (SNVs)

13. Which is not a step in the next-generation sequencing wet bench process?

    a. DNA extraction

    b. library prep

    c. fragmentation

    d. gel electrophoresis

14. Target enrichment for NGS can be accomplished by?

    a. capture-based hybridization

    b. RNA extraction

c. column purification
d. all of the above

15. The next-generation sequencing chemistry most resembling Sanger sequencing is?
    a. pyrosequencing
    b. sequencing by synthesis
    c. multivariant analysis
    d. ion Torrent

16. All of the following can be considered a single nucleotide variant (SNV) except:
    a. nonsense mutation
    b. indels
    c. missense mutations
    d. transversions

17. Micro RNAs are:
    a. variant of transfer RNA
    b. single strand coding small RNA
    c. positively regulate their target genes at the posttranscription level
    d. protected from plasma RNase digestion through protein complexation and incorporation into exosomes
    e. 50–100 nucleotides in length

18. Which best describes digital PCR?
    a. Measures PCR amplification during the exponential phase
    b. Measures the amount of accumulated PCR products at the reaction plateau
    c. The sample is diluted and placed into several thousand small compartments
    d. Melting curves are digitized
    e. Results are compared to standards

19. What are the major advantages of digital PCR?
    a. subject to inhibitors
    b. high analytical precision
    c. only PCR method that is quantitative
    d. only PCR method that measures micro RNA
    e. qPCR enables droplet technology for further improvement in sensitivity

20. Which of the following is not a clinical advantage for next-generation sequencing?
    a. Detection of rare genetic mutations that can cause disease, e.g., cystic fibrosis
    b. Detection of resistant infectious disease microorganisms
    c. Pharmacogenomic analysis of genes that are highly polymorphic, e.g., CYP2D6

    d. Circulating cell-free DNA

    e. Copy number variants (CNVs) with use of short DNA read lengths

21. Which of the following is false of telomeres?

    a. caps at the end of chromosomes that protect loss of DNA during replication

    b. short telomeres are associated with aging

    c. useful in predicting treatment efficacy for bone marrow failure

    d. can be measured using flow cytometry FISH analysis

    e. can be caused by telomerase genetic variants

22. Molecular point-of-care testing (POCT) devices employ either polymerase chain reaction (PCR) or isothermal nucleic acid amplification techniques. Isothermal methods are often faster than traditional PCR. What is the underlying reason for isothermal methods to be faster?

    a. lack of thermocycling

    b. faster polymerases

    c. simple assay/instrument design

    d. more sensitive detection of amplicons

    e. none of the above

23. Given the four phases of a polymerase chain reaction (PCR) in the PCR curve of the following figure, what is the phase indicated by C.

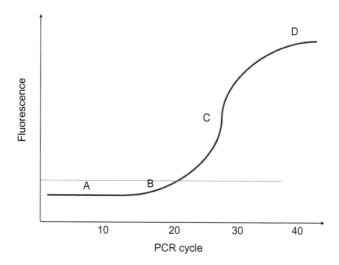

    a. baseline

    b. plateau

    c. linear

    d. growth

24. What is the likely reason for Region D flattening out over time in the polymerase chain reaction curve in the following figure?

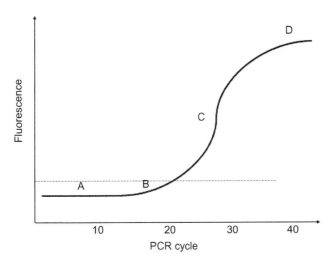

a. PCR thermocycler stopped at 40 cycles
b. depletion of PCR reagents
c. inhibition of polymerase function from amplification products
d. achieved the upper limit of detection
e. none of the above

25. Identify the appropriate complementary DNA strand for the following sequencing 5′-ATTCAGCTAC-3′.
a. 5′-TAAGTCGATG-3′
b. 5′-GTAGCTGAAT-3′
c. 5′-UAAGUCGAUG-3′
d. 5′-GUAGCUGAAU-3′
e. 5′-GTTGCTGAAT-3′

26. Which molecular probe will have the highest melting point?
a. 5′-TAAGTCGATG-3′
b. 5′-GTAGCTGAAT-3′
c. 5′-AAAGACGATG-3′
d. 5′-GCAGCGGCCU-3′
e. 5′-GTTGCTGAAT-3′

27. Loop-mediated isothermal amplification (LAMP) is one form of isothermal nucleic acid amplification that is used in certain point-of-care molecular diagnostics. Other than being isothermal, which is NOT attribute of traditional LAMP.
a. lower instrument cost
b. faster analytical turnaround time

    c. qualitative results only

    d. simplified instrument design

    e. none of the above

28. Assuming a starting concentration of a single target DNA in a patient sample, and consider single-strand initiates and double-strand counts at the end, what is the estimated concentration of DNA following 10 cycles of polymerase chain reaction?

    a. 512 copies

    b. 1 million copies

    c. 1024 copies

    d. 2048 copies

    e. unable to determine

Answers

1. c. The exon is the DNA sequence that can encode the amino acid sequence of proteins. Introns are noncoding regions between exons of a gene. NCI Dictionary of Cancer Terms. *https://www.cancer.gov/publications/dictionaries/genetics-dictionary/def/exon.*

2. b. Chromosomal microarrays are useful for detecting prenatal aneuploidy, other large changes in chromosomal structure, and microscopic abnormalities for screening for some genetic diseases. *Microarrays and next-generation sequencing technology: the use of advanced genetic diagnostic tools in obstetrics and gynecology. Am Coll Obstet Gynecol 2016, https://www.acog.org/Clinical-Guidance-and-Publications/Committee-Opinions/Committee-on-Genetics/Microarrays-and-Next-Generation-Sequencing-Technology?IsMobileSet=false.*

3. d. Whole genome sequencing includes exons and introns. *NCI Dictionary of Cancer Terms. https://www.cancer.gov/publications/dictionaries/genetics-dictionary/def/whole-genome-sequencing.*

4. c. According to the National Cancer Institute, sporadic cancer occurs in people without a *family history of that cancer or an inherited change in their DNA that would increase their risk. NCI Dictionary of Cancer Terms. https://www.cancer.gov/publications/dictionaries/cancer-terms/def/sporadic-cancer.*

5. a. Barcoding of DNA is a method to identify specimens and detect undescribed species, and is a necessary step for Sanger and next-generation sequencing. *Shokralla S, et al. Next-generation DNA barcoding: using next-generation sequencing to enhance and accelerate DNA barcode capture from single specimens. Mol Ecol Resour 2014;14:892–901.*

6. b. CGAT is a tool to compare closely related genomes or mutated genes. *Agrawal N, et al. Comparative genomic analysis of esophageal adenocarcinoma and squamous cell carcinoma. Cancer Discov 2012;2:899–905.*

7. b. Each of these is involved with DNA repair except the removal of the deoxyribose phosphate group. *Chatterjee N, et al. Mechanisms of DNA damage, repair and mutagenesis. Environ Mol Mutagen 2017;58:235–63.*

8. d. DNA damaging agents stimulate the upregulation of PARP. Inhibition of PARP activity may be useful for cancer treatment. *Morales JC, et al. Review of poly (ADP-ribose) polymerase (PARP) mechanisms of action and rationale for targeting in cancer and other diseases. Crit Rev Eukaryot Gene Exp 2014;24:15–28.*

9. c. The primary analysis produces the genomic sequence. The secondary analysis produces variant sequences. The tertiary analysis produces validation of variances. *Oliver GR, et al. Bioinformatics for clinical next generation sequencing. Clin Chem 2015;61:124–35.*

10. a. Fastq is a text-based format for storing nucleic acid sequence along with a quality score. FASTQ format guide. BAM may also be correct. *https://en.wikipedia.org/wiki/FASTQ_format.*

11. c. Read depth is the number of unique reads that include a given nucleotide in a reconstructed sequence. Aiming for a high number of unique reads in each region is referred to as deep sequencing. Coverage genetics. https://en.wikipedia.org/wiki/Coverage_(genetics).

12. a. Monoallelic gene expression enables the transcription of one of the two genes, while the other is silent. *Eckersley-Maslin M, et al. Random monoallelic expression: regulating gene expression one allele at a time. Trends Gen 2014;30:237–44.*

13. d. Gel electrophoresis is needed to separate fragments using Sanger sequencing. The other steps listed are necessary for NGS. Next-generation sequencing. Basic steps of NGS method. *https://www.slideshare.net/USDBioinformatics/basic-steps-of-ngs-method.*

14. a. Hybrid capture occurs when the target DNA is hybridized specifically to preprepared DNA fragments complementary to the regions of interest, either in solution or on a solid support. Other methods include selective circularization and PCR amplification. *Mertes F, et al. Targeted enrichment of genomic DNA regions for next-generation sequencing. Brief Funct Gen 2011;10:374–86.*

15. b. Both NGS and Sanger sequencing DNA fragments, except it is performed in a massively parallel manner. *Garrido-Cardenas JA, et al. DNA sequencing sensors: an overview. Sensors (Basel) 2017;17:588. doi: 10.3390/s17030588.*

16. b. SNVs and SNPs are similar terms. SNPs are well characterized, occur at an expected frequency, and are listed in a dbSNP database. SNVs are not as well characterized. *Katsonis P, et al. Single nucleotide variations: biological impact and theoretical interpretation. Protein Sci 2014;23:1650–66.*

17. d. miRNAs are noncoding RNAs of 17–25 nucleotides and negatively regulate their target genes. *Wang J et al. MicroRNA as biomarkers and diagnostics. J Cell Physiol 2016;231:25–30.*

18. c. Dilution of the sample into thousands of reaction cells enables each cell to have either one or zero DNA molecules. The presence/absence of a signal defines the digital signal. *Huggett JF et al. Considerations for digital PCR as an accurate molecular diagnostic tool. Clin Chem 2015;61:79–88.*

19. b. Digital PCR has higher precision than quantitative PCR. Droplet technology is an enhancement of dPCR. *Huggett JF et al. Considerations for digital PCR as an accurate molecular diagnostic tool. Clin Chem 2015;61:79–88.*

20. e. The error rate for an estimated 1.2% of a single genome differs from the reference human genome when considering CNVs is higher than with single nucleotide polymorphisms, although this may be changing. *Teo SM, et al. Statistical challenges associated with detecting copy number variations with next-generation sequencing. Bioinform 2012;28:2711–8.*

21. c. Aplastic anemia, acute myeloid leukemia, and pulmonary fibrosis are associated with inherited mutations in telomere genes. To date, *however, there are no values in measuring telomere length to access treatment efficacy. Mangaonkar AA, et al. Short telomere syndromes in clinical practice: bridging bench and bedside. Mayo Clinic Proceed 2018;93:904–16.*

22. a. Polymerase chain reaction requires heating and cooling cycles to help denature, anneal, and amplify target nucleic acids. Isothermal techniques such as loop-mediated isothermal amplification (LAMP) eliminate the need for thermocycling. *Keikha M. Lamp as one of the best candidates for replacing with PCR method. Malay J Med Sci 2018;25:121–3.*

23. d. There are four phases of PCR: baseline, exponential, linear, and plateau. The C phase is the exponential phase. *Rice J, et al. Assay reproducibility in clinical studies of plasma miRNA. PLOS One 2015;10:e0121948.*

24. c. Recent studies suggest the PCR curve plateau phase use due to polymerase inhibition by its own amplification products. These include the high concentrations of target amplicons themselves. Jansson L, et al. Challenging the proposed causes of the PCR plateau phase. Biomol Detect Quant 2019;17:1 7. *Rice J, et al. Assay reproducibility in clinical studies of plasma miRNA. PLOS One 2015;10:e0121948.*

25. b. Orientation of all nucleic acids should be 5′–3′. The complementary strand will also need to correctly pair A to T and G to C. The antisense strand must also be aligned correctly to the sense strand. *Watson JD, et al. Molecular structure of nucleic acids: a structure of deoxyribose nucleic acid. Nature 1953;171:737–8. Garibyan L, et al. Research techniques made simple: polymerase chain reaction. J Invest Dermatol 2013;133:e6.*

26. d. The more G-C content, the higher the melting point. G-C base pairing requires three hydrogen bonds, whereas A-T pairing requires two hydrogen bonds. *Watson JD, et al. Molecular structure of nucleic acids: a structure of deoxyribose nucleic acid. Nature 1953;171:737–8. Garibyan L, et al. Research techniques made simple: polymerase chain reaction. J Invest Dermatol 2013;133:e6.*

27. c. Original LAMP was considered to be only capable of "end point" analysis only by detecting amplicons. Quantitative LAMP is now feasible [28]. *Noden BH, et al. Development of a loop-mediated isothermal amplification (LAMP) assay for rapid screening of ticks and fleas for spotted fever group rickettsia. PLOS One 2018;13:e0192331.*

28. a. The number of double-stranded DNA is doubled after each cycle. Therefore a single double-stranded piece of DNA becomes 2 after one cycle (i.e., 4 single strands), 4 after two cycles, etc. *Garibyan L, et al. Research techniques made simple: polymerase chain reaction. J Invest Dermatol 2013;133:e6.*

Chapter 21

Molecular diagnostics—Genetic diseases

Greg Tsongalis[a] and Alan H.B. Wu[b]

[a]Dartmouth Hitchock Hospital, Hanover, NH, United States, [b]University of California, San Francisco, CA, United States

1. Cancer therapy applied after surgery is known as:
   a. cytotoxic
   b. adjuvant
   c. transcriptomic
   d. neo-surgical

2. Colon tumors that display chromosomal instability tend to have mutations in which gene:
   a. proto-oncogene (BRAF)
   b. mutS homolog 6 (MSH6)
   c. anaplastic lymphoma kinase (ALK)
   d. adenomatous polyposis coli (APC)

3. Cell-free DNA can be used for:
   a. noninvasive prenatal testing
   b. monitoring of drug response
   c. testing of resistance for inhibitors
   d. all of the above

4. The most commonly affected gene responsible for homologous recombination repair deficiency (HRD) is
   a. BRCA1
   b. RAD51C
   c. BARD1
   d. ATM

5. Which PARPi is useful in high-grade serous ovarian cancers?
   a. trastuzumab
   b. imatinib
   c. olaparib
   d. keytruda

Self-assessment Q&A in Clinical Laboratory Science, III. https://doi.org/10.1016/B978-0-12-822093-1.00021-1

6. The NTRK gene family encodes:
   a. a receptor tyrosine kinase family
   b. a DNA repair complex
   c. immune checkpoint regulators
   d. a MMR pathway
7. Aberrant expression of NTRK results from:
   a. homologous recombination
   b. in frame fusion
   c. point mutation
   d. out-of-frame deletion
8. Aberrant NTRK expression occurs most commonly in:
   a. Infantile fibrosarcoma
   b. NSCLC
   c. CRC
   d. Glioma
9. Testing for which gene is not recommended in lung adenocarcinoma:
   a. ROSI
   b. RET
   c. ALK
   d. BRAF
10. Which is the most clinically relevant mutated gene in nonsmall cell lung cancer?
    a. BRAF
    b. MEK
    c. PIK3CA
    d. TP53
11. KRAS Q61H in colon cancer is indicative of:
    a. Increased sensitivity to EGFR tyrosine kinase inhibitors.
    b. Decreased response to anti EGFR antibody therapy.
    c. No response to tyrosine kinase inhibitors.
    d. Nothing, it is a variant of unknown significance.
12. Nonsmall cell lung cancers (NSCLC) with ROS1fusions are:
    a. exclusive of other mutations in NSCLC
    b. not sensitive to MET inhibitors
    c. responsive to Crizotinib
    d. all of the above
13. Gastrointestinal stromal tumors with a PDGFRA exon 12 mutation have:
    a. increased sensitivity to imatinib
    b. resistant to sunitinib
    c. resistance to crizotinib
    d. responds the same as those with KIT exon 13 mutations

14. Which type of cancer gene, when mutated, most likely results in activation of the gene product?
   a. oncogene
   b. tumor suppressor gene
   c. MMR
   d. TP53

Match the following tumor types with the most common somatic mutation found in them**15.** lung squamous carcinoma

16. invasive ductal breast carcinoma

17. colorectal cancer

18. melanoma

19. neuroblastoma
   a. KRAS
   b. MAPK
   c. PIK3CA
   d. BRAF
   e. ALK

20. Which targeted therapy is associated with BRAF V600E mutation?
   a. alectinib
   b. cetuximab
   c. vemurafenib
   d. imatinib

21. For which mutation is there an FDA-approved liquid biopsy assay?
   a. KRAS G12D
   b. ALK rearrangement
   c. TPS53 R116D
   d. EGFR T790M

22. What area of lab diagnostics does circulating miRNA analysis appear to hold the most promise?
   a. cancer
   b. diabetes
   c. cardiovascular
   d. liver disease
   e. infectious disease

23. Cell-free DNA analysis has potential for in the diagnosis of human diseases. To predict the potential medical utility, cell-free DNA can cross most readily from
   1. the placenta
   2. the blood brain barrier
   3. solid tumors
   4. organs that have capsules such as thyroid or glioma

    a. 1,2,3
    b. 1,3
    c. 2,4
    d. 4 only
    e. all of the above
24. Which of the following is false regarding BRCA1 and BRCA2 mutation testing?
    a. There are only a few genetic mutations that have been identified for both.
    b. Sporadic inactivation of this gene is a common cause of breast cancer.
    c. The genes are autosomal recessive.
    d. Also associated with a high risk for cervical cancer.
    e. Mutations are somatic and not germline.
25. Which is true regarding DNA methylation?
    a. occurs through environmental factors only
    b. addition of methyl group to adenosine
    c. has no medical significance
    d. methylation can be a mediator, modifier, or consequence of the disease
    e. can provide additional information to genome-wide association studies (GWAS)

Answers

  1. b. According to the National Cancer Institute, the goal of adjuvant treatment is to lower the risk of cancer recurrence. It can include chemotherapy, radiation therapy, hormone therapy, targeted therapy, or biological therapy. Tamoxifen is an example of adjuvant therapy for breast cancer. *NCI Dictionary of Genetic Terms. https://www.cancer.gov/publications/ dictionaries/cancer-terms/def/adjuvant-therapy.*

  2. d. Fodde R. The APC gene encodes a protein that is involved with cell adhesion. Loss of APC protein function allows for clonal expansion and genetic instability enabling tumor progression and malignant transformation. *Fodde R. The APC gene in colorectal cancer. Eur J Cancer 2002;38:867–71.*

  3. d. The utility in prenatal screening is well described. But cell-free DNA can also be used for cancers. *Bronkhorst AJ. The emerging role of cell-free DNA as a molecular marker for cancer management. Biomol Detect Quant 2019;17: https://doi.org/10.1016/j.bdq.2019.100087.*

  4. a. DNA breaks are repaired by homologous recombination. While germline mutations in the BRCA 1/2 genes are the well-known mechanisms of HRD, such as seen in ovarian cancer, other mechanisms, such as germline and somatic mutations in other homologous recombination genes and epigenetic modifications, are known. *Bonadio RRCC, et al. Homologous recombination deficiency in ovarian cancer: a review of its epidemiology and management. Clinics (Sao Paulo) 2018;73 (Suppl. 1):e450s.*

5. c. Poly (ADP-ribose) polymerase has a role in cellular growth, regulation, and cell repair which helps the cancer cells repair themselves and survive. PARP inhibitors include olaparib, niraparib, rucaparib, talazoparib, and Veliparib. *Mittica G, et al. PARP inhibitors in ovarian cancer. Recent Pat Anticancer Drug Discov 2018;13:392–410.*

6. a. NTRK is neurotrophic-tropomyosin receptor tyrosine kinase which promotes the proliferation and survival of neuronal cells. *Okamura R. Analysis of NTRK alterations in pan-cancer adult and pediatric malignancies: implications for NTRK-targeted therapies. JCO Precis Oncol 2018; https://doi.org/10.1200/PO.18.00183.*

7. b. *NTRK* transcript fusions drive overexpression of mRNA and hence protein production. This leads to activation of downstream signaling. *Okamura R, et al. Analysis of NTRK alterations in pan-cancer adult and pediatric malignancies: implications for NTRK-targeted therapies. JCO Precis Oncol 2018; https://doi.org/10.1200/PO.18.00183.*

8. a. IFS is a soft-tissue sarcoma that arises in the 12 months of life, is the most common nonrhabdomyosarcoma and typically manifests as a nontender, rapidly enlarging, circular, erythematous mass over the distal extremities. *Bender J, et al. Refractory and metastatic infantile fibroscaroma harboring LMNA-NTRK1 fusion shows complete and durable response to crizotinib. Cold Spring Harb Mol Case Stud. 2019;5(1):a003376. https://doi.org/10.1101/mcs.a003376.*

9. b. RET mutations are seen in patients with medullary thyroid carcinoma and pheochromocytomas. *Kato S, et al. RET aberrations in divers cancers: next generation sequencing of 4,871 patients. Clin Can Res 2017;23:1988–97.*

10. a. The BRAF protein is part of a signaling pathway that helps transmit chemical signals from outside the cell to the nucleus. *Auliac JB, et al. Patients with non-small-cell lung cancer harbouring a BRAF mutation: a multicentre study exploring clinical characteristics, management, and outcomes in a real-life setting: EXPLORE GFPC 02-14. Current Oncol 2018;25: https://doi.org/10.3747/co.25.3945.*

11. b. Epidermal growth factor receptor (EGFR) blockage is an effective treatment for patients with metastatic colorectal cancer. Unfortunately, resistance to treatment occurs due to mutations in the KRAS and BRAF genes. *Gong J, et al RAS and BRAF in metastatic colorectal cancer management. J Gastrointest Oncol 2016;7: https://doi.org/10.21037/jgo.2016.06.12.*

12. d. ROS1 rearrangements occur in 1%–2% of patients with NSCLS which can be treated with MET prototype oncogene receptor tyrosine kinase and Crizotinib, a c-ros oncogene 1 inhibitor. *Sehgal K, et al. Targeting ROS1 rearrangements in non-small cell lung cancer with crizotinib and other kinase inhibitors. Transl Cancer Res 2018;7:S779–S786. Transl Cancer Res. 2018; 7(Suppl. 7):S779–S786.*

13. a. Patients with Gi stromal tumors have this and variants in other exons benefit to imatinib treatment. *Oppelt PJ, et al. Gastrointestinal stromal tumors (GISTs): point mutations matter in management, a review. J Gastrointest Oncol 2017;8:466–73.*

14. a. An oncogene can transfer a cell into a tumor. *National Cancer Institute Dictionary of Cancer Terms. https://www.cancer.gov/publications/ dictionaries/cancer-terms/def/oncogene.*

15. b. *Kim C, et al.* MEK inhibitors *under development for treatment of non-small-cell lung cancer. Expert Opin Investig Drugs 2018;27:17–30.*

16. c. *Agahozo MC, et al. PIK3CA mutations in ducal carcinoma in situ and adjacent invasive breast cancer. Endocr Relat Cancer 2019; https://doi. org/10.1530/ERC-19-0019.*

17. a. *Dinu D, et al.* Prognostic *significance of KRAA gene mutations in colorectal cancer–preliminary study J Med Life 2014;7:581–7.*

18. d. *Cheng L, et al.* Molecular *testing for BRAF mutations to inform melanoma treatment decisions: a move toward precision medicine. Mod Pathol 2018;31:24–38.*

19. e. *Della Corte CM, et al. Role* and *targeting of anaplastic lymphoma kinase in cancer. Mol Cancer 2018;17. https://doi.org/10.1186/s12943-018-0776-2.*

20. c. BRAF is an intracellular kinase involved in the mitogen-activated protein kinase (MAPK) pathway and regulates cell growth, division, differentiation, and apoptosis. Vemurafenib is approved by the FDA and a BRAF kinase inhibitor used in the treatment of patients, shown to be effective on patients' unresectable or metastatic melanoma. *Dean L. Vemurafenib therapy and BRAF and NRAS Genotype. 2017 Aug 15. In: Pratt V, et al., editors. Medical Genetics Summaries [Internet]. Bethesda (MD): National Center for Biotechnology Information (US); 2012.*

21. d. Osimertinib as a therapeutic was coapproved by the FDA with a laboratory test for the epidermal growth factor receptor mutation T790M. *Odogwu L, et al. FDA benefit-risk assessment of osimertinib for the treatment of metastatic non-small cell lung cancer harboring epidermal growth factor receptor T790M mutation. Oncologist 2018;23:353–9.*

22. a. Dysregulation of miRNAs occurs in different forms of cancer. They can be useful for early diagnosis, differentiate cancer subtypes, monitoring tumor metastasis, and response to therapy. *Wang H, et al. Circulating microRNAs as potential cancer biomarkers: the advantage and disadvantage. Clin Epigen 2018;10:59. https://doi.org/10.1186/s13148-018-0492-1.*

23. b. Cell-free DNA testing can be used for diagnosing pregnancy disorders and tumors from the colon, breast, pancreas, and bladder. *Khier S, et al. Kinetics of circulating cell-free DNA for biomedical applications: critical appraisal of the literature. Future Sci OA 2018;4: https://doi.org/10.4155/ fsoa-2017-0140.*

24. e. There are dozens of BRCA1/2 somatic mutations as a cause of hereditary breast cancer and ovarian cancers. *Kotsopoulos J. BRCA mutations and breast cancer prevention. Cancers 2018;10:524. https://www.ncbi. nlm.nih.gov/pmc/articles/PMC6315560/pdf/cancers-10-00524.pdf.*

25. e. Epigenetic changes have been seen in cancer, autoimmune disease, neurologic diseases, and many others. *Jin YL, et al. DNA methylation in human diseases. Genes Diseases 2018;5:1–8. http://ees.elsevier.com/gendis/ default.aspGenes&Diseases (2018)5,1e8.*

Chapter 22

Molecular microbiology

Nam Tran[a] and Jeffrey Whitman[b]
[a]University of California, Davis, CA, United States, [b]University of California, San Francisco, CA, United States

1. An elderly patient presenting to the Emergency Department with respiratory symptoms was screened using a point-of-care (POC) molecular influenza A/B and respiratory syncytial virus (RSV) assay. He indicated symptom onset was two days ago. Results showed the patient was influenza A positive, and influenza B and RSV negative. The patient was started on oseltamivir, admitted for observation due to being in a higher risk category, and placed under isolation precautions. The Infectious Disease service ordered a subsequent multiplex respiratory panel three days later to determine if the patient can be released from isolation. Results came back as rhinovirus positive and influenza A/B negative. The Infectious Disease service was concerned the POC result was a false positive. The Molecular Laboratory reran the patient sample collected from the third day and showed rhinovirus positive, influenza A positive, and influenza B negative. What is the likely reason for the discordant results?
   a. Oseltamivir therapy reduced viral nucleic acid loads at or near the limit of detection.
   b. Incorrect patient for the POC test.
   c. Rhinovirus RNA interfering with the influenza probes for the multiplex respiratory pathogen panel.
   d. The patient contracted rhinovirus in the hospital.
   e. Environmental contamination of the PCR sample.

2. A patient with a suspected *Staphylococcus aureus* skin and soft tissue infection was admitted to the Emergency Department. Wound swabs were collected for microbiological culture and paired with a rapid polymerase chain reaction (PCR) assay performed at the point of care. The PCR assay has the ability to differentiate between methicillin-resistant *Staphylococcus aureus* (MRSA) and methicillin-sensitive *Staphylococcus aureus* (MSSA). PCR reported MSSA and the patient was prescribed antibiotics (nafcillin). Two days later, the patient presents with worsening signs of

infection and previous microbiological culture results report MRSA and requiring a treatment change to vancomycin therapy to target the pathogen. What is the likely cause of the discrepant PCR result?

a. The pathogen was a variant or did not contain the *mec*A gene for methicillin resistant despite presenting phenotypically MRSA.

b. MRSA concentrations were below the detection limit of the PCR test and picked up the more common MSSA microbe.

c. The PCR reagents failed to amplify the *mec*A gene causing an incorrect MSSA result.

d. An interfering substance on the skin inhibited the detection of MRSA by PCR.

e. none of the above

3. Point-of-care molecular Group A *Streptococcus* (GAS) assays are now commercially available as a CLIA-waived test. In contrast to rapid antigen tests for GAS, negative molecular results should be followed by:

a. microbiological culture to confirm results

b. empiric antimicrobial therapy targeting GAS

c. no antimicrobial therapy targeting GAS

d. confirmatory molecular testing for GAS

e. none of the above.

4. What is the Center for Disease Control and Prevention recommendation for molecular influenza testing during "Flu Season"?

a. Molecular influenza testing should be performed on all suspected patients to optimize antiviral therapy.

b. Molecular influenza testing is not recommended during high prevalence settings and influenza can be diagnosed based on clinical presentation alone.

c. Multiplex respiratory panels should be performed to identify patients with other viral infections and support antimicrobial stewardship practices.

d. Molecular influenza testing is not required since most patients show up to the hospital >48 days after onset of symptoms and may not benefit from antiviral therapy.

e. none of the above

5. The American College of Obstetrics and Gynecology (ACOG) recommends the following regarding molecular influenza testing in pregnant women:

a. Pregnant women should be screened using a molecular influenza test if presenting with suspicion of influenza.

b. Molecular testing is not required, and patients should be empirically treated with antivirals.

c. Targeted molecular respiratory pathogen testing in high-risk obstetrics cases

d. Targeted molecular influenza testing for high-risk obstetrics cases

e. none of the above

6. The primary specimen type for CLIA-waived molecular influenza A/B point-of-care testing is?
   a. nasopharyngeal swab
   b. throat swab
   c. bronchoalveolar lavage
   d. sputum
   e. none of the above

7. The cycle threshold (Ct) value can be used to troubleshoot clinical polymerase chain reaction (PCR) assays performance issues including those used at the point of care. Ct-values are influenced by the starting concentration of target nucleic acids. Which statement best describes the relationship of Ct-value and initial concentration of target nucleic acids?
   a. Ct-value is proportional to the initial concentration of target nucleic acids
   b. Ct-value is inversely proportional to the initial concentration of target nucleic acids.
   c. Ct-value is equal to the initial concentration of nucleic acids.
   d. Ct-value is an arbitrary value defined by the assay manufacturer.
   e. none of the above

8. Given a point-of-care molecular *Staphylococcus aureus* assay that can differentiate between methicillin-sensitive *Staphylococcus aureus* (MSSA) and methicillin-resistant *Staphylococcus aureus* (MRSA). The molecular assay detects the gene for Protein A (*spa*), methicillin resistance gene (*mec*A), and a mobile genetic element called Staphylococcal cassette chromosome *mec* (SCC*mec*). Which genes are detected for *Staphylococcus aureus* positive results?
   a. *spa* positive, *mec*A negative, SCC*mec* negative
   b. *spa* positive, *mec*A positive, SCC*mec* positive
   c. *spa* positive, *mec*A negative, SCC*mec* positive or negative
   d. *spa* negative, *mec*A negative, SCC*mec* positive
   e. *spa* negative, *mec*A positive, SCC*mec* positive or negative

9. A patient who was screened negative for the *Clostridioides difficile* toxin A/B by enzymatic immunoassay (EIA) prior to being transferred to an academic medical center. Upon admission at the new facility, the patient was rescreened for *C. difficile* by a point-of-care polymerase chain reaction (PCR) assay. The PCR result was positive for *C. difficile*. The patient was placed under isolation and treatment started with metronidazole,

but never developed signs/symptoms of *C. difficile* enterocolitis. Stool samples from the PCR specimen were then sent for toxigenic culture and were found to be negative. What is the likely reason for this discrepant result?

a. False positive PCR result—detected *C. difficile* when the microorganism was not present.

b. False positive PCR result—a subset of patients serves as carriers of nontoxin producing *C. difficile*.

c. False negative EIA result at the initial facility.

d. True positive result—patient acquired *C. difficile* shortly before being transferred.

e. False negative EIA due to not measuring glutamate dehydrogenase (GDH).

10. Multiplex molecular meningitis panels are now available at the point of care. Which pathogen is the most common cause of meningitis for both adults and children in the United States?
   a. *Haemophilus influenzae*
   b. *Neisseria meningitidis*
   c. *Streptococcus agalactiae*
   d. *Streptococcus pneumoniae*
   e. *Escherichia coli*

11. Influenza A and B, as well as respiratory syncytial virus (RSV) are common targets for molecular point-of-care testing. These three viruses can be classified as the following based on their genome:
   a. RNA viruses
   b. influenza A and RSV are RNA viruses, while Influenza B is a DNA virus
   c. influenza A and B are RNA viruses, and RSV is a DNA virus
   d. influenza A and B are DNA viruses, and RSV is an RNA virus
   e. none of the above

12. Other than specific primers and probes, what is a key fundamental difference between polymerase chain reaction testing for Group A Streptococcus (GAS) versus Influenza A?
   a. Reverse transcriptase is required for influenza A PCR assays.
   b. Reverse transcriptase is required for GAS PCR assays.
   c. *Taq* polymerase is required for influenza A PCR assays.
   d. *Taq* polymerase is required for GAS PCR assays.
   e. Thermocycling is required for GAS due to the circular genome.

13. In 2009 the H1N1 "swine flu" pandemic revealed certain rapid antigen tests were not able to detect the variant pathogen. However, molecular assays were able to detect this H1N1 variant. What is the likely reason for this discrepancy between immunoassays vs molecular tests?

a. antigenic shift
b. antigenic drift
c. single nucleotide polymorphisms
d. H1N1 viruses were more virulent
e. H1N1 infections typically present with lower viral titers

14. *Chlamydia trachomatis* and *Neisseria gonorrhea* can now be detected by polymerase chain reaction (PCR) including at the point of care using cartridge-based platforms. Other molecular methods available include transcription-mediated amplification (TMA). Which of the following best describes the difference between PCR vs TMA?
    a. PCR requires thermal cycling, TMA is isothermal.
    b. PCR produces DNA amplicons from DNA, while TMA produces RNA amplicons from DNA.
    c. TMA amplification products are more liable which minimizes potential for carryover contamination.
    d. TMA produces 100–1000 copies per cycle vs PCR which only doubles the nucleic concentration per cycle.
    e. all of the above

15. Molecular testing has been shown to provide rapid and objective means to detect fastidious pathogens causing bacterial vaginosis (BV). What is the primary pathogen causing BV?
    a. *Gardnerella vaginalis*
    b. *Candida* species
    c. *Bacteroides* species
    d. *Atopobium vaginae*
    e. *Prevotella* species

16. Multiplex molecular meningitis panels are now available at the point of care. Which pathogen is the most common cause of meningitis in neonates in the United States?
    a. *Haemophilus influenzae*
    b. *Neisseria meningitidis*
    c. *Streptococcus agalactiae*
    d. *Streptococcus pneumoniae*
    e. *Escherichia coli*

17. Which of the following is true regarding testing for covid-19 virus?
    a. The preferred specimen is serum.
    b. The preferred testing methodology is real-time reverse transcriptase polymerase chain reaction for the virus's RNA.
    c. The preferred testing methodology is ELISA for the antibody.
    d. The preferred testing methodology is ELISA for the protein antigen.
    e. The preferred testing methodology is western blot for the virus protein.

18. Which of the following sexually transmitted diseases (STD) should be screened for by nucleic acid amplification testing (NAAT) versus immunoassay?
    a. human Immunodeficiency virus (HIV)
    b. syphilis
    c. hepatitis C virus (HCV)
    d. chlamydia and gonorrhea
    e. all of the above

19. What routine nucleic acid amplification testing (NAAT) is commonly performed in cerebrospinal fluid for the initial workup of acute encephalitis in adults?
    a. herpes simplex virus (HSV)–1/2
    b. enterovirus
    c. varicella zoster virus (VZV)
    d. cryptococcus
    e. syphilis
    f. a., b., c.
    g. all of the above

20. Testing for drug resistance mutations is regularly performed for which of these viral infections?
    a. human immunodeficiency virus (HIV)
    b. hepatitis A Virus (HAV)
    c. West Nile Virus (WNV)
    d. tuberculosis
    e. a. and d.

21. Why is sequencing for HIV-1 drug resistance usually performed in patient plasma over peripheral blood mononuclear cells (PBMCs)?
    a. Plasma testing is more sensitive than PBMC testing.
    b. PBMCs may contain HIV DNA archived from past timepoint in the infection.
    c. PBMC sequencing is not technically feasible in most clinical laboratories.
    d. HIV does not infect PBMCs.

22. Methicillin-resistant *Staphylococcus aureus* (MRSA) can be tested for in clinical settings by nucleic acid amplification testing (NAAT). What gene confers resistance to methicillin in this organism?
    a. ampCerm
    b. mecA
    c. vanA/B
    d. cfr

23. Polymerase chain reaction (PCR) amplification and sequencing of the 16S rRNA gene is used to identify which types of microorganisms?
   a. acid-fast bacilli (AFB)/mycobacteria
   b. bacteria
   c. protozoa
   d. yeast
   e. a. and b.
   f. all of the above

24. NS5A inhibitors are used in the treatment of hepatitis C virus (HCV). What genotype(s) of HCV is most associated with a high prevalence of NS5A mutations that confer resistance to this class of antivirals?
   a. 1a
   b. 1b
   c. 3
   d. 6
   e. a. and b,
   f. a. and c.

25. What is the methodological principle behind metagenomic next-generation sequencing for microbial identification?
   a. Sequencing of PCR amplicons from primers covering hypervariable regions of conserved genes within a class of microorganisms.
   b. Using multiple targeted PCR primers to identify a large proportion of clinically relevant infectious agents.
   c. Identification of microorganisms based on epigenetic methylation pattern.
   d. Sequencing the entirety of genetic material within a clinical sample followed by informatic separation of human and microbial hits.

26. *Mycobacterium tuberculosis* (MTB) infections are regularly screened for rifampicin resistance by nucleic acid amplification testing (NAAT) by looking for *rpoB* gene mutations. What other drug resistance is highly associated with rifampicin resistance?
   a. amikacin/kanamycin
   b. isoniazid
   c. quinolones
   d. ethambutol
   e. pyrazinamide

27. *KatG* is a mutated gene that confers isoniazid resistance in *M. tuberculosis* (MTB) infections. What is the function of the wild-type *katG* gene product?
   a. biosynthesis of mycolic acids
   b. transcribing RNA (RNA polymerase subunit)

    c. activation of prodrug
    d. unwinding of DNA (DNA gyrase)
    e. prokaryotic ribosomal activity (16s rRNA)

28. Commercial nucleic acid amplification tests (NAATs) for *Clostridium difficile* infection contain primers for identifying both *C. difficile* and its toxin genes (*tcdA/tcdB*). What is the importance of testing for the presence of toxin in these cases?
    a. It is important to quantify the toxin with quantitative PCR for clinical management.
    b. The presence of toxin without identification of *C. difficile* is suggestive of another infectious agent.
    c. The ratio of toxin to *C. difficile* is used for antibiotic dosing.
    d. The absence of toxin confirms effective antibiotic therapy.
    e. Nontoxigenic *C. difficile* is known to colonize humans and not associated with diarrheal disease.

29. Nucleic acid amplification testing is recommended for the diagnosis of which of the following parasitic infections?
    a. *Trichomonas vaginalis*
    b. *Toxoplasma gondii*
    c. *Leishmania* spp.
    d. *Plasmodium* spp.
    e. *Strongyloides stercoralis*

Answers
  1. a. Oseltamivir was started at admission and was within 4 days following onset of influenza symptoms. Studies show oseltamivir administration to be independently associated with an accelerated decrease in viral RNA concentration with viral RNA clearance by the first week if given in this 4-day window. This rapid decrease in viral RNA would reach the limits of detection of molecular assays. At the detection limit, like any other assay, exhibits high imprecision. The initial negative result was the result of this imprecision and after retesting, the positive result was recovered. Modern molecular POCT performance has been shown to be comparable to laboratory-based molecular assays. Coinfection with rhinovirus is not uncommon and there is no evidence to say the patient acquired the pathogen before or after admission since the POC assay is not designed to detect this virus. *Lee N, et al. Viral loads and duration of viral shedding in adult patients hospitalized with influenza. J Infect Dis 2009;492–500. Ling L, et al. Parallel validation of three molecular devices for simultaneous detection and identification of influenza A and B and respiratory syncytial viruses. J Clin Microbiol 2018;56:e01691–17.*

  2. a. PCR-based assays can only identify what their primers and probes are designed to detect. Genetic changes may be sufficient to prevent detection by PCR. Therefore novel variants of MRSA may be missed by PCR tests.

Alternately, the *mec*A gene alone is not the only mechanism that can impart resistance to methicillin or oxacillin. Borderline oxacillin-resistant *Staphylococcus aureus* (BORSA) strains have been identified which can be phenotypically identified as "MRSA" and lack the *mec*A gene. *Sakoulas G, et al. Methicillin-resistant Staphylococcus aureus: comparison of susceptibility testing methods and analysis of mecA-positive susceptible strains. J Clin Microbiol 2001;39:3946–51.*

3. c. The clinical sensitivity and specificity of modern point-of-care GAS testing is superior to rapid antigen tests. In fact, the clinical sensitivity is sufficiently high to not require negative result follow-up confirmation by microbiological culture as recommended by the Infectious Disease Society of America. Therefore antimicrobial therapy is not indicated, and in fact, discouraged in this scenario. *Shulman ST, et al. Clinical practice guideline for diagnosis and management of Group A Streptococcal pharyngitis: 2012 update by the Infectious Disease Society of America. Clin Infect Dis 2012;55:e86–e102.*

4. b. The pretest probability for diagnosing influenza based on clinical presentation alone during "flu season" is sufficiently high to not require molecular testing as recommended by the Centers for Disease Control and Prevention. *Centers for Disease Control and Prevention website: https://www.cdc.gov/flu/professionals/diagnosis/overview-testing-methods.htm, Accessed on January 29, 2020.*

5. b. Pregnant women are considered a high-risk population, therefore empiric treatment for influenza is recommended by the American College of Obstetrics and Gynecology (ACOG). *American College of Obstetrics and Gynecology (ACOG) Guidelines: https://www.acog.org/Clinical-Guidance-and-Publications/Committee-Opinions/Immunization-Infectious-Disease-and-Public-Health-Preparedness-Expert-Work-Group/Assessment-and-Treatment-of-Pregnant-Women-With-Suspected-or-Confirmed-Influenza?IsMobileSet=false, Accessed on January 29, 2020.*

6. a. CLIA-waived molecular influenza A/B devices are approved for testing using nasopharyngeal swab. Nasopharyngeal swabs are appropriate for patients with suspected upper respiratory tract infections. Throat swabs are inadequate. Bronchoalveolar lavage and sputum samples are appropriate for assessing lower respiratory tract infections. CLIA waiver typically excludes invasive samples such as BAL. *Irving SA, et al. Comparison of nasal and nasopharyngeal swabs for influenza detection in adults. Clin Med Res 2012;10:215–18.*

7. b. The cycle threshold (Ct) value is defined as the number of cycles required for a fluorescent signal be detectable above the background noise of a polymerase chain reaction assay. To this end, the starting concentration of target nucleic acids has an inverse relationship with the Ct-value. The higher the concentration of initial target nucleic acid translates to

needing less cycles to amplify nucleic acids by PCR to reach the detection threshold. *Caraguel CG, et al. Selection of a cut-off value for real-time polymerase chain reaction results to fit diagnostic purpose: analytical and epidemiological approaches. J Vet Diagn Invest 2011;23:2–15.*

8. c. Methicillin-sensitive *Staphylococcus aureus* must have the *spa* gene. The SCc*mec* region is optional and only confers methicillin resistance if the *mec*A gene is present. *Cepheid GXMRSA/SA-SSTI product insert 301-0190, Rev D September 2012.*

9. b. Studies show a subset of patients with *C. difficile* PCR positive results may not have toxin-producing pathogens. These patients are potential carriers of nonpathogenic *C. difficile*. Combination of *C. difficile* PCR results with toxicogenic culture helps arbitrate discordant EIA and PCR results. Glutamine dehydrogenase (GDH) assays can be used to screen for *C. difficile*; however, GDH is expressed in both toxin-producing and nontoxin-producing strains of the organism. *Polage CR, et al. Over diagnosis of Clostridium difficile infection in the molecular test era. JAMA Intern Med 2015;175:1792–1801.*

10. d. *S. pneumoniae* accounts for about 61% of cases in the United States. *Brower MC, et al. Epidemiology, diagnosis, and antimicrobial treatment of acute bacterial menigitis. Clin Microbiol Rev 2010;23:467–92.*

11. a. Influenza A/B and RSV are RNA viruses. *Pleschka S. Overview of influenza viruses. Curr Top Microbiol Immunol 2013;370:1–20. Meng J, et al. An overview of respiratory syncytial virus. PLOS Pathog 2014;10:e1004016.*

12. a. Influenza A is an RNA virus, while GAS are bacteria and contain circular DNA. To this end, RNA assays require reverse transcriptase (RT) to convert RNA back into DNA for amplification. Thus, influenza A assays are RT-PCR assays. *Center for Disease Control and Prevention website: https://www.cdc.gov/flu/professionals/diagnosis/molecular-assays.htm, Accessed on February 1, 2020.*

13. a. Antigenic shift occurs when there is genetic reassortment across species—allowing the pathogen to cross species boundaries. First described in April 2009, the variant virus appeared to be a new strain of H1N1 which was produced following a previous triple reassortment of bird, swine, and human flu viruses further combined with a Eurasian pig flu virus, leading to the term "swine flu." *Trifonov V, et al. Geographic dependence, surveillance, and origins of the 2009 Influenza A (H1N1) virus". N Engl J Med 2009;361:115–9. Herzum I, et al. Diagnostic performance of rapid influenza antigen assays in patients infected with the new influenza A (H1N1) virus. Clin Chem Lab Med 2010:48:53–6.*

14. e. TMA differs from PCR in that it produces RNA from DNA, produces more amplicons per cycle, RNA is more labile than DNA, and the technique is isothermal in nature. *Wroblewski JK, et al. Comparison of transcription-mediated amplification and PCR assay results for various genital specimen types for detection of Mycoplasma genitalium. J Clin Microbiol 2006;44:3306–3312. Gaydos CA, et al. Performance of the*

APTIMA Combo 2 assay for detection of Chlamydia trachomatis and Neisseria gonorrhea in female urine and endocervical swab specimens. J Clin Microbiol 2003;41:304–309.

15. a. *G. vaginalis* is the most common cause of BV. *Coleman J, et al C. Molecular diagnosis of bacterial vaginosis: an update. J Clin Microbiol. 2018;56:e00342–18.*

16. c. *S. agalactiae* or Group B *streptococcus* is the most common cause of meningitis in neonates—accounting for 66% of meningitis events in the first three months of life in the United States. *Phares CR, et al. Epidemiology of invasive group B streptococcal disease in the United States, 1999–2005. JAMA 2008;299:2056–65.*

17. d. Covid-19 is a member of the corona family of RNA viruses. The preferred specimen is a nsasal or throat swab, and the preferred methodology is RT-PCR. *Wang D, et al. Clinical characteristics of 138 hospitalized patients with 2019 novel coronavirus-infected pneumonia in Wuhan, China. JAMA. 2020;323(11):1061–9*

18. d. Screening for STDs is recommended by the CDC based on risk factors, including age, sex, pregnancy status, HIV status, and men who have sex with men. Recommended diseases for screening can include chlamydia, gonorrhea, syphilis, trichomoniasis, herpes simplex virus (HSV), HIV, oncogenic human papilloma virus (HPV)/cervical cancer, hepatitis B virus (HBV), and HCV. Laboratory tests for STDs are generally broken into two methodologies, NAAT, such as PCR, or immunoassays for microbial antigens or human antibodies. As of January 2020, CDC recommendations for availability of laboratory testing for STDs include NAAT for the following infectious diseases: *Chlamydia trachomatis* and *Neisseria gonorrhea* (urogenital and extragenital), HPV (with Pap smear), and HSV (viral culture is an acceptable NAAT substitute). Availability of serological testing is recommended for hepatitis A, B, and C, HIV (4th generation antigen/antibody), syphilis, and HSV (in addition to HSV NAAT or culture). *Sexually Transmitted Diseases: Summary of 2015 CDC Treatment Guidelines. J Miss State Med Assoc 2015;56(12):372–5. https://www.ncbi.nlm.nih. gov/pubmed/26975162. Barrow RY, et al. Recommendations for Providing Quality Sexually Transmitted Diseases Clinical Services. MMWR Recomm Rep 2020;68(5):1–20. http://doi.org/10.15585/mmwr.rr6805a1.*

19. f. The differential diagnosis for acute encephalitis is very broad but contains many infectious agents. Testing can be broken up by bodily fluid compartment (CSF and peripheral blood) and routine versus conditional/secondary studies. For CSF specifically, routine NAAT (e.g., PCR) is recommended for enterovirus, HSV-1/2, VZV. Immunoassay testing for syphilis and cryptococcal antigen (D and E) is routinely done in CSF, but is not NAAT as the question asked. *Venkatesan A et al. Diagnosis and management of acute encephalitis: a practical approach. Neurol Clin Pract 2014;4(3),206–15. http://doi.org/10.1212/CPJ.0000000000000036.*

20. a. For newly acquired or therapy failing HIV-1 infections, genes of drug targets are regularly sequenced to evaluate for mutations, which may confer resistance. HAV and WNV (b. and c.) are not tested for drug resistance considering their management is primarily supportive without antiviral therapy. Tuberculosis (d.) caused by *Mycobacterium tuberculosis* is regularly tested for resistance-associated mutations, but this infection is mycobacterial, not viral. *Gunthard HF, et al. Human immunodeficiency virus drug resistance: 2018 Recommendations of the International Antiviral Society-USA Panel. Clin Infect Dis 2019;68(2):177–87. http://doi.org/ 10.1093/cid/ciy463.*

21. b. HIV infects CD4 T-cells, which are a component of PBMCs along with CD8 T-cells, B-cells, and NK cells. Because of this, sequencing the HIV genome is actually more sensitive in PBMCs than plasma (a.). However, since HIV integrates into the T-cell genome, the sequencing results may reflect previous timepoints in infection and could be discordant with plasma testing, which represents the current active virus. *Gunthard HF, et al. Human immunodeficiency virus drug resistance: 2018 Recommendations of the International Antiviral Society-USA Panel. Clin Infect Dis 2019;68(2):177–87. http://doi.org/10.1093/cid/ciy463.*

22. c. In *Staphylococcus* spp., *mecA* encodes Penicillin Binding Protein 2a (PBP2a). This protein confers resistance to all beta-lactams except ceftaroline. Molecular detection of *mecA* combined with detection of a *Staphylococcus aureus*-specific gene target (e.g., Staphylococcal protein A (*spa*)) are available in multiple commercial assays for quick and accurate detection. The other answers include various other resistance genes in bacteria not seen in *Staphylococcus*. *Marlowe EM, et al. Conventional and molecular methods for the detection of methicillin-resistant staphylococcus aureus. J Clin Microbiol 2011;49(9 Suppl.):S53–6. http://doi.org/10. 1128/jcm.00791-11.*

23. e. The 16S ribosomal RNA (16S rRNA) gene found in prokaryotes contains both highly conserved functional regions and nonconserved hypervariable regions which have had sequence drifts with evolution. PCR primers can be made to the highly conserved regions to amplify all prokaryotic DNA within a specimen, while sequencing of the hypervariable regions can be specific enough to identify the organism to species in most cases. Because 16s rRNA is only found in prokaryotes, it is only useful for identifying AFB and bacteria. Additional targets for identification in AFB include 65kDa heat shock protein (hsp65) and RNA polymerase beta (rpoB) subunit genes, which are more variable than the 16S rRNA in rapid growing mycobacterium species. To identify yeast (d.), the 26S rRNA and Internal Transcribed Spacer 1 and 2 (ITS1 and 2) regions can be amplified and sequenced. No universal gene targets are widely used for identifying protozoan parasites. *Doern CD. Pocket Guide to Clinical Microbiology, Fourth Edition 2018:*

American Society of Microbiology. Lefterova MI et al. Next-generation sequencing for infectious disease diagnosis and management: a report of the Association for Molecular Pathology. J Mol Diagn 2015;17(6): 623–34. http://doi.org/10.1016/j.jmoldx.2015.07.004.

24. f. HCV is divided into clinically relevant genotypes 1–6, with genotype 1 subdivided into 1a and 1b. HCV genotypes 1a and 3 are associated with a high prevalence of NS5A mutations at baseline, which necessitate evaluation depending on planned NS5A-inhibitor treatment regimen and clinical presentation. Next-generation sequencing is most commonly used to evaluate for the major Y93H variant as well as any other minor variants that may not be picked up by targeted nucleic acid amplification testing. Other important genes involved with HCV antiviral therapy resistance include NS3 and NS5B. *Pawlotsky JM. Hepatitis C virus resistance to direct-acting antiviral drugs in interferon-free regimens. Gastroenterol 2016;151(1):70–86. http://doi.org/10.1053/j.gastro.2016.04.*

25. d. Metagenomic next-generation sequencing (mNGS) refers to the sequencing of all genetic material within a sample, which can include both host and pathogen. This has shown great success for the diagnosis of infectious diseases considering it does not require targeted amplicons for specific organisms. Answer a. refers to targeted sequencing, which identifies microorganisms through the sequences of hypervariable regions within highly conserved genes, such as 16S rRNA, 26S rRNA, hsp65, rpoB. The major limitation of targeted sequencing is the inability to identify viruses. Answer b. refers to multiplex PCR, which is a targeted approach, which can accurately identify dozens of clinically relevant organisms per test. Commercial FDA-cleared multiplex PCR assays are widely available, where targeted sequencing and mNGS are only available at a limited number of reference laboratories. Lastly, epigenetic methylation c. is a not a recognized method for microbial identification. *Gu W, et al. Clinical metagenomic next-generation sequencing for pathogen detection. Annu Rev Pathol 2019;14:319–38. doi:10.1146/annurev-pathmechdis-012418-012751. Lefterova MI et al. Next-generation sequencing for infectious disease diagnosis and management: A report of the Association for Molecular Pathology. J Mol Diagn 2015;17(6):623–34. http://doi.org/10.1016/j.jmoldx.2015.07.004.*

26. b. The presence of rifampicin resistance is highly correlated to isoniazid resistance and often serves as a proxy for multidrug resistant MTB (MDR-TB). Screening for rifampicin/isoniazid resistance is important considering these are part of the first-line drug treatment regimens. Commercial FDA-cleared NAAT tests for MTB/rifampicin resistance have been endorsed and recommended by the World Health Organization as high priority resistance testing options. *World Health Organization.Companion Handbook to the WHO Guidelines for the Programmatic*

Management of Drug-Resistant Tuberculosis. In Companion Handbook to the WHO Guidelines for the Programmatic Management of Drug-Resistant Tuberculosis. Geneva, 2014.

27. c. The *katG* gene product is a catalase enzyme that converts isoniazid to its active form in MTB. Activated isoniazid inhibits mycolic acid biosynthesis via inhibition of the *inhA* gene product. *InhA* mutations can also confer resistance to isoniazid; however, these are frequently in the promoter region and cause overexpression of mycolic acid biosynthetic machinery (a.). The major mediator of rifampicin resistance in MTB infection is *rpoB* mutations, which encode the B-subunit of RNA polymerase (b.). Fluoroquinolone resistance is caused by mutations in gyrA/gyrB encoding DNA gyrase (d.). Amikacin/kanamycin resistance can be seen with mutations in *rrs*, which encodes 16S rRNA (e.). *Almeida Da Silva PE, et al. Molecular basis and mechanisms of drug resistance in Mycobacterium tuberculosis: classical and new drugs. J Antimicrob Chemother, 2012;66(7):1417–30. http://doi.org/10.1093/jac/dkr173.*

28. e. *Clostridium difficile* can be broken down into toxigenic (toxin A- or B-producing) and nontoxigenic strains. Studies have shown healthy adults can be carriers of nontoxigenic strains, which do not warrant treatment. Toxigenic *C. difficile* infection (CDI), on the other hand, can be a cause of significant diarrheal disease requiring specialized antibiotic therapy. Nucleic acid amplification testing (NAAT) provides a sensitive assessment for CDI when used in appropriate clinical algorithms. *McDonald LC, et al. Clinical practice guidelines for Clostridium difficile infection in adults and children: 2017 Update by the Infectious Diseases Society of America (IDSA) and Society for Healthcare Epidemiology of America (SHEA). Clin Infect Dis 2018;66(7):e1–e48. http://doi.org/10.1093/cid/cix1085.*

29. a. Parasitic infections are often diagnosed by microscopy or serological testing. Nucleic acid amplification tests are rarely employed except under certain circumstances (e.g., acute infection). One infection that does benefit from NAAT is trichomoniasis, caused by *Trichomonas vaginalis*; a sexually transmitted parasite that most commonly infects the urogenital track of men and women. Symptoms of infection can include irritation, itching or burning, and discharge. NAAT in urogenital swabs and urine has shown superior sensitivity and specificity and is currently the recommended method for screening by the CDC. *Barrow RY, et al. Recommendations for providing quality sexually transmitted diseases clinical services. MMWR Recomm Rep 2020;68(5):1–20. http://doi.org/10. 15585/mmwr.rr6805a1.* Hobbs MM et al. Modern diagnosis of Trichomonas vaginalis infection. Sex Transm Infect 2013;89(6):434–8. *http://doi. org/10.1136/sextrans-2013-051057.*

Chapter 23

Infectious disease serology and molecular diagnostics for infectious diseases

Yvette McCarter[a] and Nam Tran[b]
[a]*University of Florida, Jacksonville, FL, United States,* [b]*University of California, Davis, CA, United States*

1. Which of the following tests is useful for the diagnosis of neurosyphilis?
   a. TP-PA
   b. RPR
   c. Darkfield
   d. VDRL
   e. FTA-abs
2. A patient with a recent travel history to New England is tested for Lyme disease with the following results:
   EIA (signal/cutoff ratio)
   Patient: 1.75
   Assay cutoff: 0.80
   Western Blot
   IgM: p23
   IgG: p18, p45, p66
   The best interpretation for these results is that the patient:
   a. has early Lyme disease
   b. has late Lyme disease
   c. does not have Lyme disease
   d. has Lyme neuroborreliosis
   e. has Lyme arthritis
3. The gold standard for the diagnosis of West Nile virus encephalitis is which of the following?
   a. antibody detection
   b. nucleic acid amplification
   c. antigen detection

Self-assessment Q&A in Clinical Laboratory Science, III. https://doi.org/10.1016/B978-0-12-822093-1.00023-5
267

    d. culture

    e. CSF glucose and protein

4. A 3-year-old child suspected of having Epstein-Barr virus infection is tested using the heterophile antibody test. The result is negative. What is the next best step to take?

    a. perform CMV serology

    b. collect throat washings for EBV viral culture

    c. perform EBV-specific serology

    d. consider the child EBV negative

    e. perform electron microscopy on throat washings

5. An HIV positive woman from Brazil is suspected of having visceral Leishmaniasis. Serology for *Leishmania* antibodies is negative. What is the most likely cause of this result?

    a. Patients with visceral disease develop antibody late in the course of disease.

    b. Patients with cutaneous disease most often develop detectable antibody.

    c. She does not have visceral leishmaniasis since all patients with visceral disease develop antibody.

    d. Immunocompromised individuals with visceral disease often do not develop detectable antibody.

    e. She most likely has mucocutaneous disease and her antibody level has already declined to undetectable levels.

6. A patient who traveled outside of the United States presents with a liver abscess. His physician wants to determine if *Entamoeba histolytica* is the cause. Which would be the best test to perform in this patient?

    a. antigen detection on stool

    b. stool O&P examination

    c. trichrome stain on abscess fluid

    d. culture of abscess fluid

    e. serum antibody detection

7. Which of the following tests is the most useful for detecting invasive aspergillosis in an immunocompromised host?

    a. serum antibody detection by complement fixation

    b. serum galactomannan antigen

    c. urine mannan antigen

    d. serum antibody detection by immunodiffusion

    e. CSF polysaccharide antibody detection

8. A pregnant woman is tested for *Toxoplasma* antibodies and is found to be positive for *Toxoplasma* IgM antibody and negative for *Toxoplasma* IgG antibody. She is retested again 2 weeks later and the same results are obtained. What is the most probable explanation for these results?

    a. both results were obtained very early in infection

    b. the patient has reactivated disease

    c. the IgM is a false positive result

    d. IgG antibody develops late in the course of infection

    e. negative IgG results are common in pregnancy

9. A patient with a history of pyoderma develops nephritis. His physician suspects that the patient has poststreptococcal disease. The best course of action is to perform which of the following tests?

    a. throat culture

    b. DNase B antibody test

    c. streptolysin O antibody test

    d. group A streptococcal antigen

    e. carbohydrate A antibody test

10. A patient with no reported history of syphilis is evaluated using the reverse testing algorithm and the following results are obtained:

    T. pallidum antibody test—reactive

    RPR—nonreactive

    TP-PA—reactive.

    How should these results be interpreted?

    a. The results indicate that the patient is not infected with *T. pallidum.*

    b. The patient has incubating syphilis.

    c. The patient has secondary syphilis.

    d. The patient has latent syphilis.

    e. The results are inconclusive and the patient should be retested with the conventional algorithm.

11. A patient with abnormal liver function tests has the following hepatitis marker profile:

    HBsAg: negative

    IgM anti-HBc: positive

    IgM anti-HAV: negative

    Anti-HCV: negative

    This patient has which of the following?

    a. chronic hepatitis B infection

    b. recent hepatitis A infection

    c. acute hepatitis B infection

    d. recent hepatitis C infection

    e. concurrent hepatitis B and C infection

12. A patient with fulminant hepatitis has the following hepatitis marker profile:

    HBsAg: positive

    IgM anti-HBc: negative

    Total anti- HBc: positive

    IgM anti-HDV: positive

    Total anti- HDV: positive

    Anti-HCV: negative

    IgM anti-HAV: negative

Total anti-HAV: positive
This patient has which of the following?
a. acute hepatitis B infection
b. hepatitis D superinfection
c. concurrent hepatitis A and B infection
d. hepatitis D coinfection
e. chronic hepatitis B infection

13. Which of the following is true regarding rapid antigen testing for group A streptococcus in children with pharyngitis?
    a. They are highly sensitive so a negative result does not require culture confirmation.
    b. They are neither sensitive nor specific and culture confirmation is always required.
    c. They are highly specific so a positive result does not require culture confirmation.
    d. They have a high sensitivity and positive predictive value so results do not require culture confirmation.
    e. They should not be used in children due to the presence of pharyngeal colonization with group A streptococcus.

14. The following *Histoplasma* urine antigen results are obtained on an HIV positive patient diagnosed with acute disseminated histoplasmosis:
    Initial diagnosis: 20 ng/mL
    1-month follow-up: 10 ng/mL
    3-month follow-up: 15 ng/mL
    What is the most probable explanation of these results?
    a. treatment failure
    b. reinfection with another strain
    c. successful therapy
    d. release of antigen from killed organisms
    e. cross reaction with *Cryptococcus* antigen

15. An elderly man presents to the Emergency Department with a 1-week history of malaise, muscle aches, rapid onset of dry cough, fever, and diarrhea. His physician suspects *Legionella* infection and orders a *Legionella* urine antigen test and culture of sputum. The antigen test is negative but at 5 days the culture is positive with *Legionella micdadei*. What is the most likely cause of this apparent discrepancy in results?
    a. The antigen test was performed too early in the course of disease.
    b. *Legionella micdadei* is a culture contaminant.
    c. The antigen test was performed too late in the course of disease.
    d. The urine antigen test only detects *Legionella pneumophila*.
    e. The urine antigen test was falsely negative due to a high antibody titer.

16. An HIV positive patient presents with suspected cryptococcal meningitis. Which of the following tests would be the best to diagnose this patient's infection?

a. antibody detection in serum
b. polysaccharide antigen detection in CSF
c. India ink stain of CSF
d. antibody detection in urine
e. Mannan antigen detection in CSF

17. A patient's serum is tested for varicella zoster virus antibody using a latex agglutination assay. The serum is tested at a 1:2 dilution and a negative result is obtained. Which of the following is the next best action?
   a. Report the result as negative (not immune).
   b. Repeat the test in duplicate before reporting.
   c. Test the specimen undiluted.
   d. Retest the specimen at 1:2 using another latex method.
   e. Dilute the specimen 1:40 and repeat testing.

18. The presence of a high antibody titer to phase I antigens of *Coxiella burnetii* is common in:
   a. Q fever endocarditis.
   b. Rocky Mountain spotted fever.
   c. asymptomatic seroconversion.
   d. Kawasaki disease.
   e. acute Q fever pneumonia.

19. Which of the following is true regarding *Helicobacter pylori* antigen testing?
   a. It should be performed on gastric biopsies.
   b. It is less sensitive than antibody testing.
   c. It can be useful for both diagnosis and monitoring therapeutic response.
   d. It is not useful for initial diagnosis of infection in the elderly.
   e. Previous treatment is likely to cause false positive results.

20. Based on current CDC recommendations, what test should be performed on a sample with the following results:
   HIV Ag/Ab EIA—Repeatedly reactive
   HIV1/2 differentiation assay—Negative
   a. perform IFA testing
   b. perform a quantitative HIV RNA
   c. perform an HIV-1 Western blot
   d. perform a qualitative HIV-1 RNA test
   e. perform an HIV-2 Western Blot

21. A patient undergoing an insurance physical has the following hepatitis marker results:
   HBsAg: negative
   Total anti-HBc: positive
   Total anti-HBs: positive
   How would you interpret these results?
   a. This patient has chronic hepatitis B and should be tested for HBeAg.
   b. This patient is immune to hepatitis B following immunization.
   c. This patient is immune to hepatitis B following natural infection.

    d. This patient may have early hepatitis B infection and should be tested for IgM anti-HBc.

    e. The testing should be repeated since total anti-HBc and total anti-HBs are both positive.

Answers

1. d. The VDRL test is the oldest test for the diagnosis of neurosyphilis. The VDRL test in CSF has high specificity but low sensitivity. A negative CSF VDRL therefore does not rule our neurosyphilis. The sensitivity of the test can be increased by testing only patients with documented infection with *Treponema pallidum. Treponemal tests are not recommended for testing CSF. Versalovic J, et al., Manual of Clinical Microbiology, 10th ed., 2011, p. 958.*

2. c. All samples that are positive or indeterminate for *B. burgdorferi* antibody by EIA should be tested further by Western immunoblot. Western immunoblot greatly improves the specificity of Lyme disease diagnosis. An IgM blot (indicative of early Lyme disease) is positive if two or more of the following bands are present: p23 (OspC), p39 (BmpA), and p41 (FlaB). An IgG Western blot (indicative of late Lyme disease) is positive if five or more of the following bands are present: p18 p23 (OspC), p28, p30, p39 (BmpA), p41 (FlaB), p45, p58, p66, and p93. Immunoblots that do not meet the criteria for positivity are considered negative. *Versalovic J, et al., Manual of Clinical Microbiology, 10th ed., 2011, p. 933.*

3. a. The detection of IgM antibody in CSF is strongly suggestive of CNS infection with West Nile virus. Thus detection of antibody is most frequently used to detect neuroinvasive disease. In serum it is important to note that cross-reactive antibodies can be present after infection with other flaviviruses. It is recommended that positive serum antibody results should be confirmed by neutralizing antibody testing of acute- and convalescent-phase serum specimens. Molecular tests are useful, when positive, for detecting early infection and infection in immunocompromised patients. However, a negative nucleic acid amplification test does not rule out infection. Only about one-half of patients with serologically confirmed West Nile virus disease will have a positive CSF molecular test. Culture is not routinely performed and antigen testing for West Nile virus is not currently available. *Mandell et al., Principles and Practice of Infectious Diseases, 9th ed., 2020, pp. 1240–2.*

4. c. False negative heterophile antibody tests occur frequently in young children and can also occur in adolescents and adults. If the heterophile antibody test is negative and EBV-associated disease is suspected, EBV-specific serologic testing is useful and should be performed. *Versalovic J, et al., Manual of Clinical Microbiology, 10th ed., 2011, pp. 1578–81.*

5. d. Serologic testing is recommended for patients with suspected visceral leishmaniasis in whom definitive diagnostic tests for the parasite

(microscopic identification, culture, and molecular testing) cannot be performed or have negative results. Tests for antileishmanial antibodies should not be performed as the sole diagnostic assay. In patients who are immunocompromised because of concurrent HIV/AIDS, serologic testing is of little value since antibodies may be undetectable or present at low levels resulting in false negative results. Antibody detection is not helpful in the diagnosis of cutaneous leishmaniasis. *Aronson N, et al. Diagnosis and treatment of Leishmaniasis: clinical practice guidelines by the Infectious Diseases Society of America (IDSA) and the American Society of Tropical Medicine and Hygiene (ASTMH). Clin Infect Dis 2016;63(12):e202–64.*

6. e. Antibody detection is the most useful test to diagnose patients with extraintestinal infections due to *Entamoeba histolytica*. In extraintestinal disease, organisms are generally not detectable in the stool by either conventional ova and parasite examinations or antigen detection. In addition, the yield of culture or staining of the abscess fluid is low since most organisms are in the wall of the abscess and not in the abscess fluid. *Mandell et al., Principles and Practice of Infectious Diseases, 9th ed., 2020, p. 3284. Versalovic J, et al., Manual of Clinical Microbiology, 10th ed., 2011, pp. 215–405.*

7. b. Antibody detection is helpful for the diagnosis of fungal infections in immunocompetent individuals, but are less useful in immunocompromised patients due to the likelihood of false negative results. Detection of galactomannan antigen in serum is effective in diagnosing invasive aspergillosis in immunocompromised patients. It often detects disease before the onset of clinical signs of infection. Detection of serum antibodies by immunodiffusion is a reliable test for diagnosing aspergillosis in immunocompetent individuals. Detection of serum antibodies by complement fixation may be useful in diagnosing aspergilloma but is insensitive in detecting invasive aspergillosis. *Mandell et al., Principles and Practice of Infectious Diseases, 9th ed., 2020, p. 3112. Versalovic J, et al., Manual of Clinical Microbiology, 10th ed., 2011, pp. 1781, 1788–9.*

8. c. This represents a false positive *Toxoplasma* IgM result. A positive IgM result in the absence of a positive IgG result should be viewed cautiously. Blood should be redrawn from the patient in three weeks and tested along with the previous sample. If the first sample was drawn very early in infection, both the IgM and IgG antibodies should be present in the second sample which would require further evaluation. *Versalovic J, et al., Manual of Clinical Microbiology, 10th ed., 2011, pp. 2132–3.*

9. b. The DNase B antibody test is more reliable than the streptolysin O test for detecting preceding streptococcal skin infection in patients with poststreptococcal sequelae. The anti-A-carbohydrate test is not used in assessing patients with poststreptococcal nephritis. Throat culture and antigen

tests are useful for the diagnosis of group A streptococcal infection and are not used in the assessment of poststreptococcal sequelae. *Mandell et al., Principles and Practice of Infectious Diseases, 9th ed., 2020, p. 2454. Versalovic J, et al., Manual of Clinical Microbiology, 10th ed., 2011, pp. 343–4.*

10. d. In recent years many laboratories have moved from the conventional testing algorithm (nontreponemal test with treponemal-specific test confirmation) to a reverse algorithm (automated treponemal test followed by a nontreponemal test if positive). Both algorithms detect patients with active disease but the reverse algorithm detects seropositive individuals with early disease or established latent infection that would be missed with the nontreponemal test. A positive TP-PA result confirms the presence of *T. pallidum* infection. *Mandell et al., Principles and Practice of Infectious Diseases, 9th ed., 2020, pp. 2883–2884. Versalovic J, et al., Manual of Clinical Microbiology, 10th ed., 2011, pp. 957–9.*

11. c. The diagnosis of acute hepatitis B is based on the detection of HBsAg and IgM anti-HBc. HBsAg is the serologic hallmark of HBV infection and is the first marker to appear after acute infection. During acute infection, IgM anti-HBc is detectable at the time symptoms appear. It can be the sole marker of HBV infection during what is called the *window period* between the disappearance of HBsAg and the appearance of anti-HBs. *Mandell et al., Principles and Practice of Infectious Diseases, 9th ed., 2020, p. 1952.*

12. b. Hepatitis D virus requires the surface antigen from hepatitis B virus for replication. The disease may occur as an acute coinfection with hepatitis B or as a superinfection of a chronic hepatitis B infection. Hepatitis D virus is diagnosed with detection of anti-HDV IgG, and because this remains positive with HDV clearance. IgM anti-HDV can be detected either during coinfection or superinfection. Coinfection is distinguished from superinfection by the presence of IgM anti-HBc in patients with coinfection and the absence in patients with hepatitis D superinfection. *Mandell et al., Principles and Practice of Infectious Diseases, 9th ed., 2020, pp. 1965–6. Versalovic J, et al., Manual of Clinical Microbiology, 10th ed., 2011, pp. 1672–3.*

13. c. *Streptococcus pyogenes* causes 15%–30% of pharyngitis in pediatric patients. Rapid antigen detection tests facilitate the early diagnosis and therapy in children with streptococcal pharyngitis. Most currently available rapid antigen tests for group A streptococcus are highly specific so a positive result does not require culture confirmation. However, the sensitivity of these tests is variable and in most cases less than the sensitivity of culture. Therefore the American Academy of Pediatrics recommends that negative rapid antigen tests be confirmed with throat culture. *Mandell et al., Principles and Practice of Infectious Diseases, 9th ed., 2020, pp. 2451–2.*

14. a. *Histoplasma* urine antigen is useful for the diagnosis of disseminated histoplasmosis, especially in AIDS patients. Antigen levels in urine and serum decline with effective treatment, becoming undetectable in most patients. Failure of antigen levels to decrease during therapy suggests therapy failure. In patients who have responded to therapy and antigen levels have decreased, increases in antigen levels suggest relapse. Accordingly, changes in antigen levels can be used as indicators for response to therapy. *Mandell et al., Principles and Practice of Infectious Diseases, 9th ed., 2020, p. 3172. Versalovic J, et al., Manual of Clinical Microbiology, 10th ed., 2011, pp. 1908, 1913–5.*

15. d. The *Legionella* urine antigen test is most sensitive for the Pontiac subtype of *Legionella pneumophila* serogroup 1 which causes approximately 90% of the cases of Legionnaire's disease. While urine antigen testing is rapid and sensitive, the advantage of culture diagnosis is that positivity is not dependent on the species of *Legionella*. Optimal test yield requires performing more than one type of test; nucleic acid amplification (when available), lower respiratory culture, and urine antigen testing are the preferred tests. *Mandell et al., Principles and Practice of Infectious Diseases, 9th ed., 2020, pp. 2814–5.*

16. b. The detection of *Cryptococcus* capsular polysaccharide in CSF and serum is very accurate for the diagnosis of cryptococcal infection with a sensitivity and specificity greater than 90%. Detection of organisms by India ink stain is less sensitive than CSF antigen detection. Serum antibody detection tests are available; however, they are less useful than antigen detection, especially in immunocompromised patients. *Mandell et al., Principles and Practice of Infectious Diseases, 9th ed., 2020, pp. 3157–8. Versalovic J, et al., Manual of Clinical Microbiology, 10th ed., 2011, pp. 1804–5.*

17. e. An important limitation of agglutination assays is the potential for false negative results due to high antibody concentrations. This is referred to as "prozone" effect. In the presence of high antibody concentrations, agglutination is inhibited because the high concentration of antibody binds to the antigenic sites on the latex in such a way that cross linking cannot occur. In these cases *antibody can be detected if the sample is diluted to decrease the antibody concentration and retested. Versalovic J, et al., Manual of Clinical Microbiology, 10th ed., 2011, pp. 63–4.*

18. a. There are two distinct antigenic phases (phase I and phase II) to which humans develop antibody responses to *Coxiella burnetii*. Antibody response to this antigenic difference is important in diagnosis. In acute cases of Q fever (such as pneumonia), antibody to phase II is usually higher than phase I. In chronic Q fever (such as endocarditis), phase I antibody is higher. Antibodies to phase I antigens of *C. burnetii* generally require longer to appear and indicate continued exposure to the bacteria.

IgM antibodies usually rise at the same time as IgG and remain elevated for months or longer and are of limited diagnostic value on their own. In addition, IgM antibodies are less specific than IgG antibodies and more likely to result in a false positive. In the past, Kawasaki disease was thought to be a variant of Q fever; however, support for this has not been found. Rocky Mountain spotted fever is caused by *Rickettsia rickettsii*. *Mandell et al., Principles and Practice of Infectious Diseases, 9th ed., 2020, pp. 2362–4. Versalovic J, et al., Manual of Clinical Microbiology, 10th ed., 2011, p. 1031.*

19. c. *Helicobacter pylori* positive patients shed antigen in their stool. The antigen assay is a noninvasive means for detecting positive patients and for monitoring therapeutic response 4 weeks after completing treatment. Many serological tests have suboptimal sensitivity and specificity. Detection of antibody to *H. pylori* is no longer a recommended method for diagnosis. The stool antigen test and the urea breath test are currently the most accurate noninvasive diagnostic tools for the diagnosis of *H. pylori* infection. *Versalovic J, et al., Manual of Clinical Microbiology, 10th ed., 2011, p. 906.*

20. d. In an effort to increase detection of HIV infections in the window period infections, fourth- and fifth-generation assays have been developed that detect HIV-1 and HIV-2 antibodies and HIV-1 p24 antigen. Current recommendations indicate that specimens reactive with a fourth- or fifth-generation assay should be tested with an FDA-approved antibody immunoassay that differentiates HIV-1 antibodies from HIV-2 antibodies. These differentiation assays have enabled whole-scale replacement of the Western Blot and other confirmatory assays. Specimens that are reactive on the initial antigen/antibody combination immunoassay and nonreactive or indeterminate on the HIV-1/HIV-2 antibody differentiation immunoassay should be tested with an FDA-approved qualitative HIV-1 nucleic acid test. There are currently no quantitative HIV-1 nucleic acid assays FDA approved for the diagnosis of HIV. Centers for Disease Control and Prevention and Association of Public Health Laboratories. *Laboratory testing for the diagnosis of HIV infection: Updated recommendations. http:// stacks.cdc.gov/view/cdc/23447. Published June 27, 2014. Mandell et al., Principles and Practice of Infectious Diseases, 9th ed., 2020, pp. 1627–30.*

21. c. Total anti-HBC antibodies remain positive indefinitely after IgM anti-HBc antibody disappears. They persist longer than anti-HBS. Total anti-HBc is the best marker for documenting prior exposure to hepatitis B. Anti-HBc is not present in vaccinated individuals unless they were infected with hepatitis B prior to vaccination. *Mandell et al., Principles and Practice of Infectious Diseases, 9th ed., 2020, p. 1953. Versalovic J, et al., Manual of Clinical Microbiology, 10th ed., 2011, pp. 1668–9.*

Chapter 24

Microbiology

Melanie L. Yarbrough and Carey-Ann D. Burnham
Washington University School of Medicine, St. Louis, MO, United States

1. A 58-year-old man presented to the emergency department with a fever of 102°F, a cough, and night sweats. The man had many animal exposures as a large animal veterinarian, and frequently walked in the woods near his home. Blood cultures were obtained and were positive for a small Gram-negative coccobacillus that grew on sheep blood agar and chocolate agar. No growth was seen on MacConkey agar. The organism was nonmotile and nonhemolytic. Biochemical testing reveals that the organism was oxidase and urease positive. Which microorganism cannot be ruled out in this scenario?
   a. *Brucella melitensis*
   b. *Burkholderia pseudomallei*
   c. *Coxiella burnetii*
   d. *Francisella tularensis*
   e. *Yersinia pestis*
2. An active 56-year-old man began to notice exercise intolerance. He became increasingly fatigued with generalized weakness and developed oral bleeding and hematuria. He presented to his physician, who ordered numerous laboratory studies including blood cultures. After 18 h of incubation, one set of blood cultures became positive with a Gram-variable bacillus. After approximately 48 h of incubation, growth was observed on the anaerobic brucella agar plate exclusively. Colony morphology showed a thin film of growth, "swarming" over the plate and Gram stain of the isolate revealed large, filamentous bacilli with rare subterminal spores. Additional testing revealed that the isolate was negative for catalase, lecithinase, lipase, indole, and urea. Considering the most likely identity of this organism, what clinical condition should this patient's physician be suspicious of?
   a. botulism
   b. gastrointestinal neoplasm
   c. contaminated blood culture
   d. cutaneous anthrax
   e. tetanus

Self-assessment Q&A in Clinical Laboratory Science, III. https://doi.org/10.1016/B978-0-12-822093-1.00024-7

3. A 32-year-old male presents to a hospital with fatigue and intermittent fevers. His past medical history is significant for congenital heart disease requiring multiple surgeries and documentation of a positive PPD 20 years prior to the present illness. Transesophageal echocardiography (TEE) demonstrated a small vegetation on the patient's aortic valve. Blood cultures were collected and signaled positive after approximately 3 days of incubation. Blood culture broth was subcultured to blood, chocolate, and MacConkey agars. Small colonies that slightly pitted the media grew on blood and chocolate agar. No growth was observed on MacConkey agar. A Gram stain of the growth is pictured as follows. The organism was oxidase positive, catalase negative, and indole positive. What is the most likely identity of the organism?

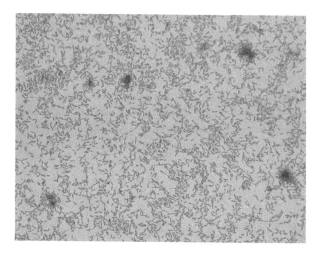

   a. *Cardiobacterium hominis*
   b. *Pseudomonas aeruginosa*
   c. *Escherichia coli*
   d. *Kingella kingae*
   e. *Mycobacterium tuberculosis*

4. The most common cause of urinary tract infections:
   a. deaminates phenylalanine
   b. hydrolyzes urea
   c. has an acid/acid reaction on triple sugar iron agar
   d. is Gram positive
   e. is oxidase positive

5. When matrix-assisted laser desorption/ionization-time of flight (MALDI-TOF) mass spectrometry is used for bacterial identification, which of the following two species cannot be differentiated without additional testing?
   a. *Escherichia coli* and *Enterobacter cloacae* complex
   b. *Escherichia coli* and *Shigella sonnei*
   c. *Klebsiella oxytoca* and *Enterobacter aerogenes*

d. *Staphylococcus aureus* and *Staphylococcus epidermidis*
e. *Staphylococcus aureus* and *Staphylococcus intermedius*
6. The most common cause of community acquired pneumonia:
    a. is bile insoluble
    b. is coagulase positive
    c. is susceptible to optochin
    d. is vancomycin resistant
    e. requires X and V factors for growth
7. A 62-year-old male with acute myeloid leukemia being treated with chemotherapy developed fever and abdominal pain. Upon admission to the hospital, he was noted to have watery diarrhea. A *C. difficile* stool antigen test was negative. Blood cultures were drawn and were positive after 24 h. The isolate was subcultured to aerobic and anaerobic media and a gram stain of the specimen revealed long, thin Gram-variable bacilli. The following day, there was growth of pinpoint, gray alpha hemolytic colonies on both the aerobic and anaerobic blood agar plates that were catalase negative. Gram stain of the colony from the aerobic media revealed Gram-variable bacilli with no spores. The isolate was subcultured to blood agar for special potency disk testing, which revealed that the organism was susceptible to kanamycin and vancomycin but resistant to colistin. What is the most likely identity of this organism?
    a. *Bacillus cereus*
    b. *Clostridium difficile*
    c. *Clostridium tertium*
    d. *Eggerthella lenta*
    e. *Lactobacillus rhamnosus*
8. A 58-year-old woman on peritoneal dialysis presented with fatigue and intermittent fevers over approximately one week. Peritoneal dialysate was collected and submitted to the microbiology laboratory for culture. After approximately 48 h incubation, small, slightly mucoid colonies were observed on blood and chocolate agar. No growth was observed on MacConkey agar. A Gram stain was performed, revealing very small, Gram-negative coccobacilli that were catalase, oxidase, and indole positive. The isolate was positive for nitrate reduction and negative for urease and cefinase. What is the most likely source of this woman's infection?
    a. bioterrorism event
    b. pet cat
    c. poor dentition
    d. running over a rabbit with the lawn mower
    e. tap water
9. Which of the following test results would require further investigation?
    a. *Enterobacter aerogenes* resistant to cefazolin
    b. *Escherichia coli* susceptible to ampicillin
    c. *Klebsiella oxytoca* susceptible to ampicillin
    d. *Proteus vulgaris* resistant to tetracycline
    e. *Stenotrophomonas maltophilia* resistant to meropenem

10. A *Salmonella* species was isolated from a blood culture and testing for antimicrobial susceptibility was set up. Which of the following antibiotics should not be reported on this isolate?
    a. ampicillin
    b. cefazolin
    c. ceftriaxone
    d. ciprofloxacin
    e. trimethoprim-sulfamethoxazole

11. Which of the following is the vector for *Borrelia burgdorferi, Borrelia miyamotoi, Anaplasma phagocytophilum,* and *Babesia*?
    a. *Aedes aegypti*
    b. *Amblyomma americanum*
    c. *Dermacentor variabilis*
    d. *Ixodes scapularis*
    e. *Ornithodoros*

12. A 64-year-old man sustained a small puncture on his bare hands while gardening. One week later he noticed a nodular lesion on his right forearm that was swollen. He treated the lesion with a topical over-the-counter antibiotic with no improvement. Within a few days, he presented to his primary care physician, who was able to aspirate pus from the abscess that was sent to the lab for routine aerobic and anaerobic culture. A Gram stain of the specimen showed abundant polymorphonuclear leukocytes and a moderate amount of a beaded Gram-positive organism with branching filaments. Organisms were negative by Kinyoun stain but demonstrated partial acid-fast characteristics by modified Kinyoun stain. Which of the following microorganisms is consistent with the patient's history and specimen gram stain?
    a. *Lactobacillus paracasei*
    b. *Mycobacterium chelonae*
    c. *Nocardia brasiliensis*
    d. *Sporothrix schenckii*
    e. *Streptococcus constellatus*

13. A 36-year-old man sustained a deep wound on his finger while cleaning and repairing the blade of his lawn mower. After approximately 5 days, he presented to the emergency department where the finger was noted to be swollen with frank pus. Cultures of the purulent material were sent to the microbiology laboratory. After approximately 5 days of incubation, the mycobacterial culture signaled positive and acid-fast organisms were noted by Kinyoun staining. This liquid culture is subcultured to solid medium. After approximately 3 days, growth of yellow colonies is observed. Which of the following Mycobacterium species is the most likely cause of infection?
    a. *Mycobacterium chelonae*
    b. *Mycobacterium gordonae*
    c. *Mycobacterium kansasii*
    d. *Mycobacterium marinum*
    e. *Mycobacterium neoaurum*

14. A 56-year-old woman with no significant past medical history sustained a minor puncture wound to her thumb while opening a wine bottle with a corkscrew. Two weeks later the thumb developed redness and swelling, and she subsequently developed extensor tenosynovitis of the associated joint. The wound was debrided, and cultures were sent to the clinical microbiology lab. After 2 weeks, growth was seen on a Lowenstein Jensen slant that was incubated at 30°C. A Kinyoun stain of the organism (Fig. 1) and colony morphology on 7H11 agar (Fig. 2) are shown. No growth was observed in cultures incubated at 35°C. What is the most likely source of this patient's wound infection?

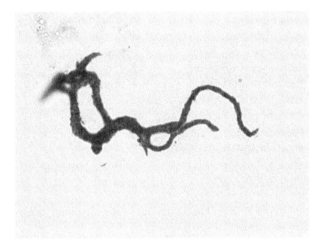

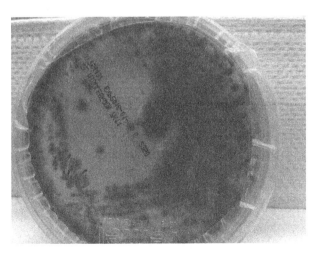

    a. the cork from the wine bottle
    b. cleaning a fish tank
    c. a lick from her pet dog
    d. working in the garden
    e. a manicure at a salon

15. A Gram stain from a blood culture specimen from a neutropenic patient reveals oval, budding yeast and pseudohyphae. Upon subculture, the organism has star-like projections or "feet." A Germ tube is performed, the isolate is Germ-tube positive. What is the most likely identity of the organism?

    a. *Aspergillus fumigatus*
    b. *Candida albicans*
    c. *Candida glabrata*
    d. *Cryptococcus neoformans*
    e. *Saccharomyces cerevisiae*

16. Match the following yeast or yeast-like organisms with the appropriate characteristics on Cornmeal Tween 80 agar.

    1. *Geotrichum candidum*
    2. *Trichosporon* spp.
    3. *Rhodotorula* spp.
    4. *Candida albicans*
    5. *Candida dubliniensis*

Choices:

    a. Pseudohyphae and blastoconidia along hyphae; single terminal chlamydospores
    b. Pseudohyphae and true hypha; blastoconidia and arthroconidia
    c. Pseudohyphae and blastoconidia along hyphae; terminal chlamydospores in pairs or clusters
    d. True hyphae that segment into arthroconidia; no blastoconidia
    e. No pseudohyphae, true hyphae, or arthroconidia

17. An infant with an indwelling catheter receiving total parenteral nutrition (TPN) becomes febrile. Blood cultures are collected and signal positive. The Gram stain of the blood culture broth reveals yeast that are round on one end, with a budlike structure with a broad base at the other end. The organism does not grow on initial subculture to SAB, blood, or chocolate agar. Sterile olive oil is added to the medium and the organism exhibits growth. What is the most likely identity of the organism?

    a. *Aspergillus fumigatus*
    b. *Blastomyces dermatitidis*
    c. *Candida tropicalis*
    d. *Malassezia furfur*
    e. *Malassezia pachydermatis*

18. A 47-year old man presented three months after a bone marrow transplant with fever, cough, chest pain, and hemoptysis. He was severely neutropenic and a chest CT revealed a left upper lobe cavitary lesion. Sputum was sent to the microbiology lab for fungal culture. While waiting for these results, serum was sent for galactomannan testing, which was positive. After 1 week, a mold was isolated from the fungal culture. The colony was greenish gray with a velvety texture. Microscopic morphology was described as septate hyphae with short conidiophores that terminated in vesicles. The upper two-thirds of the vesicles were covered in uniseriate phialides with columnar conidia. What organism is most likely the cause of this patient's infection?
  a. *Aspergillus clavatus*
  b. *Aspergillus flavus*
  c. *Aspergillus fumigatus*
  d. *Aspergillus niger*
  e. *Aspergillus terreus*
19. A 34-year-old woman, status postbone marrow transplant presents with acute onset nausea, vomiting, and diarrhea. The physician caring for the patient is suspicious of Norovirus infection. What is the preferred method of making the diagnosis of Norovirus infection?
  a. ELISA for Norovirus antigen performed on serum
  b. ELISA for Norovirus antigen performed on stool
  c. reverse-transcriptase PCR on stool
  d. serology for Norovirus antibodies
  e. viral culture of stool

20. The causative agent of cysticercosis is:
  a. *Ascaris lumbricoides*
  b. *Naegleria fowleri*
  c. *Taenia saginata*
  d. *Taenia solium*
  e. *Trichuris trichiura*
Answers
1. a. *Brucella melitensis.* Bacteria such as *Brucella* spp., *Yersinia pestis,* and *Francisella tularensis* are classified as select agents (SA) and have the potential to be used as bioterrorism agents (BA). Once a SA is suspected, further workup to rule out the agent should be completed under a biological safety cabinet. In the setting of a sentinel laboratory, if a SA cannot be ruled out, timely notification to public health officials is required and the isolate should be referred to an appropriate tier laboratory for further workup. *Brucella* spp. are small, Gram-negative coccobacilli that grow on blood and chocolate agar with poor to no growth on MacConkey agar. Colonies are nonhemolytic and organisms are nonmotile and oxidase and urease positive.

Burkholderia pseudomallei, which causes melioidosis, is a motile Gram-negative rod that forms small, nonhemolytic colonies that are oxidase positive. Growth can be seen on blood, chocolate, and MacConkey agar plates. *Coxiella burnetii* is a pleomorphic, Gram-negative, intracellular coccobacillus that cannot be cultured using routine bacteriological media but can be cultivated using cell culture. *Francisella tularensis* is a small, fastidious, pleomorphic, Gram-negative coccobacillus that is oxidase and urease negative. *Yersinia pestis*, which causes plague, is oxidase and urease negative and the gram stain of this Gram-negative rod may show a characteristic "safety pin" appearance due to bipolar staining. *Snyder JW, et al. Biothreat Agents. In: Carroll KC, et al. eds. Manual of Clinical Microbiology, Vol. 1. 12th ed., 2019, pp. 243–53.*

2. b. Gastrointestinal neoplasm. The description of the organism isolated from this patient's blood culture is consistent with *Clostridium septicum*, which has a very characteristic "swarming" morphology when grown on solid agar medium. It is an obligate anaerobic Gram-positive or Gram-variable large bacillus that frequently chain or give the appearance of long filaments. *C. septicum* does not make abundant spores, but when they are seen they are subterminal and frequently lemon shaped. *C. septicum* is not normal fecal microbiota of humans, although it is occasionally found in the appendix of healthy individuals. *C. septicum* should not be dismissed as a contaminant when isolated in the bloodstream. Most patients who have *C. septicum* isolated from their blood have an anomaly of the gastrointestinal system, such as diverticular disease, neutropenic enterocolitis involving the terminal ileum or cecum, or an underlying malignancy or gastrointestinal neoplasm, with colon cancer being the most common. *Khalid M, et al. C. Clostridium septicum sepsis and its implications. BMJ Case Rep 2012;2012.*

3. a. *Cardiobacterium hominis* is a small, pleomorphic Gram-negative bacillus with a distinctive Gram stain morphology of pairs and short chains that frequently forms rosettes. *C. hominis* is considered one of the HASCEK organisms (*Haemophilus* spp., *Aggregatibacter* spp., *Suttonella indologenes*, *Cardiobacterium* spp., *Eikenella corrodens*, *Kingella kingae*), which are fastidious Gram-negative bacilli that are frequently part of the normal microbiota of the nasopharynx and/or oral cavity of humans. The classic disease manifestation associated with HASCEK organisms is endocarditis. In general terms, HASCEK organisms are oxidase positive, require carbon dioxide for growth, and do not grow on MacConkey agar, and growth is enhanced in environments with increased humidity. The Gram-stain of *S. indologenes* is a plump Gram-negative bacillus that can occasionally form rosettes. *S. indologenes* may be misidentified as *C. hominis*, as this organism is oxidase and indole positive and catalase variable. *Zbinden R. Aggregatibacter, Capnocytophaga, Eikenella, Kingella, Pasteurella, and other fastidious or rarely encountered gram-negative rods. In: Carroll KC, et al., eds. eds. Manual of Clinical Microbiology, Vol. 1. 12th ed., 2019, pp. 656–69.*

4. c. Has an acid/acid reaction on triple sugar iron agar. *Escherichia coli* is the most common cause of urinary tract infections and produces an acid/acid reaction on triple sugar iron (TSI) agar. TSI can be helpful in determining the carbohydrate utilization pattern of an isolate. TSI agar has two parts: an aerobic slant and anaerobic butt. A colony of the test organism is touched with a sterile inoculating needle, stabbed into the agar butt, and struck over the surface of the slant. The tube is then inoculated for 18–24 h. TSI contains three sugars: glucose, lactose, and sucrose in addition to ferrous sulfate and a pH indicator. The lactose and sucrose concentrations are 10x that of glucose. The fermentation pattern is assessed by evaluating the reaction of the slant and then the butt. *Enterobacterales* preferentially ferment glucose (yellow, acid), but glucose is limited in the reaction so the aerobic slant will revert to alkaline conditions (red) due to oxidative utilization of peptone. Fermentation of sucrose and/or lactose results in a sustained yellow color in the slant. If an organism can reduce sulfur, H_2S production can be detected by a black precipitate in the media that forms when H_2S combines with ferric salts in the agar. In addition, the absence or presence of gas (bubbles or cracks in the media) should also be noted. *E. coli* are usually lactose fermenters (acid/acid on TSI) but may appear as nonlactose fermenters. They are H_2S negative and some isolates produce gas. *Salmonella, Citrobacter,* and *Edwardsiella* species are nonlactose fermenters, reduce sulfur, and produce gas on TSI (alkaline/acid + H_2S + gas). *Shigella* and *Plesiomonas* species are nonlactose fermenters that do not produce H_2S (alkaline/ acid). *Pseudomonas* species are nonfermenters and produce an alkaline/ alkaline reaction on TSI. *Buchan BW, et al. Escherichia, Shigella, and Salmonella. In: Carroll KC, et al. eds. Manual of Clinical Microbiology, Vol. 1. 12th ed. Washington, D.C.: ASM Press, 2019, 688–723. The Enterobacteriaceae. In: Carroll KC, et al. eds. Konemann's Color Atlas and Textbook of Diagnostic Microbiology, Vol. 12th ed., 2017, pp. 213–315.*

5. b. Matrix-assisted laser desorption/ionization-time of flight mass spectrometry (MALDI-TOF MS) is a rapid and accurate technique for bacterial identification. The method detects highly abundant bacterial proteins of bacteria, with ribosomal proteins being the primary peptides detected. The mass spectra generated by MALDI-TOF MS are compared with a reference spectra database to facilitate identification of a variety of bacteria. *Shigella* species and *Escherichia coli* are very closely related bacteria belonging to the *Enterobacterales* family. Because they are so closely related, methods such as 16S rRNA gene sequencing and MALDI-TOF MS are unable to reliably differentiate between *Shigella* species and *E. coli*. Thus, biochemical testing is critical for differentiation of these organisms. While both organisms are oxidase negative, *E. coli* are typically indole positive, motile, and are able to ferment lactose (pink colonies on MacConkey agar or alkaline slants on triple sugar iron agar). *Shigella sonnei* are most commonly indole negative, nonmotile, and nonlactose fermenters.

However, nonlactose-fermenting isolates of *E. coli* have been described, which can pose a significant challenge to identification. *Buchan BW, et al. Escherichia, Shigella, and Salmonella. In: Carroll KC, et al. eds. Manual of Clinical Microbiology, Vol. 1. 12th ed., 2019, pp. 688-723; Richter SS, et al. Identification of Enterobacteriaceae by matrix-assisted laser desorption/ionization time-of-flight mass spectrometry using the VITEK MS system. Eur J Clin Microbiol Infect Dis 2013;32:1571–8.*

6. c. *Streptococcus pneumoniae* are Gram-positive cocci that are the most common etiologic agent of community acquired pneumonia (CAP). Because some MALDI-TOF MS systems cannot reliably differentiate *S. pneumoniae* from other viridans group streptococci (VGS), biochemical testing remains an important part of the identification of this important pathogen. Testing for optochin susceptibility may be used to differentiate between *S. pneumoniae* (susceptible) and VGS (resistant). Alpha-hemolytic colonies that are suspicious for *S. pneumoniae* may also be subjected to bile solubility testing. *S. pneumoniae* colonies will dissolve upon contact (bile soluble) with a bile salt such as sodium deoxycholate while VGS colonies will remain intact (bile insoluble). Of note, *S. pneumoniae* is not associated with vancomycin resistance. *Haemophilus influenzae* and *Staphylococcus aureus* are less common causes of CAP. Growth of organisms in the presence or absence of X and V factors can be used to differentiate among different *Haemophilus* species, while coagulase positivity is associated with *S. aureus*. *Spellerberg B, et al. Streptococcus. In: Carroll CK, et al. eds. Manual of Clinical Microbiology, Vol. 1. 12th ed., 2019, pp. 399–417.*

7. c. *Clostridium* species are a group of Gram-positive spore-forming bacilli that may exhibit Gram-variable staining patterns. While most *Clostridium* species are obligate anaerobes, *Clostridium tertium* can grow aerobically. *C. tertium* is found in soil and is a commensal organism of the mouth and gastrointestinal tract of humans and animals. It is an infrequent cause of human infection but has been associated with bacteremia in patients with neutropenia or mucosal injury, peritonitis, septic arthritis, and brain abscesses. Because of its ability to grow aerobically, *C. tertium* is sometimes confused with *Bacillus* spp. However, *C. tertium* can be differentiated from *Bacillus* spp. based on the fact that it does not produce catalase and forms terminal spores only under anaerobic conditions. The susceptibility pattern to special potency antibiotic disks is useful for the identification of many anaerobes. In general, *Clostridium* spp. are colistin resistant and usually susceptible to vancomycin and kanamycin. *Eggerthella lenta* is an anaerobic, nonspore forming, catalase positive, Gram-positive bacillus that is a cause of bacteremia associated with gastrointestinal tract abnormalities. Similar to *C. tertium*, *Lactobacillus rhamnosus* is an aerotolerant Gram-positive bacillus that is catalase negative but is resistant to vancomycin. *L. rhamnosus* bacteremia is rare and has been linked to the use of probiotics. *Miller DL, et al. Significance of Clostridium tertium bacteremia in neutropenic and*

nonneutropenic patients: review of 32 cases. Clin Infect Dis 2001;32:975–8. Butler-Wu S, et al. Non-spore-forming anaerobic gram-positive rods. In: Carroll KC, et al. eds. Manual of Clinical Microbiology, Vol. 1. 12th ed., 2019, pp. 938–67.

8. b. Pet cat. The colony morphology, Gram-stain and biochemical characteristics are consistent with *Pasteurella* species, which are most commonly associated with normal flora of dog and cat mouths. Infections due to *Pasteurella* are commonly soft tissue infections secondary to bite wounds. However, *Pasteurella* can cause a variety of disease manifestations, such as bloodstream infections, meningitis, peritonitis, endocarditis, and osteomyelitis. It is important to keep *Pasteurella* in mind when a small Gram-negative coccobacillus that does not grow on MacConkey agar is isolated in the laboratory. Agents of laboratory-acquired infections that fit this description are ruled out prior to manipulation of the culture outside of the biological safety cabinet. The acronym "COIN" is helpful to remember the biochemical characteristics of *Pasteurella multocida*, that is, catalase, ornithine/oxidase, indole, and nitrate positive. *Francisella* is weakly catalase positive, oxidase negative, urease negative, and cefinase positive. *Brucella* is urease positive. *Snyder JW, et al. Biothreat Agents. In: Carroll KC, et al. eds. Manual of Clinical Microbiology, Vol. 1. 12th ed. Washington, D. C.: ASM Press, 2019, pp. 243–53. Zbinden R. Aggregatibacter, Capnocytophaga, Eikenella, Kingella, Pasteurella, and other fastidious or rarely encountered gram-negative rods. In: Carroll KC, et al. eds. Manual of Clinical Microbiology, Vol. 1. 12th ed., 2019. pp. 656–69.*

9. c. *Klebsiella oxytoca* is intrinsically resistant to ampicillin. Knowledge of intrinsic resistance, which is defined as antimicrobial resistance that is not acquired by an organism, is helpful for evaluating accuracy of microorganism identification and/or antimicrobial susceptibility testing (AST) results and to ensure proper reporting of AST results. If a drug for which an organism is intrinsically resistant tests as susceptible, further investigation is necessary to ensure that the correct isolate was tested and that the AST reagents are working properly. Intrinsic resistance patterns for commonly isolated organisms are published by the Clinical and Laboratory Standards Institute (CLSI) in the M100 document and by the European Committee on Antimicrobial Susceptibility Testing (EUCAST). *CLSI. Performance Standards for Antimicrobial Susceptibility Testing; Twenty-ninth Informational Supplement. CLSI document M100-S30. Vol. Wayne, PA: Clinical and Laboratory Standards Institute, 2020; The European Committee on Antimicrobial Susceptibility Testing. Intrinsic Resistance and Unusual Phenotypes. Version 3.2. http://www.eucast.org/fileadmin/src/media/PDFs/EUCAST_files/Expert_Rules/2020/Intrinsic_Resistance_and_Unusual_Phenotypes_Tables_v3.2_20200225.pdf (Accessed March 24 2020).*

10. b. Cefazolin. *Salmonella* is a nonfermenting gram-negative rod in the *Enterobacterales* family. Nontyphoidal *Salmonella* cause intestinal infections and occasional urinary tract infections, wound infections, and bacteremia. Intestinal infections with *Salmonella* are usually self-limiting, especially in an immunocompetent host. Treatment of acute intestinal infections may prolong carriage of the organism and is generally not recommended in healthy individuals. Thus, routine susceptibility testing of *Salmonella* isolates from intestinal specimens is not recommended. Susceptibility testing is recommended for extraintestinal *Salmonella* isolates and should include ampicillin, a fluoroquinolone, trimethoprim-sulfamethoxazole, and a third-generation cephalosporin. First and second generation cephalosporins such as cefazolin and aminoglycosides should never be reported for *Salmonella* isolates. While these drugs may test susceptible *in vitro*, they are not clinically effective for treatment of *Salmonella* infections. *Buchan BW, et al. Escherichia, Shigella, and Salmonella. In: Carroll KC, et al. eds. Manual of Clinical Microbiology, Vol. 1. 12th ed., 2019, pp. 688–723; CLSI. Performance Standards for Antimicrobial Susceptibility Testing; Twenty-ninth Informational Supplement. CLSI document M100-S30. Vol. Wayne, PA: Clinical and Laboratory Standards Institute, 2020.*

11. d. *Ixodes scapularis.* Vectors are organisms that transmit disease from one host to another. *Ixodes scapularis*, known commonly as the blacklegged tick, is located throughout the upper Midwest and northeast of the United States and Canada. This tick species serves as the vector that transmits Lyme Disease (*B. burgdorferi*) and several other infections, including babesiosis (*Babesia*), tickborne relapsing fever (*B. miyamotoi*), and human granulocytic anaplasmosis (*Anaplasma phagocytophilum*). *Dermacentor variabilis*, the American dog tick, is found across the midwestern and eastern United States and Canada and is the vector for tularemia (*Francisella tularensis*) and Rocky Mountain spotted fever (Rickettsia rickettsia). *Amblyomma americanum*, the Lone Star tick, is found in the southeastern and eastern United States and is a vector of human monocytic ehrlichiosis (*Ehrlichia chaffeensis*) and tularemia. *Aedes aegypti* is a mosquito that can transmit yellow fever, Dengue fever, and a number of other diseases. *U.S. Department of Health and Human Services. Tickborne Diseases of the United States. https://www.cdc.gov/lyme/resources/TickborneDiseases.pdf (Accessed March 24 2020).*

12. c. The presence of a branching, filamentous Gram-positive organism on Gram stain is consistent with aerobic actinomycetes such as *Nocardia* species. *Nocardia brasiliensis* is associated with cutaneous abscesses in immunocompetent individuals that are often due to traumatic inoculation. Disseminated infections are most likely to occur in immunocompromised patients. Of note, *Nocardia* species stain partially acid fast using modified Kinyoun or Ziehl-Neelsen acid fast stains. *Mycobacterium* species would be expected to be fully acid fast and most other organisms would not exhibit acid-fast characteristics. While *Streptococcus constellatus* is associated with abscess formation, this organism is a Gram-positive cocci. *Mycobacterium marinum* and *Sporothrix schenckii* may also cause cutaneous abscesses that spread along the lymphatics. However, neither organism would exhibit a branching, filamentous appearance on gram stain. *Conville PS, et al. Nocardia, Rhodococcus, Gordonia, Actinomadura, Streptomyces, and other aerobic actinomycetes. In: Carroll KC, et al. eds. Manual of Clinical Microbiology, Vol. 1. 12th ed., 2019:525–57.*

13. e. *Mycobacterium neoaurum.* Many species of mycobacteria are associated with infections following traumatic wounds. Mycobacteria are commonly classified by growth rate and pigmentation. Rapidly growing *Mycobacterium* (RGM) are those with the ability to grow within 7 days of subculture, while slowly growing *Mycobacteria* require more than 7 days for growth upon subculture. Pigmented species may produce pigment only when exposed to light (photochromogens) while others are pigmented regardless of light exposure (scotochromogens). *M. neoaurum* is an RGM that produces a yellow pigment. It is an uncommon cause of human infection but has been found to cause skin and soft tissue infection following traumatic inoculation and line-associated infection. *M. chelonae* is an RGM that is associated with traumatic wound infections, but does not have a yellow pigment. *Mycobacterium marinum* is a slowly growing photochromogen associated with traumatic injuries and exposure to contaminated water or fish tanks. *M. marinum* requires special growth temperatures (28–30°C) for primary isolation. *Mycobacterium gordonae* is a slowly growing scotochromagen that is commonly encountered in clinical laboratories and is almost never a cause of human infection. *Mycobacterium kansasii* is a slowly growing photochromogen associated with pulmonary infections. *Brown-Elliott BA, et al. Clinical and taxonomic status of pathogenic nonpigmented or late-pigmenting rapidly growing mycobacteria. Clin Microbiol Rev 2002;15:716–46. Brown-Elliott BA, et al. Mycobacterium neoaurum and Mycobacterium bacteremicum sp. nov. as causes of mycobacteremia. J Clin Microbiol 2010;48:4377–85.*

14. b. Cleaning a fish tank. *Mycobacterium marinum* is a slow-growing environmental mycobacterium that is distributed widely in aquatic environments, particularly fish tanks and naturally occurring bodies of water. *M. marinum* causes cutaneous infections that are a result of skin trauma and subsequent exposure to freshwater fish tanks or salt water. *M. marinum* has a restricted growth temperature of 28–30°C for primary isolation. Because of the requirement for nonroutine culture techniques and the relative rarity of *M. marinum* as a cause of infection, diagnosis of this organism is often delayed. Incubation of cultures of specimens from the head, neck, extremities, and other superficial sites at 30°C facilitates growth of *M. marinum.* Of note, *M. marinum* is a photochromogen, which means

that colonies of this organism appear buff colored but produce a yellow pigment upon exposure to light. *Caulfield AJ, et al. Mycobacterium: Laboratory characteristics of slowly growing mycobacteria other than Mycobacterium tuberculosis. In: Carroll KC, et al. eds. Manual of Clinical Microbiology, Vol. 1. 12th ed., 2019, pp. 595–611.*

15. c. *Candida albicans. Candida* spp. are yeasts that are commensal organisms found in the gastrointestinal tract and on skin and mucocutaneous membranes of humans. *Candida* spp. cause opportunistic infections in patients with risk factors such as immune suppression, diabetes, IV drug abuse, and in patients with catheters. *Candida* yeast are typically seen as oval budding yeast on wet preparation. Colonies are glabrous (without aerial hyphae) and smooth and *C. albicans* in particular may produce projections on certain media types that are often referred to as "feet." *Candida* spp. may appear as budding yeast and many species produce pseudohyphae. *C. albicans* can produce germ tubes (the beginning of true hyphae) when placed in a nutritionally rich calf serum, which differentiates this species from other species within the genus. *Borman AM, et al. Candida, Cryptococcus, and Other Yeasts of Medical Importance. In: Carroll KC, et al. eds. Manual of Clinical Microbiology, Vol. 2. 12th ed., 2019, pp. 2056–86.*

16. 1. d; 2. b; 3. e; 4. a; 5. c. Cornmeal agar with Tween 80 is used to differentiate yeasts and yeast-like organisms on the basis of morphological characteristics. The addition of the surfactant Tween 80 stimulates rapid and abundant chlamydospore formation, which is useful in the differentiation of *Candida* spp. Placement of a coverslip over the yeast inoculum generates a microaerophilic environment that further encourages the production of chlamydospores. On cornmeal Tween 80 agar, *Candida* spp. form pseudohyphae and some true hyphae, with clusters of round blastoconidia at the septa. *C. albicans* forms a single terminal, thick-walled chlamydospore, while *C. dubliniensis* usually forms pairs or clusters of chlamydospores. *Rhodotorula* spp. are round to oval budding yeasts that do not produce pseudohyphae. Colonies usually appear pink to coral on the plate. *Geotrichum candidum* and *Trichosporon* spp. are yeast-like organisms that produce arthroconidia from hyphae on cornmeal Tween 80 agar. *Trichosporon* also produce pseudohyphae and blastoconidia along the hyphae, which helps differentiate this organism from *Geotrichum candidum. Borman AM, et al. Candida, Cryptococcus, and other yeasts of medical importance. In: Carroll KC, et al. eds. Manual of Clinical Microbiology, Vol. 2. 12th ed. Washington, D.C.: ASM Press, 2019, pp. 2056–86. Walsh TJ, et al. Larone's Medically Important Fungi: A Guide to Identification, Vol. 6th ed., 2018.*

17. d. *Malassezia* species are members of human skin microbiota that can cause skin infections (tinea versicolor) and bloodstream infections. While there are many species in the genus, *Malassezia furfur* is the organism that is most often associated with central line infections secondary to lipid supplementation. Many species, with the exception of *M. pachydermatis*, require long-chain fatty acid supplementation to grow on microbiologic media. *Malassezia* produce yeast-like cells that are actually phialides with small "collarettes" or a scarring where the daughter cell buds from the parent cell. These yeast-like organisms are usually single-celled but can produce hyphal elements that may be directly observed in tissue specimens. *Walsh TJ, et al. Larone's Medically Important Fungi: A Guide to Identification, Vol. 6th ed., 2018.*

18. c. *Aspergillus fumigatus*. Invasive aspergillosis (IA) is a life-threatening infection in patients with prolonged and profound neutropenia. Knowledge of the species of *Aspergillus* can facilitate proper treatment of IA. *A. fumigatus*, the most common cause of invasive aspergillosis, has greenish gray colonies with uniseriate phialides that cover the upper two-thirds of the vesicle. *A. flavus* colonies are yellowish green to brown with rough conidiophores and phialides that cover the entire vesicle. *A. niger* colonies start out light but turn black with age and have biseriate phialides that cover the entire vesicle. Colonies of *A. clavatus* are green with a white border. Phialides are densely packed on a large, elongated vesicle. Galactomannan is a carbohydrate found in the cell wall of *Aspergillus* species. Galactomannan antigen detection by enzyme immunoassay in serum and bronchoalveolar lavage is most useful in the setting of patients with hematological malignancy or who have undergone hematopoietic cell transplantation, where the selection of high-risk patients maximizes positive predictive value. Thus, it may be useful as an aid in the early diagnosis of invasive aspergillosis but is not a substitute for a thorough diagnostic workup that includes imaging, histology, and culture. *Patterson TF, et al. Practice guidelines for the diagnosis and management of aspergillosis: 2016 update by the Infectious Diseases Society of America. Clin Infect Dis 2016;63:e1–e60. Chen SC-A, et al. Aspergillus, Talaromyces, and Penicillium. In: Carroll KC, et al. eds. Manual of Clinical Microbiology, Vol. 2. 12th ed., pp. 2103–3131.*

19. c. Reverse-transcriptase PCR on stool. The majority of acute gastroenteritis cases are caused by viruses. Norovirus is a segmented double stranded RNA virus in the *Caliciviridae* family that is associated with acute moderate to severe gastroenteritis and hospital-acquired infections. Immunocompromised patients are at risk for chronic infections and may shed the virus for prolonged periods. Infections with this virus often peak in the winter months but outbreaks may occur year-round. Norovirus does not grow using routine viral culture methods employed by clinical laboratories. In addition, serology is not a good choice for diagnosis of enteric viruses, as the need for acute and convalescent sera is not practical for diagnosis of these mainly self-limiting infections. Nucleic acid amplification testing is useful for detection of viruses causing gastroenteritis,

including Norovirus. Nucleic acid extraction from feces can be a challenge due to presence of PCR inhibitors. Additionally, due to the presence of nonviral nucleic acids in stool, assays use virus-specific probes to eliminate nonspecific PCR products. *Pang X, et al. Gasteronteritis Viruses. In: Carroll KC, et al. eds. Manual of Clinical Microbiology, Vol. 2. 12th ed., 2019, pp. 656–73.*

20. *d.* *Taenia solium* is the cause of cysticercosis, a parasitic infection that occurs when larval cysts infect brain, eyes, muscles, or other tissues of an intermediate host, which is a host that harbors sexually immature parasitic forms. Swine are usually the intermediate host for *T. solium*, but humans may serve as incidental intermediate hosts if infectious eggs are ingested from contaminated food or water. If this occurs, the larvae hatch and penetrate the intestinal wall, whereupon larva circulate and invade musculature and other tissues. Humans are normally the definitive hosts for *T. solium* and may acquire intestinal taeniasis upon ingestion of the larval form (cysticerci) in undercooked pork. Definitive hosts serve as the vector in which a parasite reaches sexual maturity and reproduces. *T. saginata*, the beef tapeworm, does not cause cysticercosis, as humans cannot serve as intermediate hosts for this parasite. *Garcia LS. Intestinal Cestodes. Diagnostic Medical Parasitology, Vol. 6th ed., 2016, pp. 418–46.*

Section D

Other clinical laboratory sections

Chapter 25

Hematology and coagulation

Zane D. Amenhotep
University of California, San Francisco, CA, United States

A care provider noted that an hours-old full-term male neonate with a normal respiratory rate had mild perioral and central cyanosis, despite an otherwise normal physical exam. The patient was immediately placed on supplemental oxygen, but remained cyanotic. Hemoglobin oxygen saturation by pulse-oximetry (SpO$_2$) ranged from 90% to 93%. Upon collection, his blood was noted to have a chocolate-brownish discoloration. His complete blood count (including hemoglobin concentration) and blood gas measurements were all normal for age. However, CO-oximetry revealed the following proportions of hemoglobin derivatives.

Oxyhemoglobin (O$_2$HB):	80.2% (reference interval: 95% to 98%)
Carboxyhemoglobin (COHb):	2.5% (reference interval: <1.5%)
Methemoglobin (MetHb):	17.3% (reference interval: <2.0%)

1. Which of the following mechanisms could explain the cause of this patient's cyanosis?
   a. maternal smoking during pregnancy
   b. decreased activity of NADH-cytochrome-b_5 reductase
   c. increased activity of NADPH-methemoglobin reductase
   d. abundance of reduced glutathione
   e. in utero exposure to high levels of ascorbic acid
2. What is the reason for the discrepancy between the pulse oximetry reading and the reading of the CO-oximeter in the preceding case?
   a. the pulse-oximeter is likely malfunctioning
   b. the CO-oximeter is likely malfunctioning
   c. the patient has an M hemoglobin
   d. the CO-oximeter reads light absorption at more wavelengths
   e. the patient has been treated with ascorbic acid
3. A 22-year-old female is referred for workup of incidental findings on a complete blood count. The report includes the following erythrocyte indices.

Self-assessment Q&A in Clinical Laboratory Science, III. https://doi.org/10.1016/B978-0-12-822093-1.00025-9

RBC	$5.98 \times 10^6/\mu L$	(Reference: $3.80–5.20 \times 10^6/\mu L$)
Hemoglobin	11.1 g/dL	(Reference: 11.7–15.7 g/dL)
MCV	76 fL	(Reference: 80–100 fL)

Which of the following diagnoses would *best* explain these RBC indices?

  a. iron deficiency anemia
  b. thalassemia trait
  c. anemia of chronic disease
  d. secondary polycythemia
  e. sideroblastic anemia

4. A 68-year-old male with a history of chronic lymphocytic leukemia/small lymphocytic lymphoma (CLL/SLL) presents with bluish purple discoloration of his hands and feet. A direct antiglobulin test is positive for complement only. A complete blood count is requested and the initial run on the analyzer generated the following red blood cell indices.

RBC	$2.0 \times 10^6/\mu L$	(Reference: $4.4–5.90 \times 10^6/\mu L$)
HGB	10.5 g/dL	(Reference: 13.3–17.7 g/dL)
MCV	127 fL	(Reference: 80–100 fL)

What is the most likely cause for the high mean corpuscular volume (MCV)?

  a. high WBC count
  b. cryoglobulin
  c. cold agglutinin
  d. large platelets
  e. in vitro hemolysis

5. Which of the following automated hematology analyzer indices is thought to provide the earliest sign of bone marrow regeneration following chemotherapy and/or engraftment in bone marrow or hematopoietic stem cell transplantation?

  a. reticulocyte count %
  b. absolute reticulocyte count
  c. immature reticulocyte fraction
  d. reticulocyte hemoglobin content (CHr)
  e. reticulocyte hemoglobin equivalent (Ret-He)

6. Which of the following conditions will tend to increase the erythrocyte sedimentation rate?

  a. decreased fibrinogen
  b. anemia
  c. the presence of sickle cells
  d. high serum albumin
  e. microcytic erythrocytes

7. Which of the following increases the transcription of erythropoietin (EPO)?

  a. red cell 2,3-diphosphoglycerate
  b. hypoxia-inducible factor 2α
  c. wild-type von Hippel Lindau gene

d. *JAK2* V617F

e. *JAK2* Exon 12

8. Which of the following hemoglobins are due to a mutation in an α-globin gene?

   a. hemoglobin S

   b. hemoglobin C

   c. hemoglobin E

   d. hemoglobin Lepore

   e. hemoglobin Constant Spring

9. Which of the following hemoglobins is composed of a tetramer of four gamma globin chains (γ_4)?

   a. hemoglobin H

   b. hemoglobin Bart's

   c. hemoglobin C Harlem

   d. hemoglobin M Boston

   e. hemoglobin G Philadelphia

10. A 20-year-old male presents for workup of asymptomatic microcytosis (MCV of 78 femtoliters). The following hemoglobin HPLC analysis was obtained; see Figure 1.

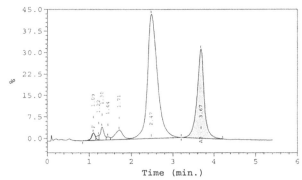

Peak Name	Calibrated Area %	Area %	Retention Time (min)
F	1.4*	- - -	1.09
Unknown	- - -	0.6	1.22
P2	- - -	2.3	1.30
Unknown	- - -	0.5	1.44
P3	- - -	3.5	1.71
Ao	- - -	60.2	2.47
A2	31.3*	- - -	3.67

Which of the following interpretations fits best with the HPLC result?

   a. alpha thalassemia

   b. beta thalassemia

      c. hemoglobin E trait

      d. iron deficiency anemia

      e. hemoglobin Lepore

11. A 23-year-old male is referred to a hematologist for a known blood disorder. His red blood cell indices are as follows...

RBC	$2.80 \times 10^6/\mu L$	(Reference: 4.4–$5.90 \times 10^6/\mu L$)
Hemoglobin	$10.5\,g/L$	(Reference: 13.3–$17.7\,g/dL$)
MCV	$109.1\,fL$	(Reference: 80–$100\,fL$)

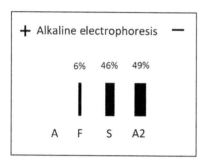

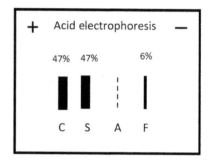

What is the best interpretation of the hemoglobin gel electrophoresis?

      a. hemoglobin S trait

      b. hemoglobin S disease

      c. hemoglobin C trait

      d. hemoglobin C disease

      e. hemoglobin SC disease

12. A 57-year-old female presents with incidentally discovered leukocytosis (WBC $> 90 \times 10^3/\mu L$). Her peripheral blood smear reveals many immature granulocytes including a relative expansion in the number of myelocytes, 5% basophils, and 2% blasts. Her hemoglobin concentration was slightly low and her platelet count was slightly increased. Which of the following diagnoses best fits the described blood smear morphology and what is the likely molecular finding?

      a. acute promyelocytic leukemia with *PML-RARA*

      b. chronic myelogenous leukemia, *BCR-ABL1* positive

      c. chronic neutrophilic leukemia with *CSF3R* mutation

      d. essential thrombocythemia with *JAK2* V617F

      e. polycythemia vera with *JAK2* Exon 12 mutation

13. A 19-year-old female college student presents with a sore throat and excessive malaise. Her complete blood count reveals moderate leukocytosis (WBC count $24 \times 10^3/\mu L$) and a review of her peripheral blood smear reveals a variety of atypical lymphocytes including some large forms with variable nuclear borders and abundant deeply basophilic cytoplasm. Both monocytoid and plasmacytoid lymphocytes are noted. Which diagnosis fits best with this morphologic pattern?

a. acute myeloid leukemia
b. diffuse large B cell lymphoma
c. chronic lymphocytic leukemia
d. Epstein-Bar virus, infectious mononucleosis
e. chronic myeloid leukemia

14. All of the following are true of G6PD deficiency *EXCEPT...*

a. Heinz bodies, which are denatured globin chains, may form in this condition.
b. Oxidative stress may provoke a hemolytic episode.
c. Antimalaria drugs are a well-described possible trigger of hemolytic episodes.
d. The optimal time for quantitative measurement of G6PD is during an acute hemolytic episode.
e. The ascorbate cyanide test is not specific for G6PD deficiency.

15. A 9-year-old male with sickle cell disease presents with progressive fatigue, pallor, and shortness of breath. A complete blood count reveals a hemoglobin measurement of 4.2 g/dL, down from 7.4 g/dL, 2 weeks ago. The patient's absolute reticulocyte count is low at $5 \times 10^3/\mu L$. What is the likely explanation for the significant drop in this patient's hemoglobin?

a. spurious result
b. myelodysplastic syndrome
c. transient erythroblastopenia of childhood
d. parvovirus infection
e. iron deficiency

16. Which of the following combinations of blood and bone marrow findings are compatible with a deficiency of cobalamin (vitamin B12)?

a. hypersegmented neutrophils, macrocytic anemia, hypercellular bone marrow
b. hyposegmented neutrophils, macrocytic anemia, hypocellular bone marrow
c. hyposegmented neutrophils, macrocytic anemia, hypercellular bone marrow
d. hypersegmented neutrophils, microcytic anemia, hypercellular bone marrow
e. hypersegmented neutrophils, microcytic anemia, hypocellular bone marrow

17. A 5-year-old female with splenomegaly, anemia, and mild jaundice presents for evaluation. Her peripheral blood smear shows many red blood cells with smaller diameter than normal RBCs and no central pallor and that stain more darkly red. Which of the following statements is true regarding this condition?

a. In an osmotic fragility test this patient's red blood cells (RBCs) would show a decreased osmotic fragility relative to normal RBCs.

b. The Ham's (acid hemolysin) test would be helpful in confirming the diagnosis.

c. Flow cytometry would have no role in making the diagnosis for this condition.

d. This condition may involve abnormalities related to ankyrin, spectrin, or band 3

e. This condition is always severe

18. Which of the following porphyrias is commonly associated with both photosensitivity and hemolytic anemia?

    a. acute intermittent porphyria (AIP)

    b. congenital erythropoietic porphyria (CEP)

    c. variegate porphyria (VP)

    d. hereditary coproporphyria (HCP)

    e. X-linked porphyria

19. A 32-year-old female presents with fatigue and no prior bleeding history. A complete blood count is performed and, although there is no evidence of anemia, the initial run on the hematology analyzer reveals unexpected thrombocytopenia.

Platelet count $75 \times 10^3/\mu L$ (Reference range $150\text{–}400 \times 10^3/\mu L$)

Review of the patient's peripheral blood smear reveals many large platelet clumps. The technologist recognizes this and requests for a new blood specimen to be sent. Which anticoagulant preservative should be used to collect the new specimen?

    a. K_2-EDTA

    b. K_3-EDTA

    c. sodium citrate

    d. lithium heparin

    e. sodium heparin

20. A 64-year-old female presented to an urgent care center with a complaint of fever and recurring bouts of severe abdominal pain accompanied by nausea, vomiting, and dark urine for the past 2 weeks. Her complete blood count reveals pancytopenia. Her absolute reticulocyte count was moderately increased. Serum studies revealed moderate transaminitis, indirect hyperbilirubinemia, increased lactate dehydrogenase, and low haptoglobin. The direct antiglobulin test was negative. Imaging studies revealed a portal vein thrombosis. What is the most likely associated underlying diagnosis?

    a. cold agglutinin disease

    b. cryoglobulinemia

    c. Coombs-positive hemolytic anemia

    d. paroxysmal cold hemoglobinuria

    e. paroxysmal nocturnal hemoglobinuria

21. Flow cytometry was performed on peripheral blood for the patient described in the previous question. Monoclonal antibodies to which of the following antigens would be most useful for evaluating RBCs, neutrophils, and monocytes for evidence of the suspected condition?
    a. CD13 and CD33
    b. CD14 and CD16
    c. CD55 and CD59
    d. CD19 and CD45
    e. CD71 and CD56

22. Which of the following could falsely shorten the clotting time of coagulation assays?
    a. underfilled collection tube
    b. polycythemia
    c. use of a discard tube
    d. heparin or iv fluid contamination
    e. improper phlebotomy tube collection order

23. What is the standard blood:anticoagulant ratio for coagulation test specimens?
    a. 1:2
    b. 2:1
    c. 4:1
    d. 9:1
    e. 20:1

24. Which of the following conditions are likely to show isolated prolongation of the prothrombin time (PT)?
    a. Factor VII deficiency
    b. unfractionated heparin therapy
    c. disseminated Intravascular coagulation
    d. hemophilia A
    e. liver failure

25. Which of the following conditions are likely to present with an isolated prolongation of the activated partial thromboplastin time (aPTT)?
    a. coumadin therapy
    b. presence of a Lupus Anticoagulant
    c. severe vitamin K deficiency
    d. intravenous fluid contamination of the sample
    e. Factor V deficiency

26. The following are the results of a dilute Russell's Viper Venom Time assay.

dRVVT Screen:	68s (Ref NPP: <44s)	Screen ratio	1.5
dRVVT 1:1 mix:	65s (Ref NPP: <46s)	Mix ratio	1.4
dRVVT Confirm:	29s (Ref NPP: <38s)	Confirm ratio	0.8

What is the best interpretation?
  a. factor deficiency
  b. a specific factor inhibitor is present
  c. no evidence of a lupus anticoagulant
  d. lupus anticoagulant is unlikely
  e. consistent with lupus anticoagulant

27. Which profile is most consistent with disseminated intravascular coagulation (DIC)?
  a. prolonged PT, normal aPTT, low platelet count, normal fibrinogen
  b. normal PT, prolonged aPTT, low platelet count, low fibrinogen
  c. prolonged PT, prolonged aPTT, high platelet count, high fibrinogen
  d. prolonged PT, prolonged aPTT, low platelet count, low fibrinogen
  e. normal PT, normal aPTT, low platelet count, low fibrinogen

28. The following tracings show tracings are from a global hemostasis assay based on viscoelastic testing. The tracing on the left is from a normal patient.

Normal   **Patient**

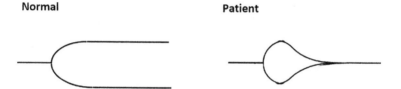

What does the tracing on the right indicate?
  a. use of antiplatelet agents
  b. fibrinolysis
  c. increased clot strength
  d. decreased clot strength
  e. anticoagulant effect

29. The following factor assays were performed at different dilutions.

Factor VIII		Factor IX		Factor XI	
1:10	2%	1:10	60%	1:10	81%
1:20	10%	1:20	65%	1:20	78%
1:40	38%	1:40	62%	1:40	76%
1:80	52%				
Ref.	50%–150%	Ref.	50%–150%	Ref.	50%–150%

What is the best interpretation?
  a. Factor VIII activity is markedly low
  b. a nonspecific inhibitor is present
  c. a specific factor inhibitor is present
  d. likely vitamin K deficiency
  e. likely von Willebrand disease

30. Each of the following tests may be affected by active or recent thrombosis except?
   a. anticardiolipin antibody
   b. antithrombin activity
   c. protein C, Functional
   d. protein S, Functional
   e. homocysteine
31. Each of the following may indicate additional risk of thrombosis except?
   a. antithrombin deficiency
   b. protein C deficiency
   c. protein S deficiency
   d. plasminogen activator inhibitor 1 (PAI-1) deficiency
   e. anticardiolipin antibody
32. Monitoring unfractionated heparin therapy with an unfractionated heparin antifactor Xa assay has advantages over using the activated partial thromboplastin time (aPTT) assay in each of the following settings except?
   a. increased level of acute phase proteins
   b. patients with DIC
   c. patients with lupus anticoagulant
   d. patients with Factor XII deficiency
   e. recent use of rivaroxaban
33. Which of the following best describes Type 2B von Willebrand disease?
   a. reduced but qualitatively normal von Willebrand factor antigen
   b. decreased binding of von Willebrand factor to collagen
   c. gain of function mutation, increased aggregation of platelet rich plasma with low dose ristocetin
   d. complete loss of all von Willebrand factor multimers
   e. deficiency of glycoprotein Ib-V-IX Complex on platelets.
34. Which of the following is most useful in providing definitive diagnosis of thrombotic thrombocytopenic purpura (TTP)?
   a. ADAMTS13
   b. LDH
   c. PTT
   d. platelet count
   e. hemoglobin

Answers
1. b. Methemoglobin (MetHb) is a hemoglobin derivative that results from oxidation of the iron in the heme molecule to the ferric state. Ferric iron in heme is unable to bind oxygen. High levels of MetHb result in cyanosis (despite normal levels of total hemoglobin) and also "chocolate-brown" blood. Methemoglobinemia may be inherited (as in this case) or acquired from exposure to various medications (such as benzocaine, dapsone, sulfonamides, and nitrate derivatives) or other chemical compounds that are

strong oxidants. Inherited forms may be due to abnormal hemoglobin chains (so-called M hemoglobins), in which heme iron is more stable in the ferric state, or due to various autosomal recessive mutations that result in decreased erythrocyte NADH-cytochrome-b_5 reductase (CYTB5R). CYTB5R is the primary component of the methemoglobin reductase system that reduces MetHb to hemoglobin. Reserve enzyme systems include reduced glutathione, (NADPH)-methemoglobin reductase and ascorbic acid. Methylthioninium chloride (new methylene blue) or high doses of ascorbic acid may be used to treat clinically symptomatic acquired methemoglobinemia or congenital methemoglobinemia due to CYTB5R deficiency. While some studies have indicated that smoking may be associated with slightly higher baseline levels of MetHb, maternal smoking would not be expected to cause this degree of methemoglobinemia in a neonate. *Kedar PS, et al. Novel mutation (R192C) in CYBR3 gene causing NADH-cytochrome b5 reductase deficiency in eight Indian patients associated with autosomal recessive congenital methemoglobinemia type I. Hematol 2018;23(8):567–73.*

2. d. Conventional pulse oximeters take absorbance readings at only two wavelengths of light that are intended to differentiate oxyhemoglobin (O_2Hb) from deoxyhemoglobin (HHb), but do not distinguish methemoglobin (MetHb), which absorbs light equally well at both wavelengths. The determination of arterial oxygen saturation (SaO2) by pulse oximetry (SpO$_2$) is based on the ratio of O_2Hb to total hemoglobin, i.e., [O_2Hb/ (O_2Hb + HHb)]. At low levels of MetHb, the decrease in the oxygen saturation measured by pulse oximetry (SpO$_2$) is less than expected for the degree of increase in MetHb. At higher levels of MetHb, that are more likely to be clinically symptomatic, SpO$_2$ tends to plateau at around 85%, which may lead to errors in the estimation of SaO$_2$. This is in contrast to CO-oximetry, which uses multiple wavelengths to distinguish different hemoglobin derivatives. M hemoglobins may absorb at different wavelengths that read differently by CO-oximetry and may require hemoglobinopathy studies to fully resolve. Both ascorbic acid and methylene blue should reduce the proportion of methemoglobin. It should also be noted that CO-oximetry generally cannot distinguish MetHb from methylene blue, which presents a challenge for monitoring after treatment. *Ward J, et al. Congenital methemoglobinemia identified by pulse oximetry screening. Pediatr 2019;143(3). https://doi.org/10.1542/peds.2018-2814.*

3. b. Microcytosis in the setting of a high RBC count are common findings in thalassemia. The mild degree of anemia suggests thalassemia trait. Additional studies would be necessary to distinguish α-thalassemia from β-thalassemia. Iron deficiency anemia, anemia of chronic disease, and sideroblastic anemia would be expected to have a low RBC count commensurate with the low hemoglobin concentration. Patients with primary polycythemia vera tend to develop iron deficiency over time, due in part to

phlebotomy to decrease erythrocytosis. This is not necessarily true of secondary polycythemia, in which both the RBC and hemoglobin concentration are typically high. Various erythrocyte indices and published formulas have been presented as a means of screening to distinguish thalassemia trait from iron deficiency. As patients may have more than one of these conditions concurrently, additional studies may be helpful in excluding each possibility. *Kar YD, et al. Erythrocyte indices as differential diagnostic biomarkers of iron deficiency anemia and thalassemia. J Pediatr Hematol Oncol 2019;23.* http://doi.org/10.1097/MPH.0000000000001597.

4. c. A variety of phenomena can interfere with the accuracy of readings from automated cell counters. Cold agglutinin disease represents an autoimmune condition in which pentameric IgM antibodies cause agglutination of red blood cells. It may be associated with medical conditions like CLL/SLL, viral infections, etc. Agglutinates of RBCs are "seen" by many hematology analyzers as single large red blood cells rather than several normal size red blood cells. This results in a falsely increased mean corpuscular volume (MCV) and a falsely low RBC concentration. This change also affects derived parameters that are calculated from the MCV and/or RBC, like hematocrit (HCT), mean corpuscular hemoglobin (MCH), and mean corpuscular hemoglobin concentration (MCHC). Warming the specimen to 37°C and repeating the run can improve the accuracy of measurements. High WBC counts can also falsely increase the MCV, though this is more likely to occur with myeloid cells than with the smaller lymphocytes seen in CLL/SLL and it would not explain the low RBC count. Cryoglobulins are immunoglobulins that precipitate in colder temperature. They do not generally bind or interact with RBCs and are not associated with hemolysis. Significant numbers of large platelets may be "seen" by some hematology analyzers as if they were RBCs. Even when this occurs, they are usually still smaller (on average) than RBCs and may falsely decrease the MCV. In vitro hemolysis can cause a falsely low RBC though would not be expected to increase the MCV. *Bessman JD, et al. Spurious macrocytosis, a common clue to erythrocyte cold agglutinins. Am J Clin Pathol 1980;74:797–800.*

5. c. Modern hematology analyzers can provide a variety of automated measurements beyond the traditional red blood cell indices like reticulocyte count % or absolute reticulocyte count, which give general information on the status of the function of the erythroid compartment of the bone marrow in the setting of anemia. Some studies have indicated that the immature reticulocyte fraction can serve as a marker of engraftment that precedes other parameters, such as conventional reticulocyte counts, absolute neutrophil count (ANC), or immature platelet fraction (IPF). Reticulocyte hemoglobin content and Reticulocyte hemoglobin equivalent can provide early indications of iron depletion. *Noronha JF, et al. Immature*

reticulocytes as an early predictor of engraftment in autologous and allo-geneic bone marrow transplantation. Clin Lab Haematol 2003;25:47–54.

6. b. Erythrocyte sedimentation rate (ESR) is a measure of the rate of sed-imentation of erythrocytes (red blood cells) in a column and is used as a nonspecific marker of underlying inflammation. Conditions that decrease the negative charge (zeta potential) of erythrocytes that repels individual RBCs, like high concentrations of fibrinogen and similar acute phase pro-teins, will tend to accelerate this rate. Anemia, which lowers the concen-tration of red blood cells (RBCs) in plasma, also accelerates the sedimentation. Irregularly shaped RBCs, or high concentrations of albu-min and microcytes, both tend to slow the ESR. *McPhersonRA, et al. eds. Henry's Clinical Diagnosis and Management by Laboratory Methods, 23rd ed., 2017.*

7. b. Hypoxia-inducible factors (HIFs) are transcription factors that target the erythropoietin (EPO) gene and result in increased EPO production. Under conditions of normal oxygen tension, HIFs are degraded by interactions between oxygen, the von Hippel Lindau tumor gene product, and other enzymes. Under hypoxic conditions, this degradation rate is decreased, resulting in increased transcription of EPO. Mutations in the VHL gene result in defects in this degradation pathway, causing an increase in HIF1α and therefore EPO, a mechanism underlying secondary familial polycythe-mias. JAK2 is a receptor-mediated tyrosine kinase that becomes autopho-sphorylated when erythropoietin (EPO) binds to its receptor (EPOR). *JAK2* mutations have been implicated in clonal myeloid neoplasms includ-ing polycythemia vera, which results in polycythemia in the setting of low erythropoietin. 2,3-Diphosphoglycerate (2,3-DPG) is an intermediate in the glycolytic pathway in erythrocytes that accumulates under hypoxic conditions (as when RBCs travel through peripheral vascular beds) and causes alterations in the conformation of globin chains that stabilize the low oxygen affinity state, thus promoting the release of oxygen to the tis-sues (i.e., a rightward shift in the hemoglobin–oxygen dissociation curve). *Farsijani NM, et al. Renal epithelium regulates erythropoiesis via HIF-dependent suppression of erythropoietin. J Clin Invest 2016;126(4): 1425–3.*

8. e. Hemoglobin Constant Spring (Hb CS) is formed by a mutation in the termination codon of the α2-globin gene, resulting in an unstable, elon-gated α-globin chain. Hb CS homozygotes may have a thalassemia inter-media phenotype. Hemoglobin S, C, and E result from mutations that affect the β-globin chain. Hemoglobins Lepore result from various δβ-globin fusions arising from unequal crossover events between δ-globin and β-globin genes. Lepore contains two normal α-globin chains and two δβ-globin chains. *Komvilaisak P, et al. Clinical course of homozygous*

hemoglobin Constant spring in pediatric patients. J Pediatr Hematol Oncol 2018;40(5):409–12.

9. b. Hemoglobin Bart's and Hemoglobin H are each associated with alpha thalassemia with multiple gene deletions and each contains tetramers of a single type of globin chain. They form in the setting of a significant imbalance in globin chain production. Hemoglobin Bart's is comprised of four gamma globin chains (γ_4) and hemoglobin H is comprised of 4 beta globin chains (β_4). Bart's hemoglobin is most notable at birth, whereas hemoglobin H becomes more prevalent with age as γ-globin chain production decreases and β-globin chain production increases. Hemoglobin C Harlem results from two mutations in the beta chain ($\beta6$ Glu > Val, $\beta73$ Asp > Asn). Hemoglobin M Boston results when distal histidines in the heme pockets of α chains are replaced by tyrosines [His E7(58) > Tyr] resulting in a hemoglobin in which iron tends to be oxidized to the ferric state, which is unable to bind oxygen. Hemoglobin G Philadelphia results from an alpha chain substitution ($\alpha68$ Asn > Lys). *Wu MY, et al. Neonatal screening for α-thalassemia by cord hemoglobin Barts: how effective is it? Int J Lab Hematol 2015;37(5):649–53.*

10. c. The chromatogram in the figure shows an abnormal hemoglobin variant that was later confirmed to represent hemoglobin E trait (Hb E), which may show mild microcytosis. Using this high-performance liquid chromatography (HPLC) method, hemoglobin E has a retention time that masks hemoglobin A2 (Hb A2). As such, the amount of Hb A2 cannot be accurately quantified. Hemoglobin A2 ($\alpha_2\delta_2$) may be increased as a compensatory response to beta thalassemia. However, the proportion of Hb A2 generally does not exceed 10% of total hemoglobin. As such, a coeluting hemoglobin should be suspected. Hemoglobin Lepore also has a similar retention time, though usually makes up a proportion that is in between that of Hb A2 and Hb E. Both mild α-thalassemia (trait) and iron deficiency would typically show a normal hemoglobin chromatogram. In the case of alpha thalassemia that is because each of the normal hemoglobins that are usually seen in adults (Hb A, HbA2, and HbF) contain α-globin and therefore would be proportionally reduced. In more severe forms of α-thalassemia, the imbalance of globin chain production may result in the formation of tetramers of the excess chains, i.e., hemoglobin Bart (γ_4) or hemoglobin H (β_4). *Joneja U, et al. The results of hemoglobin variant analysis in patients revealing microcytic erythrocytosis on complete blood count. Lab Med 2018;49(2):147–53.*

11. e. The pattern on the alkaline gel shows two nearly equal size bands one in the hemoglobin S position and one in the hemoglobin A2 position. The quantity of the band in the A2 position is too high to be hemoglobin A2, which rarely exceeds 10% of total hemoglobin. If present, hemoglobins C, E, and O would also migrate in the same position as hemoglobin A2

on alkaline gels. The acid gel resolves that the two equal size bands occupy the position for hemoglobin S and hemoglobin C, supporting an interpretation of hemoglobin SC disease (Hb SC). Hemoglobin F is also present and increased as may be seen with a compensatory response or with the administration of hydroxyurea. Hemoglobin A2 adds slightly to the band on alkaline electrophoresis and is barely visible on the acid gel. *Huisman TH, et al. Chemical heterogeneity of fetal hemoglobin in subjects with sickle cell anemia, homozygous Hb-C disease, SC disease, and various combinations of hemoglobin variants. Clin Chim Acta 1972;38(1):5–16.*

12. b. Of the listed diagnoses, the morphologic findings described including the high white count, expansion in myelocytes and, in particular, the presence of basophilia are most suggestive of chronic myelogenous leukemia, which can be confirmed by RT-PCR for *BCR-ABL1* mutation. This mutation must be excluded prior to considering the other myeloproliferative neoplasms that are listed, several of which may share some features of CML. There is no mention of bilobed cells or abnormal promyelocytes or Auer rods that would raise suspicion for acute promyelocytic leukemia with *PML-RARA. World Health Organization Classification of Tumors of the Haematopoietic and Lymphoid Tissues, 2017.*

13. d. The morphologic findings described including variant lymphocyte forms and some with abundant deeply basophilic cytoplasm are most compatible with a reactive lymphocytosis as may be seen with Epstein-Bar virus (EBV) infection, the causative agent of infectious mononucleosis. Other viral infections can cause a similar reactive picture. Choice a. and e. are myeloid malignancies and would not have variant lymphocytes as the predominant feature. The lymphocytes of chronic lymphocytic leukemia would typically have a more monotonous appearance including fissured chromatin with interspersed "smudge" cells. *McPherson RA, et al. eds., Henry's Clinical Diagnosis and Management by Laboratory Methods, 23rd ed., 2017.*

14. d. Glucose-6 phosphate dehydrogenase (G6PD) is an enzyme in the pentose phosphate pathway and present in virtually all cells. It plays a prominent role in the protection of red blood cells. It is in highest concentration when red blood cells are immature and depletes as red cells age. In individuals with G6PD deficiency, upon exposure to oxidative stress due to a variety of potential causes, red blood cells have few defenses from the damaging effects of reactive oxygen species. During hemolytic episodes the most susceptible cells are cleared. As such, quantitative measurement of G6PD during an acute hemolytic episode may detect near normal levels of G6PD due to the remaining younger cells. *Sagiv E, et al. Glucose-6-phosphate-dehydrogenase deficient red blood cell units are associated with decreased posttransfusion red blood cell survival in children with sickle cell disease. Am J Hematol 2018;93(5):630–4.*

15. d. Individuals with anemia due to a variety of conditions, such as sickle cell disease, are at risk of a severe but transient aplastic crisis when infected with parvovirus B19. The sharp drop in hemoglobin can lead to congestive heart failure or stroke or may associate with an acute splenic sequestration crisis. When symptomatic, parvovirus infection typically results in a pure red cell aplasia though white blood cell and platelet counts may also decline in some instances. Myelodysplastic syndrome is not typical in this patient's age group. Transient erythroblastopenia of childhood is another form of pure red cell aplasia that typically affects children who are between 6 months and 6 years old. It is most often asymptomatic and self-limited. Iron deficiency would be more slowly progressive. The patient is clinically symptomatic and there is no indication that the result is spurious. *Blauel ER. Unexpected anemia and reticulocytopenia in an adolescent with sickle cell anemia receiving chronic transfusion therapy. J Pediatr Hematol Oncol 2015;37(7):e438–40.*

16. a. Significant vitamin B12 deficiency may present with a macrocytic anemia with hypersegmented neutrophils and a hypercellular bone marrow due to the increased, but ineffective, erythropoiesis. The maturing erythroid precursors show nuclear cytoplasmic asynchrony in that the cytoplasm matures but the nucleus (which should become smaller as the cell matures) lags behind and remains large with fenestrated chromatin, creating a megaloblastic appearance. *Farrelly SJ et al. Hypersegmented neutrophils and oval macrocytes in the setting of B12 deficiency and pancytopaenia. BMJ Case Rep. 2017:bcr-2016-218508.*

17. d. The condition described is compatible with hereditary spherocytosis, a red blood cell membrane disorder that is typically due to defects in ankyrin, band 3, or spectrin. The clinical severity varies. Osmotic fragility is usually increased. Another technique for diagnosis includes flow cytometry dye binding assessment using eosin-5′ maleimide (EMA) dye, which has greater than 90% sensitivity and specificity. The Ham's (acid hemolysin) test is used to assess for paroxysmal nocturnal hemoglobinuria. *Lumori BAE, et al. Diagnostic and treatment challenges of paroxysmal nocturnal hemoglobinuria in Uganda. Case Rep Hematol 2019;* http://doi.org/10.1155/2019/7897509.

18. b. Porphyrias are disorders of enzymes in the heme biosynthesis pathway that lead to accumulation of heme intermediates in red blood cells and other tissues. They may manifest as neurovisceral symptoms, photosensitivity, hemolytic anemia, or a combination of these conditions. Congenital erythropoietic porphyria (CEP) is a rare severe form of porphyria that is associated with both photosensitivity and hemolytic anemia. The other porphyrias listed do not typically have both. *Xu W, et al. Molecular basis of congenital erythropoietic porphyria: mutations in the human uroporphyrinogen III synthase gene. Hum Mutat 1996;7:187–92.*

19. c. Platelet clumping can cause what has been termed, "pseudothrombocy-topenia," where a hematology analyzer "sees" each platelet clump as a single large platelet rather than many normal-size platelets. This in vitro phenomenon occurs infrequently and is related to collection of blood in EDTA (the usual anticoagulant used for complete blood counts). Repeating the blood collection in citrate can often prevent this from recurring and allow for a more accurate platelet count. *Bai M, et al. Transient EDTA-dependent pseudothrombocytopenia phenomenon in a patient with anti-phospholipid syndrome. Clin Lab 2018;64(9):1581–3.*

20. e. Paroxysmal nocturnal hemoglobinuria occurs due to a mutation in the PIG-A gene in a bone marrow stem cell. While small clones of these cells may be present in normal individuals, clonal expansion may populate the bone marrow and result in marrow failure. Cells with this mutation are deficient in the GPI-anchored proteins that regulate complement, which leads to hemolytic anemia, thrombosis, and increased susceptibility to infection. *Huang Y, et al. Correction to: prediction of thrombosis risk in patients with paroxysmal nocturnal hemoglobinuria. Ann Hematol 2019;98(12):2857.*

21. c. CD55 also known as decay-accelerating factor (DAF) and CD59, also known as membrane inhibitor of reactive lysis (MIRL) are complement-regulating proteins. These proteins may be lost from the surface of red blood cells and white blood cells in patients with paroxysmal nocturnal hemoglobinuria. Other antigens may also be lost from specific subtypes of cells. *Alfinito F, et al. Blood cell flow cytometry in paroxysmal noctur-nal hemoglobinuria: a tool for measuring the extent of the PNH clone. Leukemia. 1996;10(8):1326–30.*

22. e. Coagulation tubes are usually collected early in the sequence of phlebot-omy and should be collected before tubes that contain clot activators as contamination can falsely shorten coagulation tests. Answers a., b., and c. would falsely prolong the clotting time due to an excess of antico-agulant or other contaminating fluid. A discard tube can prevent contam-ination of coagulation samples that are collected through a vascular access device. *Mcpherson RA, et al. eds., Henry's Clinical Diagnosis and Man-agement by Laboratory Methods, 23rd ed., 2017.*

23. d. Incorrect ratios of blood to anticoagulant can falsely prolong or falsely shorten clotting times. Most coagulation testing uses light blue–top tubes containing 3.2% or 3.8% sodium citrate in a blood to anticoagulant ratio of 9:1. McPherson and Pincus, Henry's Clinical Diagnosis and Management by Laboratory Methods, 23rd ed., Saunders, 2017.

24. a. Of the options listed, only Factor VII deficiency is likely to present as an isolated prolonged PT. Factor VII deficiency is a rare congenital bleeding disorder. The degree of prolongation of PT and the level of factor VII activity generally do not correlate well with the severity of the clinical presentation. Unfractionated heparin primarily affects the aPTT assay. Disseminated intravascular coagulation would affect both PT and aPTT. Hemophilia A is a deficiency of factor VIII and would primarily affect aPTT. The other options would generally prolong aPTT alone or both PT and PTT. *Tripathi P. Factor VII deficiency - an enigma; clinicohematological profile in 12 cases. Hematol 2019;24(1):97–102.*

25. b. Of the options listed only the presence of a Lupus Anticoagulant is likely to present as an isolated prolonged aPTT. Lupus anticoagulant is an antiphospholipid antibody that inhibits in vitro coagulation reactions in a phospholipid-dependent manner. Paradoxically, patients who acquire a lupus anticoagulant are at increased risk of thrombosis. As many other factors can prolong aPTT, additional assays are required to confirm the presence of a lupus anticoagulant. Coumadin primarily affects PT but in higher doses may also affect aPTT. Severe Vitamin K deficiency would affect both PT and aPTT since the synthesis of certain coagulation factors in both arms of the coagulation cascade depends on vitamin K (in particular, factors II, VII, IX, and X and proteins C and S). Similarly, the dilution effect of intravenous fluid contamination (e.g., when collecting a sample from a central venous access device through which fluids are administered, without first drawing a discard tube) would likely falsely prolong both PT and aPTT. Factor V is a factor in the final common pathway and deficiency would prolong both PT and aPTT. *Tcherniantchouk O, et al. The isolated prolonged PTT. Am J Hematol 2013;88(1):82–5.*

26. e. Testing for lupus anticoagulant requires three elements including (1) an initial screening test showing prolongation; (2) a mixing study to demonstrate persistence of prolongation in the presence of normal plasma, as evidence of inhibition; and (3) a confirmatory test that demonstrates phospholipid dependence (by showing correction with excess phospholipid). Selecting more than one test to assess different aspects of the coagulation cascade enhances the sensitivity. The dilute Russell viper venom time (dRVVT) relies on Russell Viper venom directly activating of factor X (in the final common pathway). The use of dilute phospholipid reagent in the screen further enhances the sensitivity. Using normalized ratios that compare to normal pooled plasma (NPP) reduces inter- and intra-assay variation. In the assay shown, prolongation in the screen (ratio >1.2) followed by persistent prolongation in the mixing study and correction in the confirm step is consistent with a lupus anticoagulant. Positive tests are usually repeated after 12 weeks to confirm that the lupus anticoagulant is persistent. *Moore GW. Recent guidelines and recommendations for laboratory detection of lupus anticoagulants. Semin Thromb Hemost 2014;40(2):163–71.*

27. d. Disseminated intravascular coagulation (DIC) involved the triggered activation of the coagulation and fibrinolysis systems and results in simultaneous formation of thrombin (thrombosis) and plasmin (fibrinolysis).

The sequence of clotting consumes coagulation factors, coagulation system regulators and platelets, which classically results in prolonged PT, aPTT, and thrombocytopenia and low fibrinogen. D-Dimer, though nonspecific is generally elevated and peripheral blood smears may show schistocytes as evidence of microangiopathic hemolytic anemia. A variety of potential triggers have been identified. *Kaneko T, et al. Diagnostic criteria and laboratory tests for disseminated intravascular coagulation. J Clin Exp Hematop 2011;51(2):67–76.*

28. b. The example on the right indicates early fibrinolysis or clot breakdown. Viscoelastic testing on whole blood (which uses some variation of a probe in a container where one or the other rotates or deflects as clot formation proceeds) permits a more global assessment of hemostasis, including the contribution from platelets and coagulation factors as well as fibrinolytic components. Devices measure the initial lag before clot formation begins, the angle of the increase as clot formation starts, the maximal deflection point, and any weakening of the clot over time as lysis ensues. These devices are often used to guide clinical determinations in the setting of surgery, trauma and certain cardiac procedures and provide indicators of whether a patient's hemostatic system is functioning adequately or whether they may benefit from transfusion of blood components, like platelets, plasma, cryoprecipitate or antifibrinolytic agents, and other hemostatic medications. *David JS. Fibrinolytic shutdown diagnosed with rotational thromboelastometry represents a moderate form of coagulopathy associated with transfusion requirement and mortality: a retrospective analysis. Eur J Anaesthesiol 2019;37:170–9.*

29. c. The increasing factor activity with increasing dilutions suggests the presence of a factor inhibitor. Given that this effect is limited to the factor VIII assay, a specific factor VIII inhibitor should be suspected. *Castellone DD, et al. Factor VIII activity and inhibitor assays in the diagnosis and treatment of hemophilia A. Semin Thromb Hemost 2017;43(3):320–30.*

30. a. Thrombophilia testing may be appropriate in patients with unprovoked venous thromboembolism to assess the risk of recurrence. Active or recent thrombosis can affect several of the related assays as factors and regulators are consumed during the process of an acute thrombotic event and the subsequent resolution. Tests that are initially deficient may show adequate levels with repeat testing. Of the listed entities, anticardiolipin antibody is the only exception that is listed. *Shen Y, et al. Analysis of thrombophilia test ordering practices at an academic center. PLoS One 2016;11(5): e0155326. doi:10.1371/journal.pone.0155326.*

31. d. Plasminogen activator inhibitor-1 (PAI-1) regulates fibrinolysis (clot breakdown). PAI-1 deficiency is rare and increases the risk of bleeding. The other answer choices are risk factors for thrombosis that may be included in thrombophilia screening for patients with unprovoked venous thromboembolism. *Stevens SM, et al. Guidance for the evaluation and treatment of hereditary and acquired thrombophilia. J Thromb Thrombolysis 2016;41(1):154–64.*

32. e. Rivaroxaban is an oral anticoagulant that directly inhibits factor Xa. As such, it can interfere in the measurement of unfractionated heparin by the antifactor Xa assay. *Faust AC, et al. Managing transitions from oral factor Xa inhibitors to unfractionated heparin infusions. Am J Health Syst Pharm 2016;73(24):2037–41.*

33. c. Type 2B von Willebrand disease (type 2B VWD) results in a gain-of-function mutation, and demonstrates increased ristocetin-induced platelet aggregation (RIPA) with low-dose ristocetin. Answer choice a. describes type 1 VWD. Answer choice b. describes type 2M VWD. Answer choice d. describes type 3 VWD. Answer choice e. describes Bernard–Soulier syndrome (BSS). *Frontroth JP, et al. Prospective study of low-dose ristocetin-induced platelet aggregation to identify type 2B von Willebrand disease (VWD) and platelet-type VWD in children. Thromb Haemost 2010;104 (6):1158–65.*

34. a. Thrombotic thrombocytopenic purpura (TTP) is an acquired blood disorder that results from low levels of ADAMTS13, a metalloproteinase that is responsible for cleaving ultralong von Willebrand factor multimers (UL-vWF). Deficiency in ADAMTS13 results in a buildup of UL-vWF multimers, which induce platelet aggregation and adhesion to the vascular endothelium. This results in platelet consumption and thrombosis in the microvasculature throughout the vascular system. Clinical signs and symptoms may include a classic pentad of microangiopathic hemolytic anemia (MAHA), thrombocytopenia, fever, renal insufficiency, and neurologic symptoms. The peripheral blood smear may show schistocytes indicating microangiopathic hemolytic anemia and, if renal insufficiency is noted, burr cells may be present. In this setting, a very high LDH should raise significant suspicion for TTP. However, definitive diagnosis requires demonstration of low ADAMTS13 activity levels. *Saha M, et al. Thrombotic thrombocytopenic purpura: pathogenesis, diagnosis and potential novel therapeutics. J Thromb Haemost 2017;15(10):1889–900.*

Chapter 26

Immunology and autoimmune disease

Lusia Sepiashvili[a] and Alan H.B. Wu[b]
[a]Hospital for Sick Children, University of Toronto, Toronto, ON, Canada, [b]University of California, San Francisco, CA, United States

1. Describe the general relationship between autoimmune disease prevalence with age and sex:
   a. Increasing with age and most predominant in females
   b. Decreasing with age and most predominant in males
   c. Increasing with age and no sex differences
   d. Increasing with age and most predominant in males
   e. Decreasing with age and most predominant in females
   f. Autoantibody prevalence is stable throughout various ages and does not vary between males and females
2. Identify the main immunoglobulin isotype of disease-associated autoantibodies
   a. IgG
   b. IgA
   c. IgM
   d. IgD
   e. IgE
   f. IgG4
3. Which of the following serve as indirect evidence of human disease caused by autoimmunity?
   a. Appearance of disease features due to maternal-fetal transfer of autoantibodies such as in Grave's disease or myasthenia gravis.
   b. Pemphigus vulgaris and bullous pemphigoid reproduced by transfer of serum to newborn mice.
   c. Experimental immunization by injecting antigen into a syngeneic recipient without any adjuvant and demonstrating that this results in production of autoantibodies and appearance of characteristic lesions of the disease in question.

d. Use of spontaneously occurring genetic models replicating human disease—adoptive immunization of recipients with T lymphocytes. Demonstrates that disease caused by autoimmune response rather being consequence of disease.

e. There is only direct evidence available proving that human disease is caused by autoimmunity.

f. a. and b.

Match autoantigens to the *organ-specific* autoimmune diseases they are associated with

4. Acetylcholine receptor	a. Celiac disease
5. Glutamic acid decarboxylase (GAD65)	b. Autoimmune thyroiditis
6. Cytochrome P450 2D6	c. Type 1 diabetes mellitus
7. Tissue transglutaminase	d. Myasthenia gravis
8. Thyroperoxidase	e. Autoimmune hepatitis (Type 2)
9. Alpha3 NC1 domain of type IV collagen	f. Primary biliary cirrhosis
10. Dihydrolipoamide acyltransferase	g. Goodpasture syndrome

Match autoantigens to the *systemic* autoimmune diseases they are associated with

11. Double-stranded DNA	a. Sjögren syndrome
12. Cardiolipin and β2 glycoprotein 1	b. Polymyositis
13. Jo 1 (histidyl tRNA synthetase)	c. Scleroderma
14. Citrullinated proteins	d. Systemic lupus erythematosus
15. Scl 70 (topoisomerase 1)	e. Antiphospholipid syndrome
16. Ro/SSA and La/SSB	f. CREST syndrome/limited cutaneous form of systemic sclerosis
17. Centromere	g. Rheumatoid arthritis

18. The following laboratory method(s) cannot be used for assessment of total complement function or the classical pathway of complement:

a. Radial immunodiffusion hemolytic assays

b. Flow cytometry

c. Solid-phase ELISA

d. Liposome immunoassay

e. c and d.

f. None of the above

19. What is the advantage of anti-dsDNA antibody detection using the *Crithidia luciliae* substrate as compared to solid-phase immunoassays employed in systemic lupus erythematosus (SLE) evaluation?

a. It is a more diagnostically sensitive technique.

b. The interpretation is less subjective.

c. This method detects low, medium, and high avidity antibodies.

d. It is a more diagnostically specific technique.

    e. Specific only toward antibodies of IgG isotype.
    f. Equal cross-reactivity with double-stranded and single-stranded DNA antibodies.
20. Therapeutic monoclonal antibodies have been shown to interfere with the following clinical laboratory assays
    a. Serum free light chains
    b. Serum total protein
    c. Serum protein electrophoresis
    d. Nephelometric immunoassays
    e. Serum immunofixation electrophoresis
    f. c and e.
21. The expression of which of the following proteins is induced by the action of interleukin 6 on hepatocytes, as part of the innate immune system function of this cytokine?
    a. Amyloid beta protein
    b. Alanine aminotransferase
    c. Immunoglobulin G
    d. C-reactive protein
    e. Albumin
    f. None of the above
22. Which of the following is an entry criterion for systemic lupus erythematosus (SLE) classification as part of the 2019 European League Against Rheumatism (EULAR) and the American College of Rheumatology (ACR) Recommendations?
    a. Positive antinuclear antibodies (ANA) at a titer of ≥1:80 on HEp-2 cells or an equivalent positive test (during active disease).
    b. Positive antineutrophil cytoplasmic antibodies (ANCA) at a titer of ≥1:80 on HEp-2 cells or an equivalent positive test (during active disease).
    c. Positive antineutrophil cytoplasmic antibodies (ANCA) at a titer of ≥1:80 on HEp-2 cells or an equivalent positive test (ever).
    d. Positive antinuclear antibodies (ANA) at a titer of ≥1:80 on HEp-2 cells (ever).
    e. Positive antinuclear antibodies (ANA) at a titer of ≥1:80 on HEp-2 cells or an equivalent positive test (ever).
23. Identify the laboratory criteria included in the 2006 International consensus statement on an update of the classification criteria for definite antiphospholipid syndrome (APS):
    a. Presence of lupus anticoagulant (LAC) in serum.
    b. Lupus anticoagulant, anticardiolipin antibodies (aCL), or anti-β2 glycoprotein 1 antibodies (ab2GP1) positive on two or more occasions at least 12 weeks apart.

    c. aCL or ab2GP1 cutoffs defined by the 95th percentile of normal controls.

    d. All of the above

    e. b. and c. only

24. The recommended screening tests for celiac disease in symptomatic individuals involve the following laboratory tests:

    a. Serum tissue transglutaminase IgG and IgA antibodies

    b. Serum tissue transglutaminase IgG antibodies and total IgA concentration

    c. Serum tissue transglutaminase IgA antibodies and endomysial IgA antibodies

    d. Serum tissue transglutaminase IgA antibodies and HLA-DQ2 and HLA-DQ8 genotyping

    e. Serum tissue transglutaminase IgA antibodies and gliadin (deamidated) IgG antibodies

25. According to the 2019 European Society for Paediatric Gastroenterology Hepatology and Nutrition (ESPGHAN) guidelines for diagnosing celiac disease, a biopsy could be avoided in symptomatic patients suspected of having celiac disease if the following laboratory criteria are met:

    a. Serum tissue transglutaminase IgA antibodies ≥ 10 times the upper limit of normal and positive endomysial IgA antibodies in a second serum sample.

    b. Positive celiac HLA risk alleles DQ2 and/or DQ8.

    c. Serum tissue transglutaminase IgA antibodies ≥ 5 times the upper limit of normal and positive endomysial IgA antibodies in a second serum sample.

    d. Serum tissue transglutaminase IgG antibodies ≥ 10 times the upper limit of normal and positive endomysial IgG antibodies in a second serum sample.

    e. Serum tissue transglutaminase IgA antibodies ≥ 10 times the upper limit of normal, confirmed by repeat testing in a second serum sample minimum 12 weeks later.

26. Identify the significance of serum ferritin concentrations in hemophagocytic lymphohistiocytosis (HLH):

    a. Ferritin $> 10,000 \, \text{ng/mL}$ has been shown to have high sensitivity and specificity for HLH, with minimal overlap with other inflammatory disorders.

    b. Ferritin $< 500 \, \text{ng/mL}$ excludes the possibility of HLH.

    c. Ferritin upregulates growth differentiation factor 15 (GDF15).

    d. In the absence of cytopenias and fevers, ferritin $> 10,000 \, \text{ng/mL}$ increases the index of suspicion for HLH.

    e. Elevated levels in neonates suggest the concomitant development of hemochromatosis.

27. A finding of an indirect immunofluorescence pattern consistent with perinuclear antineutrophil cytoplasmic antibody (ANCA) staining on ethanol-fixed neutrophils can be associated with any of the following findings:
  a. Reflection of duodenal disease localization in case of suspected inflammatory bowel disease.
  b. More likely to be positive for Proteinase-3 antibodies than Myeloperoxidase antibodies.
  c. May be a false positive due to the presence of an antinuclear antibodies, necessitating careful pattern review, interpretation, and reporting by a skilled laboratorian.
  d. Excludes the diagnosis of small vessel vasculitis (e.g., microscopic polyangiitis).
  e. Diagnostic for acute hemorrhagic edema of infancy (i.e., benign leukocytoclastic vasculitis).

28. Which diseases are characterized by the presence of antibodies to glomerular basement membrane?
  a. Goodpasture syndrome
  b. Crohn's disease
  c. Celiac disease
  d. Hypocomplementemic urticarial vasculitis syndrome
  e. Wegener's granulomatosis

29. Which of the following autoantibodies are not found in high titers in vasculitis?
  a. Antineutrophil cytoplasmic antibodies
  b. Anti-C1q antibodies
  c. Anti-myeloperoxidase antibodies
  d. Anti-SSA and anti-SSB
  e. Cryoglobulins

30. Anti-double stranded DNA antibodies are found in which of the following autoimmune diseases?
  a. Chronic inflammatory demyelinating polyneuropathy
  b. Lupus erythematosus
  c. Rheumatoid arthritis
  d. Polymyositis
  e. Guillain-Barre syndrome

31. Which autoantibodies are expressed in scleroderma (positive in at least 20% of cases)?
  1. antitopoisomerase I (anti-scl-70) antibody
  2. anticentromere antibody
  3. anti-topoisomerase antibody
  4. anti-PM/Scl antibody
    a. 1,3
    b. 1,2,3

 c. 2,4
 d. 4 only
 e. All of the above
Answers
 1. a. Although there are exceptions, the prevalence of autoimmune disease generally increases with age. Female sex is also linked to higher prevalence of autoimmune disease, and the ratio of females to males with autoimmune disease tends to increase with age. *Leffell M, et al. Handbook of human immunology (1st ed.). CRC 1997.*
 2. a. Most disease-associated autoantibodies are present at high titer and are of IgG isotype, most naturally occurring autoantibodies are low titer IgMs. *Leffell M, et al. Handbook of human immunology (1st ed.). CRC 1997.*
 3. d. Most evidence for human diseases classified as autoimmune is indirect. Options A-B are examples of direct evidence that these diseases are caused by immunity, specifically by circulating antibodies in the cases demonstrated here. Option C is incorrect since adjuvant such as bacterial lipopolysaccharide would be required for disease reproducibility by experimental immunization. *Leffell M, et al. Handbook of human immunology (1st ed.). CRC 1997.*
 4. d. Acetylcholine receptor—myasthenia gravis
 5. c. Glutamic acid decarboxylase (GAD65)—Type 1 diabetes mellitus
 6. e. Cytochrome P450 2D6—autoimmune hepatitis (Type 2)
 7. a. Tissue transglutaminase—celiac disease
 8. b. Thyroperoxidase—autoimmune thyroiditis
 9. g. Alpha3 NC1 domain of type IV collagen—Goodpasture syndrome
10. f. Mitochondria—primary biliary cirrhosis
11. d. Double stranded DNA—systemic lupus erythematosus.
12. e. Cardiolipin and β2 glycoprotein 1—Antiphospholipid syndrome
13. b. Jo 1 (histidyl tRNA synthetase) —polymyositis.
14. g. Cyclic citrullinated peptide—rheumatoid arthritis
15. c. Scl 70—scleroderma.
16. a. Ro/SSA and La/SSB—Sjögren syndrome
17. f. Centromere—CREST syndrome/limited cutaneous form of systemic sclerosis

 For questions 4–17, comprehensive review of the autoimmune diseases and important antigenic targets can be found in the following. *Leffell M, et al. Handbook of human immunology (1st ed.). CRC 1997. Rich RR, et al. Clinical immunology: Principles and practice: Fourth edition. In Clinical Immunology: Principles and Practice, 4th ed., 2012.*

18. b. Radial immunodiffusion hemolytic assays, solid-phase ELISAs, and liposome immunoassays have all been used for assessment of total complement functional status. Frazer-Abel A, et al. Overview of laboratory testing and clinical presentations of complement deficiencies and dysregulation. Adv Clin Chem. 2016. https://doi.org/10.1016/bs.acc.2016.06.001.

19. d. Statements a, b, and e apply to solid-phase immunoassays when compared to dsDNA autoantibody detection *via* indirect immunofluorescence on *Crithidia luciliae*. Solid-phase immunoassays are more sensitive, less subjective, and generally designed to detect autoantibodies of IgG isotype. When compared to solid-phase immunoassays, anti-dsDNA antibody detection using *Crithidia luciliae* is more specific to SLE but has lower sensitivity. *Riboldi P, et al. Anti-DNA antibodies: A diagnostic and prognostic tool for systemic lupus erythematosus? Autoimmun 2005.* https://doi.org/10.1080/08916930400022616.

20. f. Therapeutic monoclonal antibodies can appear as monoclonal protein bands on serum protein electrophoresis and serum immunofixation electrophoresis. Serum free light chain assays have been shown to be unaffected. *McCudden CR, et al. Interference of monoclonal antibody therapies with serum protein electrophoresis tests. Clin Chem 2010.* https://doi.org/10.1373/clinchem.2010.152116. *Willrich MA, et al. Monoclonal antibody therapeutics as potential interferences on protein electrophoresis and immunofixation. Clin Chem Lab Med 2016.* https://doi.org/10.1515/cclm-2015-1023. *Rosenberg AS, et al. Investigation into the interference of the monoclonal antibody daratumumab on the free light chain assay. Clin Biochem 2016.* https://doi.org/10.1016/j.clinbiochem.2016.07.016.

21. d. As part of the innate immune system, interleukin 6 acts on hepatocytes to induce expression of C-reactive protein (CRP), fibrinogen, and serum amyloid A, also known as the acute phase response. *Depraetere S, et al. Stimulation of CRP secretion in HepG2 cells: Cooperative effect of dexamethasone and interleukin 6. Agents Actions 1991.* https://doi.org/10.1007/BF01988730. *Sproston NR, et al. Role of C-reactive protein at sites of inflammation and infection. In Frontiers in Immunology 2018.* https://doi.org/10.3389/fimmu.2018.00754.

22. e. Positive ANA at a titer of ≥1:80 by immunofluorescence on HEp-2 cells was shown to have a sensitivity of 97.8% (95% CI 96.8–98.5%) in a recent meta-analysis, thereby supporting its use as an entry criterion. Since HEp-2 ANA testing is not widely available and in light of ongoing advances in the field, equivalent positive antinuclear antibody test results by solid-phase immunoassay screening with at least equivalent performance to HEp-2 ANA was included in the entry criterion recommendation. *Leuchten N, et al. Performance of antinuclear antibodies for classifying systemic lupus erythematosus: A systematic literature review and meta-regression of diagnostic data. Arth Care Res 2018.* https://doi.org/10.1002/acr.23292. *Aringer M et al. 2019 European League Against Rheumatism/American College of Rheumatology classification criteria for systemic lupus erythematosus. Ann Rheumat Dis 2019.* https://doi.org/10.1136/annrheumdis-2018-214819.

23. b. Option A is incorrect since the appropriate specimen type for lupus anticoagulant determination is plasma. Option C is incorrect since aCL or ab2GPI cutoffs are defined by the 99th percentile of normal controls, not the 95th percentile. *Devreese KMJ, et al. Laboratory criteria for antiphospholipid syndrome: communication from the SSC of the ISTH. J Thromb Haemost 2018.* https://doi.org/10.1111/jth.13976. *Miyakis S, et al. International consensus statement on an update of the classification criteria for definite antiphospholipid syndrome (APS). J Thrombo Haemost 2006.* https://doi.org/10.1111/j.1538-7836.2006.01753.

24. b. The recommended screening approach for celiac disease in symptomatic individuals involves testing for tissue transglutaminase IgA antibodies and total IgA. In patients with selective IgA deficiency, IgG-based testing should be performed (deamidated gliadin, tissue transglutaminase, or endomysial antibodies). *Husby S, et al. European society for pediatric gastroenterology, hepatology, and nutrition guidelines for the diagnosis of coeliac disease. J Pediatr Gastroenterol Nutr. 2012.* https://doi.org/10.1097/MPG.0b013e31821a23d0.

25. a. According to the 2019 ESPGHAN guidelines for diagnosing celiac disease, a biopsy could be avoided in symptomatic patients suspected of having celiac disease if the serum tissue transglutaminase IgA antibodies exceed 10 times the upper limit of normal and positive endomysial IgA antibodies in a second serum sample. *Husby S, et al. European society for pediatric gastroenterology, hepatology, and nutrition guidelines for the diagnosis of coeliac disease. J Pediatr Gastroenterol Nutr. 2012.* https://doi.org/10.1097/MPG.0b013e31821a23d0.

26. a. Hemophagocytic lymphohistiocytosis (HLH) is a life-threatening syndrome of excessive immune activation. It is most common in infants and young children. Ferritin > 10,000 ng/mL has been shown to have high sensitivity and specificity for HLH, with minimal overlap with other inflammatory disorders. However, normal or mildly elevated levels do not exclude the possibility of HLH. Ferritin can also be elevated in the other clinical scenarios such as neonatal hemochromatosis and fulminant liver failure, therefore, the potential causes contributing to extremely high ferritin should be considered during the interpretation. Growth differentiation factor 15, a protein responsible for modulation of iron homeostasis, is upregulated in HLH and has a role in modulating the dramatic increase in serum ferritin. *Allen CE, et al. Highly elevated ferritin levels and the diagnosis of hemophagocytic lymphohistiocytosis. Pediat Blood Cancer 2008.* https://doi.org/10.1002/pbc.21423. *Otrock, Z. et al. Elevated serum ferritin is not specific for hemophagocytic lymphohistiocytosis. Ann Hematol 2017.* https://doi.org/10.1007/s00277-017-3072-0. *Wu JR, et al. GDF15-mediated upregulation of ferroportin plays a key role in the development of hyperferritinemia in children with hemophagocytic lymphohistiocytosis. Pediatr Blood Cancer 2013.* https://doi.org/10.1002/pbc.24373.

27. c. Assessment of ANCA may be warranted in cases of suspected autoimmune vasculitis (primarily Wegener granulomatosis and microscopic polyangiitis) or in attempt to identify colonic disease localization in inflammatory bowel disease. Perinuclear ANCA is typically associated with myeloperoxidase antibodies. When antinuclear antibodies are present, they may appear as positive perinuclear ANCA requiring careful interpretation by a skilled laboratory professional. ANCA does not have known clinical utility in the assessment of acute hemorrhagic edema of infancy. *Bossuyt X, et al. Revised 2017 international consensus on testing of ANCAs in granulomatosis with polyangiitis and microscopic polyangiitis. Nature Rev Rheumatol 2017.* https://doi.org/10.1038/nrrheum.2017.140.

28. a. Known as antiglomerular basement membrane disease, this is a rare autoimmune disorder that affects the lungs and kidneys. *McAdoo SP. Anti-glomerular basement membrane disease. Clin J Am Soc Nephrol 2017;12:1162–72.*

29. d. Anti-SSA and anti-SSB antibodies are often seen in Sjogren's syndrome. Vasculitis involves inflammation of the blood vessels, including arteries, veins, and capillaries. It usually reduces blood flow. *de Souza AWS. Autoantibodies in systemic vasculitis. Front Immunol 2015;6:184–96.*

30. b. Anti-ds-DNA antibodies are found in 80% of patients with lupus. There are many other autoantibodies that are expressed in this disease. *Wang X, et al. Anti-double stranded DNA antibodies: origin, pathogenicity, and targeted therapies. Front Immunol 2019. doi.org/10.3389/fimmu.2019. 01667.*

31. b. Scleroderma is characterized by skin thickening. Anti-PM100 is found in a lower percentage of scleroderma patients. *Domsic RT, et al. Autoantibodies and their role in scleroderma clinical care. Curr Treatment Opt Rheumatol 2016;2:239–51.*

Chapter 27

Liquid biopsy

Erika M. Hissong, Priya D. Velu, Zhen Zhao, and Hanna Rennert
Weill Cornell Medicine, New York, NY, United States

1. What is the preferred method for collection of blood for circulating tumor DNA (ctDNA) testing?
   a. serum blood collection tubes
   b. heparin anticoagulated tubes
   c. EDTA anticoagulated tubes
   d. cell stabilizing or EDTA anticoagulated tubes
2. What is the optimal sample type for analyzing circulating tumor DNA (ctDNA)?
   a. whole blood
   b. plasma
   c. serum
   d. peripheral blood mononuclear cells (PBMC)
3. Which of the following is an important characteristic of circulating cell-free DNA?
   a. fetal cell-free DNA can be detected in maternal blood for months after birth
   b. represents only limited areas of the genome
   c. fragments are typically less than 300 DNA base pairs
   d. fragments are protected from further degradation in the circulation
4. Which method is commonly used to achieve high specificity in next-generation sequencing (NGS) cell-free DNA testing?
   a. employing ultra-deep sequencing methods
   b. employing multiple PCR cycles during library preparation
   c. using unique molecule identifiers (UMI) to tag individual DNA molecules
   d. assigning specific reads to gene variants
5. What is noninvasive prenatal testing (NIPT)?
   a. A screening genetic test of fetal DNA in maternal blood to determine the probability of certain chromosomal abnormalities in the fetus.
   b. A diagnostic genetic test of fetal DNA in maternal blood to determine the probability of certain chromosomal abnormalities in the fetus.
   c. A diagnostic genetic test to determine carrier status in the fetus.

    d. A diagnostic genetic test to determine carrier status in both the mother and the fetus.

6. Noninvasive prenatal testing (NIPT) is most commonly used to detect which of the following genetic abnormalities?

    a. trisomy 21, trisomy 18, and trisomy 16

    b. trisomy 21, trisomy 18, trisomy 13, and sex chromosome imbalances

    c. balanced chromosomal translocations and large deletions

    d. rare microdeletion syndromes

7. What is the minimum fraction of fetal cell-free DNA generally required for noninvasive prenatal testing (NIPT) testing?

    a. 1%

    b. 4%

    c. 10%

    d. 15%

8. If a tissue sample is unavailable, which other clinical test should be considered for determining whether or not a patient with lung adenocarcinoma should be treated with tyrosine kinase inhibitors (TKIs?)

    a. EGFR mutation analysis of peripheral blood

    b. EGFR mutation analysis of plasma circulating cell-free DNA

    c. KRAS mutation analysis of peripheral blood

    d. KRAS mutation analysis of plasma circulating cell-free DNA

9. Spiking plasma or extracted cell-free DNA with synthetic or exogenous DNA can help with:

    a. identifying extraction efficiency of cell-free DNA

    b. determining DNA fragment size bias of extraction

    c. identifying presence of PCR inhibitors

    d. all of the above

10. A false positive rate reported in a noninvasive prenatal test (NIPT) is approximately 0.2%. This means:

    a. One in 5 normal fetuses will have a positive DNA test.

    b. One in 200 normal fetuses will have a positive DNA test.

    c. One in 200 affected individuals will have a negative DNA test.

    d. One in 500 normal fetuses will have a positive DNA test.

11. What is a principal mechanism for release of cell-free DNA into the circulation?

    a. DNA glycosylase activity

    b. DNase hydrolysis

    c. apoptosis

    d. DNA strand breakage

12. Which of the following nucleic acids can be present in the circulation?

    a. exosomal DNA

    b. microbial cell-free DNA

    c. fetal cell-free DNA

    d. all of the above

13. Which of the following test(s) is most suitable for detecting genetic alterations in plasma?
   a. quantitative real-time PCR
   b. digital droplet PCR (ddPCR)
   c. next-generation sequencing (NGS)
   d. all of the above

14. A 50-year-old woman presents with metastatic liver cancer of unknown primary tumor and a circulating cell-free DNA test showing a KRAS mutation. This result is indicative of:
   a. colorectal cancer
   b. lung cancer
   c. pancreatic cancer
   d. inconclusive

15. For which of the following applications is liquid biopsy testing most useful?
   a. monitoring of treatment response and therapy selection in cancer patients
   b. screening for cancer
   c. diagnosis of hereditary cancer disorders
   d. all of the above

16. An EGFR L858R mutation analysis result of cell-free DNA plasma from a lung cancer patient is most commonly associated with:
   a. squamous cell carcinoma of lung
   b. nonsmall cell lung cancer
   c. small cell lung cancer
   d. large cell lung cancer

17. What is the approximate length of cell-free DNA fragments?
   a. 85 base pairs
   b. 170 base pairs
   c. 340 base pairs
   d. 680 base pairs

18. What is the primary source of circulating cell-free fetal DNA in maternal blood?
   a. fetal white blood cells
   b. fetal skin cells
   c. excreted fetal cells
   d. placental cells

19. What is the most common cause of a false positive test result in noninvasive prenatal testing (NIPT)?
   a. confined placental mosaicism
   b. maternal tumor
   c. vanishing twin
   d. maternal copy number variation

20. Commonly used methodologies for separation of circulating tumor cells include:
    a. tumor cell capture with anti-EpCAM or cytokeratin antibodies for epithelial cells
    b. density gradient centrifugation
    c. leukocyte (CD45) depletion techniques
    d. all of the above

21. How quickly is cell-free DNA cleared from circulation?
    a. minutes to 1–2 h
    b. 6–24 h
    c. 24–72 h
    d. days to weeks

22. High levels of plasma cell-free DNA in circulation correlate with all of the following except:
    a. metastatic disease
    b. large tumor size
    c. increased patient survival
    d. increased cellular turnover

23. What methodology is used to determine whether genomic DNA from lysed white blood cells has contaminated true circulating cell-free DNA in a sample?
    a. Measuring DNA fragment size by quantitative real-time PCR.
    b. Quantifying total amount of DNA in a sample and comparing to reference standard.
    c. Quantifying amount of DNA with sequence specific to white blood cells.
    d. Measuring relative amount of double-stranded to single-stranded DNA present in sample.

24. For noninvasive prenatal testing (NIPT), what follow-up confirmatory testing should be performed?
    a. none necessary
    b. amniocentesis if positive
    c. amniocentesis if negative
    d. follow-up testing is always performed

25. What is the earliest stage of pregnancy that cell-free DNA testing can be performed to detect fetal abnormalities?
    a. 10 weeks
    b. 13 weeks
    c. 16 weeks
    d. 21 weeks

26. Which of the following is not an advantage of liquid biopsy over traditional tissue biopsy?
    a. accounts for tumor heterogeneity

b. convenient and noninvasive sample acquisition

c. increased negative predictive value

d. ability to monitor patients over time

27. Which would not be considered a preanalytical variable for sensitivity and accuracy of liquid biopsy?

    a. DNA extraction methodology

    b. storage conditions during transportation

    c. the time from sample draw to centrifugation

    d. waccurate detection of mutations at very low variant allele frequencies

28. How is the cell-free fetal DNA fraction enriched from cell-free maternal DNA?

    a. DNA sequence

    b. DNA fragment length

    c. G-C content

    d. single-stranded vs double-stranded DNA

29. When designing a next-generation sequencing (NGS) test for detection of resistance mutations to targeted cancer therapies from cell-free DNA, what is the minimum allele frequency that should be detected?

    a. 0.1%

    b. 1%

    c. 5%

    d. 10%

30. When designing a cfDNA test for detection of mutations in several different tumor types, what is the best choice of methodology?

    a. digital droplet PCR (ddPCR)

    b. quantitative real-time PCR (qPCR)

    c. next-generation sequencing (NGS)

    d. DNA microarray

Answers

1. **d.** Testing for circulating tumor DNA (ctDNA) is optimally performed on plasma samples collected in cell stabilization tubes (Streck tubes), or K2EDTA tubes (lavender top tubes) processed within 6 h of collection to avoid lysis of white blood cells and ctDNA dilution with leukocyte DNA. In cell stabilization tubes the use of leukocyte and ctDNA stabilization reagents prevent leakage of cell genomic DNA from leukocytes and degradation of ctDNA, respectively, allowing up to several days of storage time between collection to processing. *Merker JD. Circulating tumor DNA analysis in patients with cancer: American Society of Clinical Oncology and College of American Pathologists Joint Review. J Clin Oncol 2018;36(16):1631–64.*

2. **b.** The optimal sample type for the analysis of ctDNA in blood is plasma. The concentration of total cell-free DNA (cfDNA) is higher in serum

compared to plasma due to leukocyte lysis and release of cell genomic DNA during clotting. The amount of normal DNA derived from leukocytes that can potentially dilute the ctDNA fraction is much lower in plasma. *Merker JD, et al. Circulating tumor DNA analysis in patients with cancer: American Society of Clinical Oncology and College of American Pathologists Joint Review. J Clin Oncol 2018;36(16):1631–64.*

3. c. Cell-free DNA (cfDNA) commonly refers to DNA fragments shed by cells into body fluids. Unlike genomic DNA, which is found in the cell's nucleus, these fragments are free floating. These small fragments are typically 150–170 bp in size and represent the entire genome of the mother and fetus. During pregnancy, the mother's bloodstream contains a mix of cfDNA that originated from her cells and from cells of the placenta. Fetal cfDNA is undetectable in maternal circulation shortly after birth. *Pös O, et al. Circulating cell free DNA nucleic acid: characteristics and applications. Eur J Human Gen 2018;26:937–45.*

4. c. NGS-based tests are highly sensitive for detecting of mutations present at ≤1% of the reads. However, errors introduced during PCR amplification steps of library preparation or during sequencing increase the false positive rate, limiting the accuracy of detecting positive variants present at very low frequency. One strategy involves the use of unique molecular identifiers (barcodes) to tag individual target DNA molecules. These barcodes allow PCR duplicates originating from the same target DNA molecule to be grouped together and collapsed into a single consensus sequence with error correction. *Bale TA, et al. Clinical use of cell-free DNA in tumor diagnostics. Adv Mol Pathol 2019;2:153–62.*

5. a. Noninvasive prenatal testing (NIPT) is a screening test that estimates the risk of a fetus having certain chromosomal abnormalities or genetic disorders. Because NIPT analyzes both fetal and maternal cfDNA, the test may detect a genetic condition in the mother. A positive NIPT test is followed by a diagnostic genetic test, such as chorionic villus sampling or amniocentesis. *Gregg AR, et al. Noninvasive prenatal screening for fetal aneuploidy, 2016 update: a position statement of the American College of Medical Genetics. Genet Med 2016;10:1056–65.*

6. b. Noninvasive prenatal testing (NIPT) is most often used to look for chromosomal disorders that are caused by the presence of an extra or missing copy (aneuploidy) of a chromosome, such as Down syndrome (trisomy 21), Edwards syndrome (trisomy 18), Patau syndrome (trisomy 13), and extra or missing copies of the X chromosome and Y chromosome (the sex chromosomes). The accuracy of the test varies by disorder. *Dondorp W, et al. Non-invasive prenatal testing for aneuploidy and beyond: challenges of responsible innovation in prenatal screening. Summary and recommendations. Eur J Hum Genet 2015;23(11):1438–50.*

7. b. There must be enough fetal cfDNA in the mother's bloodstream to be able to identify fetal chromosome abnormalities. The proportion of cfDNA in maternal blood that comes from the placenta is known as the fetal fraction. Generally, the fetal fraction must be above 4%, which typically occurs around the tenth week of pregnancy. Low fetal fractions can lead to an inability to perform the test or a false negative result. Reasons for low fetal fractions include testing too early in the pregnancy, sampling errors, maternal obesity, and fetal abnormality. *Dondorp W, et al. Noninvasive prenatal testing for aneuploidy and beyond: challenges of responsible innovation in prenatal screening. Summary and recommendations. Eur J Hum Genet 2015;23(11):1438–50.*

8. b. Specific mutations in the epidermal growth factor receptor (EGFR) tyrosine kinase in patients with nonsmall cell lung cancer have been shown to confer sensitivity to certain EGFR tyrosine kinase inhibitors (TKIs). Studies have demonstrated that cfDNA testing of EGFR mutations can facilitate rapid cancer genotyping testing at diagnosis and after the development of drug resistance to assist with tumor targeted therapy. Currently, next-generation sequencing (NGS) or digital droplet PCR methods (ddPCR) are clinically used for tumor mutation detection in cfDNA. *Bale TA, et al. Clinical use of cell-free DNA in tumor diagnostics. Adv Mol Pathol 2019;2:153–62.*

9. d. Preanalytical variables such as extraction efficiency and DNA fragment size bias can be assessed by spiking blood plasma with known fragment lengths and concentrations of synthetic or exogenous DNA prior to cell-free DNA extraction. Higher Cq values or lower read counts for synthetic or exogenous DNA spiked into extracted cell-free DNA compared to values from pure reference material can indicate the potential presence of PCR inhibitors. *Geeurickx E, et al. Targets, pitfalls, and reference materials for liquid biopsy tests in cancer diagnostics. Mol Asp Medi 2019; https://doi.org/10.1016/j.mam.2019.10.005.*

10. d. These results indicate a predicted increased risk for a genetic abnormality when the fetus is actually unaffected (false positive). Noninvasive prenatal testing (NIPT) results can also indicate a decreased risk for a genetic abnormality when the fetus is actually affected (false negative). False positive results may be test related or due to biological causes such as maternal or placental mosaicism. False negative results may be due to low fetal cfDNA fraction in the maternal blood. *Palomaki GE, et al. The clinical utility of DNA-based screening for fetal aneuploidy by primary obstetrical care providers in the general pregnancy population. Genet Med 2017;19:778–86.*

11. c. Cell-free DNA is thought to be released from cells during apoptosis and necrosis. Enzymatic cleavage of DNA during apoptosis results in formation of DNA fragments of an average of 150–170 bp, corresponding to how

DNA wraps around nucleosomes. Other mechanisms such as active secretion of DNA molecules in the circulation are also possible. *Ammerlaan W, et al. Biospecimen science of blood for cfDNA genetic analyses. Curr Pathobiol Rep 2019;7:9–15.*

12. d. A variety of biomarkers have been described that are released directly into blood or body fluids and detected via liquid biopsy. These include circulating tumor cells, proteins, cell-free DNA and RNA, extracellular vesicles, exosomal DNA, and microRNA. The nucleic acids can be derived from patient genomic DNA, tumor DNA, fetal cell DNA, and even microbial DNA. *Campos C, et al. Molecular profiling of liquid biopsy samples for precision medicine. Cancer J 2018;24(2):93–103.*

13. d. Various approaches to testing cfDNA can be utilized depending on the application and specific clinical scenario, from quantitative real-time PCR or ddPCR for small targeted panels that detect a few variants to NGS which aims for much broader coverage of the genome for simultaneous detection of many variants. *Ammerlaan W. Biospecimen science of blood for cfDNA genetic analyses. Current Pathobiology Reports 2019;7:9–15. Li Z, et al. The cornerstone of integrating circulating tumor DNA into cancer management. Biochem Biophys Acta Rev. 2019;Canc. 2018;* https://doi.org/ 10.1016/j.bbcan.2018.11.002.

14. d. Although cfDNA may have some utility in screening and determining origin of metastatic disease of unknown primary in the future, specific alterations detected in plasma, particularly those in common driver genes, such as KRAS, are unlikely to be of benefit in determining tissue of origin. KRAS can be a driver mutation in numerous cancer types, including colorectal, lung, and pancreatic carcinoma. *Conway, AM. Molecular characterization and liquid biomarkers in carcinoma of unknown primary (CUP): taking the 'U' out of 'CUP'. Brit J Can 2018;120:141–153. Rassy E, et al. Liquid biopsy: a new diagnostic, predictive and prognostic window in cancers of unknown primary. Eur J Cancer 2018;105:28–32.*

15. a. While many potential applications of liquid biopsy are being investigated, testing is still limited to a restricted number of clinical applications. Currently, one of the most useful and well-understood applications for liquid biopsy is for monitoring treatment response and therapy selection in cancer patients. *Castro-Giner F, et al. Cancer diagnosis using a liquid biopsy: challenges and expectations. Diagn 2018;31.* https://doi.org/ 10.3390/diagnostics8020031.

16. b. EGFR mutations, primarily deletions in exon 19 affecting the amino acid motif LREA (delE746-750) or substitution of arginine for leucine at position 858 (L858R) in exon 21, are most common in people with non-small cell lung cancer. These mutations lead to constitutive activation of the EGFR tyrosine kinase and are associated with high response rates to EGFR tyrosine kinase inhibitor (TKI) therapy. Although tissue biopsy is

the gold standard for molecular analysis, noninvasive or minimally invasive liquid biopsy methods may be also considered in certain clinical situations. *Castro-Giner F, et al. Cancer diagnosis using a liquid biopsy: challenges and expectations. Diagn 2018;31; https://doi.org/10.3390/diagnostics8020031.*

17. b. Cell-free DNA is usually released during apoptosis of cells and is composed of small DNA fragments in the range of 140–170 bp, corresponding to the length of nucleosome-protected DNA. *Fan HC, et al. Analysis of the size distributions of fetal and maternal cell-free DNA by paired-end sequencing. Clin Chem 2010;56:1279–86. Mouliere F, et al. High fragmentation characterizes tumour-derived circulating DNA. PLoS ONE 2011;6(9);e23418. Zheng YWL, et al. Nonhematopoietically derived DNA is shorter than hematopoietically derived DNA in plasma: a transplantation model. Clin Chem 2012;58(3):549–8.*

18. d. Circulating cell-free fetal DNA in maternal blood is thought to arise predominantly from apoptosis of placental cells, specifically the cytotrophoblasts of the chorionic villi. One limitation of this testing is that the findings may not be representative of the actual fetal DNA, such as in cases where a genetic abnormality is only present in the placenta and not in the fetus, also known as confined placental mosaicism. *Tjoa ML, et al. Trophoblastic oxidative stress and the release of cell-free feto-placental DNA. Am J Pathol 2006;169:400–4. Faas BH, et al. Non-invasive prenatal diagnosis of fetal aneuploidies using massively parallel sequencing-by-ligation and evidence that cell-free fetal DNA in the maternal plasma originates from cytotrophoblastic cells. Expert Opin Biol Ther 2012;12:S19–26.*

19. a. Confined placental mosaicism is when a chromosomal abnormality occurs only in the placenta but not in the fetus and can occur in up to 2% of high-risk pregnancies. As a result, there are general recommendations to confirm a positive NIPT result by invasive prenatal testing, preferably by amniocentesis. *Grati FR, et al. Fetoplacental mosaicism: potential implications for false-positive and false-negative noninvasive prenatal screening results. Genet Med 2014;16:620–24.*

20. d. A variety of approaches have been utilized both in the research and clinical setting for the isolation of circulating tumor cells. These include physical properties such as size-based filtration approaches or density gradient centrifugation, antibodies to capture epithelial cells (EpCAM), depletion of leukocytes with CD45 antibodies, or even use of functional properties (protein secretion, migratory properties, etc.). *Cristofanilli M, et al. Circulating tumor cells, disease progression, and survival in metastatic breast cancer. N Engl J Med 2004;351:787–91. Vona G, et al. Isolation by size of epithelial tumor cells. Am J Pathol 2000;156(1):57–63. Riethdorf S, et al. Detection of circulating tumor cells in peripheral blood of patients with*

metastatic breast cancer: a validation study of the CellSearch system. Clin Cancer Res 2007;13:920–8.

21. a. The estimated half-life of cfDNA in circulating blood ranges from several minutes to 1–2 h. This relatively short half-life is convenient as it allows for evaluation of treatment response and postsurgical evaluation. *Moreira G, et al. Increase in clearance of cell-free plasma DNA in hemodialysis quantified by real-time PCR. Clin Chem Lab Med 2006;44 (12):1410–5. Gauthier VJ, et al. Blood clearance kinetics and liver uptake of mononucleosomes in mice. J Immunol 1996;156(3);1151–6. Yu SCY, et al. High-resolution profiling of fetal DNA clearance from maternal plasma by massively parallel sequencing. Clin Chem 203;59(8):1228–37.*

22. c. The variability in cell-free DNA levels is thought to be attributable to stage of disease, tumor burden, cellular turnover, and accessibility of tumor cells to blood vessels. High levels of cell-free DNA have been associated with worse prognosis in some studies. *Bettegowda C, et al. Detection of circulating tumor DNA in early- and late-stage human malignancies. Sci Transl Med 2014;6:224ra24.*

23. a. The average size of circulating cell-free DNA is 150–170 bp. However, lysis of white blood cells can cause an increase in overall DNA content secondary to the release of genomic DNA. Genomic DNA is composed of larger DNA fragments than that seen with circulating cfDNA, and an increased fraction of fragments >300 bp is indicative of a compromised blood sample in which nucleated cells have released genomic DNA. *Mehrotra M, et al. Study of preanalytic and analytic variables for clinical next-generation sequencing of circulating cell-free nucleic acid. JMD 2017;19(4):514–4.*

24. b. Noninvasive prenatal testing (NIPT) is considered a screening testing, and while it offers less overall risk compared to other invasive testing modalities, it is not as accurate as chorionic villous sampling or amniocentesis. While designed to be highly sensitive, it suffers from lower specificity, and all positive results should be further investigated with invasive testing methods. *Gregg AR, et al Noninvasive prenatal screening for fetal aneuploidy, 2016 update: a position statement of the American College of Medical Genetics and Genomics. Genet Med 2016;18:1056–65.*

25. a. Studies have shown that cell-free DNA can be detected in maternal plasma at around 10 weeks, which offers the ability to screen earlier than other, invasive testing methodologies such as chorionic villus sampling (CVS) and amniocentesis. *Wang E, et al. Gestational age and maternal weight effects on fetal cell-free DNA in maternal plasma. Prenat Diagn 2013;33:662–6.*

26. c. The advantages of liquid biopsy include the convenience of sample acquisition with minimal risk to the patient, the ability to perform serial testing to monitor response to therapy or minimal residual disease, and the provision of more complete information regarding overall tumor burden and tumor heterogeneity that cannot be obtained from sampling at one

site. However, liquid biopsy sequencing has a higher overall false negative rate and negative predictive value, since alterations present in tissue may not always be present or identifiable in circulating cell-free DNA. *Merker JD, et al. Circulating tumor DNA analysis in patients with cancer: American Society of Clinical Oncology and College of American Pathologists Joint Review. J Clin Oncol 2018;36(16):1631–4.*

27. d. Preanalytical variables for cfDNA testing include specimen type (plasma, serum, other bodily fluids), type of test tube utilized, (K2EDTA vs cell stabilization tube), storage conditions (temperature), processing conditions, and DNA extraction and quantification methods. The ability to detect low variant allele frequencies (the limit of detection of the test) would be considered an analytical variable. *Geeurickx E, et al. Targets, pitfalls, and reference materials for liquid biopsy tests in cancer diagnostics. Molr Asp Med 2019; https://doi.org/10.1016/j.mam.2019.10.005.*

28. b. Cell-free fetal DNA is under 300 bp in length since it is released through apoptosis during physiologic placental cell turnover throughout pregnancy, while maternal DNA has a much more varied distribution of fragment length size. This property can be utilized to increase fetal-to-maternal DNA ratio with size fractionization methods that select for DNA fragments less than 300 bp in length. *Taglauer ES, et al. Review: Cell-free fetal DNA in the maternal circulation as an indication of placental health and disease. Placenta 2014;35(Suppl):S64–8. Sayres LC, et al. Cell-free fetal nucleic acid testing: A review of the technology and its applications. CME Review Article 2011;66(7):431–42.*

29. a. The fraction of cfDNA that is tumor derived can be as low as <0.1% and up to >10% of total DNA. Conventional NGS has a limit of detection around 1%, but biochemical and bioinformatic optimization can improve the limit of detection from 1% to up to 0.1%. *Castro-Giner F. Cancer Diagnosis using a liquid biopsy: challenges and expectations. Diagn 2018;31; https://doi.org/10.3390.* Salk JJ, et al. Enhancing the accuracy of next-generation sequencing for detecting rare and subclonal mutations. Nat Rev Genet 2018;19(5):269–85.

30. c. For identification of a few variants that are associated with an individual tumor or with resistance to targeted molecular therapies, ddPCR and quantitative real-time PCR strategies can be used. More targeted approaches such as ddPCR and real-time PCR can also offer more sensitive detection of variants at the expense of breadth of coverage. For broader coverage of the genome and detection of many variants associated with several different tumor types, NGS is the preferred sequencing strategy. *Merker JD, et al Circulating tumor DNA analysis in patients with cancer: American Society of Clinical Oncology and College of American Pathologists Joint Review. J Clin Oncol 2018;36(16):1631–64.*

Chapter 28

Urinalysis and stool testing

Alan H.B. Wu
University of California, San Francisco, CA, United States

1. Identify the urine crystal from patient with a history of diuretic use.[1]

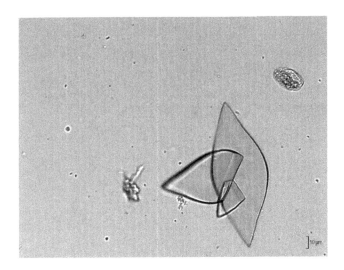

---

1. Images from Wikipedia Commons.

Self-assessment Q&A in Clinical Laboratory Science, III. https://doi.org/10.1016/B978-0-12-822093-1.00028-4
337

2. This patient has a metabolic acidosis and an elevated anion gap.[1]

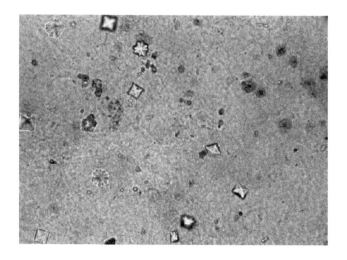

3. This patient has excretion of this amino acid.[1]

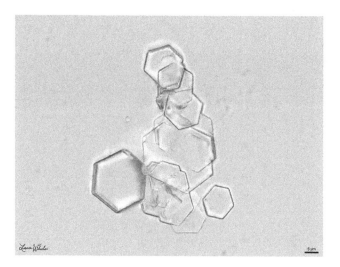

4. This patient has acidic urine and a history of nephrolithiasis.[1]

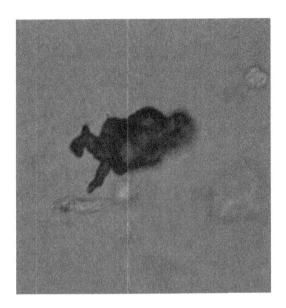

5. This patient has a bacterial infection and alkaline urine.[1]

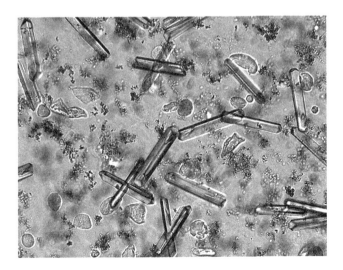

6. This crystal is seen in conjunction with lipid secretion secondary to nephrotic syndrome.[1]

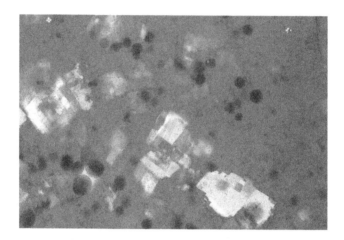

7. This patient has a metabolic disease associated with amino acid metabolism dysfunction.[2]

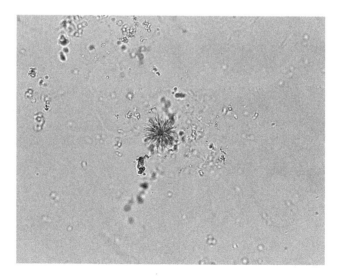

2. Images from Shutterstock.

8. This patient has obstructive liver disease.[2]

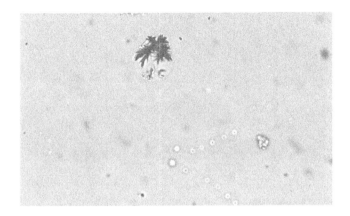

9. There is no clinical significance for finding these casts.[2]

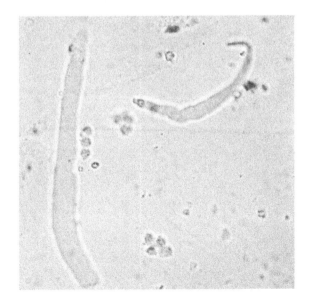

10. This cast is seen in urinary stasis.[2]

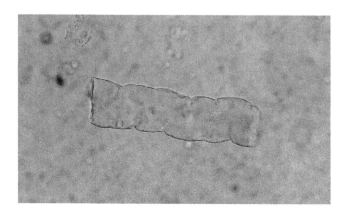

11. These casts can form from incorporation of white cells.[2]

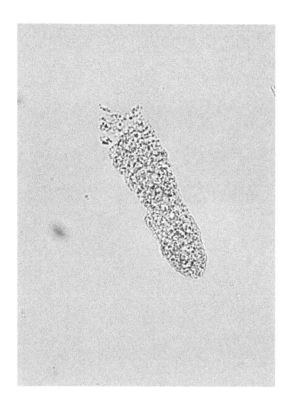

12. A normal occasional finding in adult males.[2]

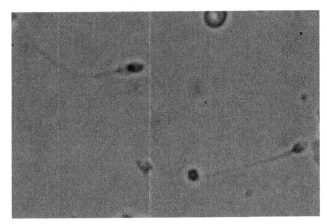

13. Presence of these cells can produce pain upon urination.[2]

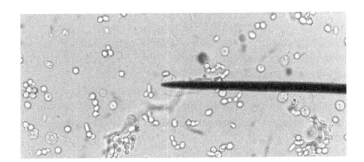

14. In postmortem urine, the presence of this structure in urine can be useful to indicate the cause of death.[2]

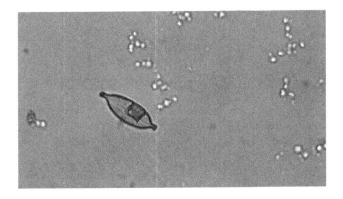

15. What is the normal threshold for excretion of glucose into the urine?
    a. 65 mg/dL
    b. 110 mg/dL
    c. 126 mg/dL
    d. 180 mg/dL
    e. 250 mg/dL
16. Which of the following disorders can produce a negative urine bilirubin and positive urobilinogen?
    a. hepatic cirrhosis
    b. hemolytic anemia
    c. urolithiasis
    d. alcoholism
    e. liver cancer
17. Which of the following is true regarding the sensitivity of the urine protein dipstick?
    a. measures all urine proteins equally
    b. most sensitive to albumin
    c. most sensitive to free light chains
    d. results correlate to microalbumin testing
    e. good agreement with quantitative urine protein analysis
18. Which of the following is false regarding hematuria and hemoglobinuria?
    a. hemoglobinuria can result from hematuria
    b. most common causes of hematuria are related to kidney diseases
    c. the dipstick is sensitive to both hemoglobin and myoglobin
    d. hemoglobinuria occurs when the transferrin binding capacity has been exceeded
    e. hematuria can be caused by drugs such as aspirin, warfarin, and heparin
19. Usually the urine and blood pH are abnormal in the same direction. For which of the following is there a discrepancy between urine pH and blood pH?
    a. lactate acidosis
    b. diabetic ketoacidosis
    c. hyperventilation
    d. chronic obstructive pulmonary disease
    e. renal tubular acidosis
20. Which of the following is not associated with a positive leukocyte with a negative nitrite in the urine?
    a. the leukocyte can be falsely positive due to skin contamination
    b. some microorganisms do not produce nitrite
    c. nitrites are produced by bacteria but are not increased due to inflammation

d. nitrites are used to adulterate urine for drug testing
e. this situation separates symptomatic from asymptomatic urinary tract infection

21. Which disorder is not associated with an increase in urine specific gravity?
    a. syndrome of inappropriate ADH
    b. use of contrast agents
    c. dehydration
    d. psychogenic polydipsia
    e. diarrhea

22. Identify this organism found in a fecal sample.

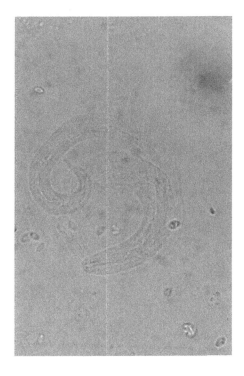

23. Identify this structure found in a child's stool sample.

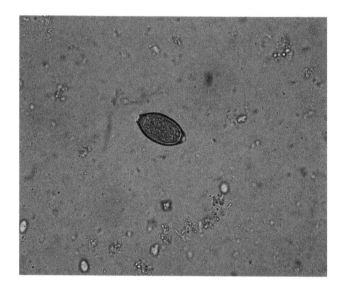

24. Which of the following is not a limitation of the fecal occult blood test (guaiac slide test) for colon cancer screening?
a. Hydrating the slide increases the sensitivity of testing.
b. Unhydrated slides produce many false positive.
c. The slide test is less expensive than immunoassays.
d. The slide test is not CLIA waived.
e. The slide test has low sensitivity for detecting polyps that do not bleed.

25. Which of the following is true regarding the fecal immunochemical test (FIT)?
a. obviates the need for colonoscopy
b. sequential testing produces similar performance to colonoscopy
c. single FIT and the 3-card guaiac offer similar performance
d. FIT testing should be conducted once every 5 years
e. quantitative offers no advantage over qualitative FIT testing

26. Which of the following disorders is the primary reason for conducting fetal fat analysis?
a. investigation of fat malabsorption
b. when high, indicates hepatic steatosis
c. determine the proper therapeutic for Shwachman–Diamond syndrome
d. determine the presence of excess small bowel bacteria
e. diagnostic of Crohn's disease

27. Which microorganism produces toxins that can be tested in stool using immunoassays?

a. COVID-19

b. *Clostridium difficile*

c. methicillin-resistant (MRSA) and methicillin-sensitive *Staphylococcus aureus*

d. Shiga toxin-producing *Escherichia coli*

e. *Salmonella*

Answers

1. Excess uric acid secretion occurs in patients with gout, an inflammatory disorder caused by certain diets, medications, and decreased renal function. Gout. MedicineNet. *https://www.medicinenet.com/gout_gouty_arthritis/article.htm.*

2. Oxalate is a major metabolite of ethylene glycol leading to calcium oxalate crystals. The acidosis is due to oxalic and glycolic acids. *Pomara C, et al. Calcium oxalate crystals in acute ethylene glycol poisoning: a confocal laser scanning microscope study in a fatal case. Clin Toxicol (Phila). 2008 Apr;46(4):322–4.*

3. Cystinuria is a genetic disease characterized by excess crystal excretion and kidney stone formation. *What is cystinuria? Healthline. https://www.healthline.com/health/cystinuria.*

4. Ammonium urate crystals are rare but can form in patients with acidic urine and is associated with ammonium urate kidney stones. *Tanaka Y, et al. Causes of presence of ammonium urate crystals in urinary sediments. Japan J Med Tech 2015;64:179–85.*

5. Triple phosphate crystals form in alkaline urine and are seen in renal disease and presence of some bacteria such as Proteus or Klebsiella. *Fogazzi G. Crystalluria: a neglected aspect of urinary sediment. Nephrol Dial Transplant 1996;11:379–87. Griffith DP, et al. Urease. The primary cause of infection-induced urinary stones. Investig Urol 1976;13:346–50.*

6. Cholesterol crystals have well-defined edges and appear as plates. *Fogazzi G. Crystalluria: a neglected aspect of urinary sediment. Nephrol Dial Transplant 1996;11:379–87.*

7. Tyrosine crystals are seen in liver diseases that are associated with an impairment of amino acid metabolism. *Fogazzi G. Crystalluria: a neglected aspect of urinary sediment. Nephrol Dial Transplant 1996;11:379–87.*

8. Excess conjugated bilirubin in urine can form crystals and seen in patients with intrahepatic and extrahepatic liver disease. *Fogazzi G. Crystalluria: a neglected aspect of urinary sediment. Nephrol Dial Transplant 1996;11:379–87.*

9. Hyaline casts contain the Tamm–Horsfall mucoprotein and outline the renal tubule. *Ringsrud KM. Casts in the urine sediment. Lab Med 2001;4 (32):191–3.*

10. Waxy casts are broad and are found in tubules when there is urinary stasis and atrophy. *Ringsrud KM. Casts in the urine sediment. Lab Med 2001;4 (32):191–3.*

11. Granular casts form from white and red cell casts as a result of aggregation of mucoproteins. *Ringsrud KM. Casts in the urine sediment. Lab Med 2001;4(32):191–3.*

12. Sperm excretion occurs after ejaculation. Retrograde ejaculation is a disorder whereby semen the bladder instead of exiting out of the penis. *Retrograde ejaculation. Mayo Clinic. https://www.mayoclinic.org/diseases-conditions/retrograde-ejaculation/symptoms-causes/syc-20354890.*

13. Candida fungus are yeast and can cause itching, pain and is caused by infections to the upper or lower urinary track. It occurs in women more often than men. The arrow indicates the presence of budding yeast. *Kauffman CA, et al. Candida urinary tract infections-diagnosis. Clin Infect Dis 2011;52(suppl. 6):S452–6.*

14. This structure is a diatom, a single cell algae found in ocean water and soils. Its presence could indicate ingestion of seawater as seen in drownings. *Ranal A, et al. Significance of diatoms in diagnosis of drowning deaths: a review. J Foren Genetic Sci 2018;1(5). doi:10.32474/PRJFGS.2018.01.000121.*

15. d. The renal threshold for reabsorption of glucose is between 160 and 180 mg/dL, above which results in glycosuria. *Ferrannini E. Learning from glycosuria. Diabetes 2011;60(3):695–6.*

16. b. High urine bilirubin is seen in obstructive liver diseases. In patients with bleeding there is an overload in the liver's ability to conjugate bilirubin and high serum bilirubin unconjugated. Due to its insolubility, conjugated must be bound to albumin and is not excreted into urine. *Kalakonda A. Physiology, bilirubin. StatPearls. https://www.ncbi.nlm.nih.gov/books/NBK470290/.*

17. b. The urine protein dipstick largely reacts to albumin but cannot measure low levels of albumin, such as in early diabetic nephropathy and is not useful for Bence Jones proteins detection. *Gaugarum R, et al. The accuracy of urine dipsticks as a screening test for proteinuria in hypertensive disorders of pregnancy. Hypertens Pregn 2005;24(2):117–23.*

18. d. Hemoglobinuria occurs when binding capacity of haptoglobin, not transferring is exceeded. *Difference between hematuria and hemoglobinuria. https://www.differencebetween.com/difference-between-hematuria-and-vs-hemoglobinuria/.*

19. c. In renal tubular acidosis, there is an inability of the kidneys to secrete acids into the tubules leading to alkaline urine and a systemic acidosis. *Mustaqeem R, et al. Renal tubular acidosis. StatPearls. https://www.ncbi.nlm.nih.gov/books/NBK519044/.*

20. c. High leukocytes indicate infection or inflammation along the urinary tract. Nitrites are produced by some bacterium. Nitrites are used as an adulterant

for urine drug testing, but values are high, not low. *What do leukocytes in the urine mean? MedicalNewsToday. https://www.medicalnewstoday.com/articles/313751.*

21. d. Contrast agents are highly dense and when excreted in urine causes high specific gravity. Diabetes insipidus and polygenic polydipsia are associated with excess water intake and a low specific gravity. What is a urine specific gravity test? MedicalNewsToday.. *https://www.medicalnewstoday.com/articles/322125.*

22. Pinworms are parasites that live in the gastrointestinal tract and can be shed into stool. Children are at risk if they have poor hygiene. Pinworm infection. Healthline. *https://www.healthline.com/health/pinworms.*

23. This is an egg of a pinworm. Female pinworms lay eggs near the skin folds of the anus. *https://www.healthline.com/health/pinworms.*

24. d. The guaiac test has been used for many years for colon cancer screening despite its limitations. The test is waived by CLIA. *Simon JB. Fecal occult blood testing: clinical value and limitations. Gastroenterol 1998;6:66–78.*

25. b. According to the US Multi-Society Task Force on Colorectal Cancer for screening colorectal cancer, FIT testing performed every year produces similar results to colonoscopy and superior results to the guaiac testing. *Robertson DJ, et al. Recommendations of fecal immunochemical testing to screen for colorectal neophasia: A consensus statement by the US Multi-Society Task Force on Colorectal Cancer. Gastroenterol 2017;52:1217–37.*

26. a. Many of the conditions listed in the question are diseases associated with fat malabsorption but the test is not diagnostic for any one in particular. Steatosis is accumulation of fat within a cell or organ, often the liver and is not due to malabsorption. *Fecal fat testing. Healthline. https://www.healthline.com/health/fecal-fat.*

27. b. Stools can be tested via enzyme immunoassay for the presence of the cytotoxin to *C. difficile*. PCR and stool culture are used for *E. coli, salmonella*, and *campylobacter*. While an ELISA is available for *E. coli*, it does not perform as well as PCR. *Merz CS, et al. Comparison of four commercially available rapid enzyme immunoassays with cytotoxin assay for detection of Clostridium difficile toxin(s) from stool specimens. J Clin Microbiol 1994;32(5):1142–7. Pulz M, et al. Comparison of a Shiga toxin enzyme-linked immunosorbent assay and two types of PCR for detection of Shiga toxin-producing Escherichia coli in human stool specimens. J Clin Microbiol 2003;41(10):4671–5.*

Chapter 29

Flow cytometry

Chuanyi Mark Lu and Alan H.B. Wu
University of California, San Francisco, CA, United States

1. Flow cytometric immunophenotyping (FCI) is used for
   1. detection of lineage-associated cluster of differentiation (CD) antigens.
   2. detection of antigens used as therapeutic targets.
   3. assessment of biologic markers associated with prognosis.
   4. detection of residual neoplastic cells following therapy.
      - **a.** 1, 2 and 3
      - **b.** 1 and 3
      - **c.** 2 and 4
      - **d.** 4 only
      - **e.** all of the above
2. Specimens that are suitable for FCI include which of the following?
   1. peripheral blood.
   2. bone marrow aspirate.
   3. lymph node.
   4. body fluid.
      - **a.** 1, 2, and 3
      - **b.** 1 and 3
      - **c.** 2 and 4
      - **d.** 4 only
      - **e.** all of the above
3. All of the following are clinical indications of FCI EXCEPT which one?
   - **a.** diagnosis and classification of acute leukemia
   - **b.** diagnosis and classification of non-Hodgkin lymphoma
   - **c.** diagnosis and classification of Hodgkin lymphoma
   - **d.** confirmation of clonal plasma cell proliferation
   - **e.** diagnosis of paroxysmal nocturnal hemoglobinuria (PNH)
   - **f.** tumor DNA content/cell cycle analysis
4. In contrast to normal maturing hematopoietic and lymphoid cells, leukemia and lymphoma cells often show which of the following?
   1. normal patterns of antigen expression.
   2. random alterations in antigen expression.

Self-assessment Q&A in Clinical Laboratory Science, III. https://doi.org/10.1016/B978-0-12-822093-1.00029-6

3. synchronous antigen expression.

4. abnormally increased or decreased levels of antigen expression, including gain or loss of antigens.

   a. 1, 2, and 3

   b. 1 and 3

   c. 2 and 4

   d. 4 only

   e. all of the above

5. All of the following statements regarding FCI are true EXCEPT which one?

   a. It is a technique that can perform multiparametric analysis on individual cells.

   b. Antigens commonly used for blast identification include CD34, CD117, HLA-DR, and terminal deoxynucleotidyl transferase (TdT).

   c. It relies on the overall pattern of antigen expression rather than expression of any single antigen for acute leukemia classification.

   d. The blast percentage obtained by flow cytometry is more accurate than morphologic blast count.

6. Which of the following is(are) definite lineage-specific antigen(s)?

   1. cytoplasmic myeloperoxidase (MPO)

   2. CD13

   3. surface CD3

   4. CD19

      a. 1, 2, and3

      b. 1 and 3

      c. 2 and 4

      d. 4 only

      e. all of the above

7. Which of the following antigens are common therapeutic targets?

   a. CD20, CD33, CD52

   b. CD10, CD34, CD19

   c. CD41, CD42, CD61

   d. CD36, glycophorin

   e. CD2, CD5, CD7

8. The immunophenotypic diagnosis of B-cell lymphomas typically relies on

   1. evidence of clonal expansion (light chain restriction).

   2. strong expression of surface immunoglobulin.

   3. alterations in intensity of antigen expression.

   4. loss of antigen expression.

      a. 1, 2, and 3

      b. 1 and 3

      c. 2 and 4

      d. 4 only

      e. all of the above

9. All of the following are analytical limitations of FCI EXCEPT which one?
   a. fibrotic or markedly "packed" bone marrow may yield too few cells for adequate analysis
   b. there is a lack of tissue architectural relationships
   c. inability of accurately defining the surface antigen profile of specific cells
   d. T-cell lymphoma that does not have an aberrant immunophenotype may not be detected
   e. a small monoclonal B-cell population may not be detected
   f. partial tissue involvement by lymphoma may be missed
   g. inability to detect/diagnose Hodgkin lymphoma
10. Aberrant CD5 expression is characteristic of which of the following?
    1. small lymphocytic leukemia/small lymphocytic lymphoma (CLL/SLL).
    2. subset of prolymphocytic leukemia.
    3. mantle cell lymphoma/leukemia.
    4. marginal zone lymphoma.
       a. 1, 2, and 3
       b. 1 and 3
       c. 2 and 4
       d. 4 only
       e. all of the above
11. CD10 is characteristically expressed in
    a. follicular lymphoma.
    b. diffuse large B-cell lymphoma.
    c. Burkitt lymphoma/leukemia.
    d. lymphoplasmacytic lymphoma.
       a. 1, 2, and 3
       b. 1 and 3
       c. 2 and 4
       d. 4 only
       e. all of the above
12. All of the following are features of hairy cell leukemia EXCEPT which one?
    a. circulating neoplastic cells with circumferential "hairy" projections
    b. strong/bright expression of CD103, CD22, and CD11c
    c. lack of surface immunoglobulin
    d. expression of B-cell antigens (CD19, CD20)
13. Which of the following may allow a rapid confirmation of a clonal T-cell malignancy?
    a. absence or down-regulation of pan-T-cell antigens
    b. flow cytometric analysis of TCR V-β
    c. an aberrant T-cell immunophenotype
    d. CD10-positive T-cell population

14. FCI is often included in the diagnostic workup of which of the following?
   a. refractory anemia
   b. Hodgkin lymphoma
   c. chronic myeloid leukemia, blast phase
   d. polycythemia vera

15. The typical immunophenotype of Sézary cells is
   a. CD3+, CD4+, CD8−, CD25+
   b. CD3+, CD4+, CD8+, CD25−
   c. CD3+, CD4+, CD8-, CD25−
   d. CD3+, CD4−, CD8+, CD25−

16. The lymphoblasts in precursor-B acute lymphoblastic leukemia (precursor B-ALL) often express all of the following antigens EXCEPT which one?
   a. terminal deoxynucleotidyl transferase (TdT)
   b. CD10
   c. HLA-DR
   d. surface immunoglobulin
   e. CD34

17. All of the following are true for plasma cell myeloma EXCEPT which one?
   a. expression of monotypic cytoplasmic immunoglobulin
   b. IgG is most common
   c. typically CD79a+, CD138+, CD19−, CD56+
   d. typically CD79a+, CD138+, CD19+, CD56−
   e. lack of surface immunoglobulin

18. Immunophenotypic features of neoplastic natural killer (NK) cells include
   1. positive for CD16 and/or CD56.
   2. positive for cytoplasmic CD3ε.
   3. variably positive for CD2.
   4. positive for surface CD3.
      a. 1, 2, and 3
      b. 1 and 3
      c. 2 and 4
      d. 4 only
      e. all of the above

19. Which of the following may be used to distinguish thymocytes from mediastinal precursor T-lymphoblastic lymphoma (T-LBL)?
   a. presence of CD4/CD8 coexpression
   b. variable (smear-like) expression of surface CD3
   c. expression of CD10
   d. expression of CD1a

20. Which of the following may be used to distinguish hematogones from residual precursor B-lymphoblastic leukemia (B-ALL) in bone marrow?
   a. coexpression of CD10/CD34
   b. expression of TdT
   c. lack of aberrant antigen expression

d. lack of immunoglobulin expression

e. discrete CD20 expression pattern

21. Flow cytometric immunophenotyping is essential for the diagnosis of which of the following?

1. acute myeloblastic leukemia with maturation.

2. acute myeloblastic leukemia without maturation.

3. acute monoblastic leukemia.

4. acute myeloblastic leukemia, minimally differentiated.

    a. 1, 2, and 3

    b. 1 and 3

    c. 2 and 4

    d. 4 only

    e. all of the above

22. All of the following are true EXCEPT which one?

    a. helper T-cell express CD3 and CD4

    b. suppressor T-cell express CD8, but not CD3

    c. B-cells express CD19, but not CD3

    d. NK-cells express CD16 and/or CD56

    e. HIV infection leads to decrease in CD4/CD8 ratio

Answers

1. e. Flow cytometry is a rapid and convenient technique for generating immunophenotypic data. The ability to perform multiparametric analysis on an individual cellular basis is a unique feature of the technique and offers distinct advantages over immunohistochemistry. Medical indications for performing flow cytometry in a clinical setting include (1) diagnosis and classification of hematopoietic and lymphoid neoplasms based on detection of lineage-specific cluster of differentiation antigens, so-called CD antigens; (2) assessment of biologic parameters associated with prognosis (e.g., CD38 and ZAP-70 for CLL/SLL); (3) detection of antigens used as therapeutic targets (e.g., CD20, CD30, CD33, CD52, etc.); and (4) detection of residual neoplastic cells following therapy. *McPherson RA et al., eds. Henry's Clinical Diagnosis and Management by Laboratory Methods, 23rd ed., 2011, p. 656.*

2. e. Any specimen from which a single cell suspension can be generated is suitable for FCI, including peripheral blood, bone marrow (aspirate), lymph node, fine needle aspirates, core biopsies, body fluids (e.g., CSF), etc. Blood and bone marrow may be anticoagulated with EDTA (lavender-top), heparin (green-top), and acid citrate dextrose (ACD) (yellow-top). Tissue specimens are best stored in tissue culture media such as RPMI 1640 and transported to the lab as soon as possible. Specimen of all types are commonly stored and transported at room temperature prior to processing. *McPherson RA et al., eds. Henry's Clinical Diagnosis and Management by Laboratory Methods, 23rd ed., 2011, pp. 659–660.*

3. c. FCI is not indicated in diagnosis and classification of Hodgkin lymphomas (HL). The neoplastic cells in HL (Reed-Sternberg cells in classical HL and "popcorn"/L&H cells in nodular lymphocyte predominant HL) are too scarce to be detected in FC sample prepared from tissue involved by HL. For patients with HL, FCI is occasionally used to exclude concurrent or secondary clonal B- or T-cell disorders, or in follow-up studies, to exclude secondary hematopoietic malignancies. Besides acute myeloid leukemia, acute lymphoblastic leukemia, non-Hodgkin lymphoma, and plasma cell myeloma, FCI is also useful in aiding the diagnosis of paroxysmal nocturnal hemoglobinuria by assessing expression of GPI-anchored antigens (e.g., CD55 and CD59) on peripheral erythrocytes and leukocytes. FCI is also used for DNA content/cell cycle analysis, which is of diagnostic and prognostic importance for a variety of malignant tumors. *McPherson RA et al., eds. Henry's Clinical Diagnosis and Management by Laboratory Methods, 23rd ed., 2011, pp. 660–672.*

4. d. The identification of hematopoietic and lymphoid neoplasia by immunophenotyping relies on the principle that neoplastic cells express patterns of antigen expression that are distinctly different from those of their normal counterparts. Antigen expression in normal cells is a tightly regulated process resulting in a characteristic pattern of antigen acquisition and loss with maturation that is cell lineage specific. Neoplastic leukemia and lymphoma cells often show nonrandom alterations in antigen expression that include (1) gain of antigens not normally expressed by cell type or lineage; (2) abnormal increase or decrease in levels of expression (intensities) including complete loss of normal antigens; (3) asynchronous antigen expression, e.g., expression of antigens normally expressed by the cell type but at an inappropriate time during maturation; and (4) abnormally homogeneous expression of one or more antigens (e.g., light chain restriction in B-cells) by a population that normally exhibits more heterogeneous expression. *McPherson RA et al., eds. Henry's Clinical Diagnosis and Management by Laboratory Methods, 23rd ed., 2011, pp. 665–671.*

5. d. The major role for flow cytometry in acute leukemia is classification. While flow cytometry is excellent at identifying blasts, there are two issues that suggest caution in utilizing flow cytometric blast percentage for the initial diagnosis of acute leukemia. First, the blast percentage obtained from bone marrow specimens is often inaccurate due to a combination of peripheral blood dilution (artifactual decrease) and partial lysis of erythroid precursors during specimen processing (artifactual increase). Second, blasts identified by immunophenotyping do not always directly correspond to blasts as identified by morphology. This is true because leukemic populations, like normal population, consist of a maturational continuum and there is no perfect concordance between specific antigen changes and the arbitrary morphologic changes that distinguish blasts from

more differentiated cells. Moreover, in some type of leukemia, early precursors (e.g., promonocytes) are considered blast equivalents and are intentionally included in morphologic blast counts. The other statements are true. *McPherson RA et al., eds. Henry's Clinical Diagnosis and Management by Laboratory Methods, 23rd ed., 2011, pp. 659–660.*

6. b. Many antigens are commonly used for lineage assignment and leukemia classification, including cytoplasmic MPO, CD13, CD33, and CD117 for acute myeloid leukemia (AML); CD19, CD22, CD79a, and TdT for precursor-B lymphoblastic leukemia (B-ALL); and CD1a, CD2, CD3, CD5, and CD7 for T-ALL. However, only MPO (myeloid) and CD3 (cytoplasmic CD3ε or surface CD3; T-lineage) are definite lineage-specific antigens. Other antigens are typically, but not definitely, associated with certain cell lineage. For example, CD19 is a very reliable B-cell marker, but it may be seen in myeloid leukemia such as AML with $t(8;21)(q22; q22)$. Likewise, CD2, CD5, and CD7 are considered T-cell markers, but it can be seen in AML and B-cell lymphomas. *Swerdlow SH, et al, eds. WHO Classification of tumours of Haematolopoietic and lymphoid tissues, 2017, p. 181.*

7. a. There are numerous commercially available therapeutic antibodies including anti-CD3 (OKT3), anti-CD20 (Rituximab), anti-CD33 (Gemtuzumab), anti-CD52 (Alemtuzamab), anti-CD22 (Epratuzamab), anti-CD30 (Brentuximab), and anti-CD38 (Daratumumab). CD41, CD42, and CD61 are *markers of megakaryocytic differentiation. Glycophorins are markers of erythroid differentiation. McPherson RA et al., eds. Henry's Clinical Diagnosis and Management by Laboratory Methods, 23rd ed., 2011, pp. 659–660.*

8. b. The flow cytometric method to diagnose B-cell lymphoma is to combine its ability to detect clonality (light chain restriction) with the ability to identify alterations in antigen intensity. Although the most important immunophenotypic evidence of B-cell lymphoma (clonal B-cell expansion) is light chain restriction, alterations in antigen intensity are often incorporated into the diagnostic interpretation. For instance, weak expression of CD20 and surface immunoglobulin (sIg) is characteristic feature of chronic lymphocytic leukemia/small lymphocytic lymphoma (CLL/SLL). Strong sIg expression is only seen in some B-cell lymphomas such as marginal zone B-cell lymphoma. Aberrant loss of certain T-cell antigens is suggestive of T-cell neoplasm (e.g., loss of CD7 is often seen in mycosis fungoides/Sézary syndrome). *McPherson RA et al., eds. Henry's Clinical Diagnosis and Management by Laboratory Methods, 23rd ed., 2011, pp. 659–660.*

9. c. One of the advantages of FCI is that multicolor (6-, 8-, 10-) and multiparametric analysis can be performed, allowing for an accurate definition of the surface antigen profile of specific cell population. Weakly expressed

surface antigens can also be detected by FCI. All other choices are limitations of FCI. Wood BL. Principles of minimal residual disease detection for hematopoietic neoplasms by flow cytometry. *Cytometry B Clin Cytom. 2016;90:47–53. PMID: 25906832.*

10. a. CD5 is a T-cell marker. Aberrant CD5 expression is characteristic of CLL/SLL and MCL, which are mature B-cell neoplasms. CLL/SLL and MCL can further be distinguished by expression of CD23 and SOX11 (CLL/SLL is typically CD23+ and SOX11 –, and MCL is CD23 – and SOX11+). B-cell PLL could be de novo or derived from underlying CLL/SLL. CD5 and CD23 are positive in 20%–30% and 10%–20% of PLL cases, respectively. MZL is typically negative for CD5. *Swerdlow SH, et al, eds., WHO Classification of tumours of Haematolopoietic and lymphoid tissues, 2017, pp. 215–44, McPherson RA et al., eds. Henry's Clinical Diagnosis and Management by Laboratory Methods, 23rd ed., 2011, pp. 665–671.*

11. a. CD10 is considered the earliest marker for lymphoid differentiation and a marker for acute lymphoblastic leukemia (ALL), for which reason it is also known as common ALL antigen (CALLA). Expression of CD10 is also a characteristic feature of germinal center B-cell derived lymphomas, including follicular lymphoma, diffuse large B-cell lymphoma, and Burkitt lymphoma/leukemia. Germinal center T-cell derived angioimmunoblastic T-cell lymphoma (AITL) is also positive for CD10. Lymphoplasmacytic lymphoma is negative for CD10. *Swerdlow SH, et al, eds., WHO Classification of tumours of Haematolopoietic and lymphoid tissues, 2017, pp. 215–244, McPherson RA et al., eds. Henry's Clinical Diagnosis and Management by Laboratory Methods, 23rd ed., 2011, pp. 665–671.*

12. c. Hairy cell leukemia (HCL) is a neoplasm of small B lymphoid cells with abundant cytoplasm with circumferential "hairy" projections, involving bone marrow, spleen, and peripheral blood. The neoplastic cells are characteristically positive for surface Ig, B-cell markers, CD22, CD25, CD11c, and CD103, and negative for CD5 and CD10. When HCL is suspected, CD103, CD22, CD25, and CD11c should be added to the flow panel. It is important not to miss HCL because it does not respond well to conventional lymphoma chemotherapy, but interferon-α, pentostatin, or cladribine can induce long-term remissions. *Swerdlow SH, et al, eds., WHO Classification of tumours of Haematolopoietic and lymphoid tissues, 2017, pp. 226–228, McPherson RA et al., eds. Henry's Clinical Diagnosis and Management by Laboratory Methods, 23rd ed., 2011, 2011, pp. 665–671.*

13. b. T-cell receptor (TCR) gene rearrangement assessment by Southern blot or PCR remains the gold-standard test for establishing a diagnosis of T-cell neoplasm, and the test turn-around time is often 7–10 days. However, flow cytometric analysis of a broad array of antibodies directed against the variable (V) region of the TCR-beta chain (TCR V-β) may allow for a rapid confirmation of a T-cell malignancy. An aberrant T-cell immunophenotype (e.g., absence or down-regulation of pan-T-cell antigens, particularly CD7) does not necessarily indicate a clonal T-cell neoplasm and may be observed in infectious mononucleosis, reactive dermatoses, and inflammatory disorders. Although demonstration of a CD10+ T-cell population may aid in the diagnosis of angioimmunoblastic T-cell lymphoma (AITL), CD10+ T-cells do exist in thymoma and a small peripheral CD10+ T-cell population may be detected in patients with reactive lymphoid proliferations as well as B-cell lymphomas. *Swerdlow SH, et al, eds., WHO Classification of tumours of Haematolopoietic and lymphoid tissues, 2017, pp. 345–421, McPherson RA et al., eds. Henry's Clinical Diagnosis and Management by Laboratory Methods, 23rd ed., 2011, pp. 665–671.*

14. c. The blast phase of chronic myeloid leukemia (CML-BP) resembles acute leukemia. In about 70% of cases, the blast lineage is myeloid. In approximately 30% of patients with CML-BP, the blast phase is due to proliferation of lymphoblasts. Rarely, patients have separate populations of myeloid and lymphoid lineage blasts simultaneously. The blasts in CML-BP are often primitive or heterogeneous, and therefore FCI is needed to define the lineage of blasts. FCI is typically not included in the diagnostic workup of low-grade myelodysplastic syndrome (e.g., refractory anemia), Hodgkin lymphoma due to the very low number of neoplastic cells, and chronic myeloproliferative disorders (e.g., polycythemia vera, essential thrombocythemia, and chronic idiopathic myelofibrosis). *Swerdlow SH, et al, eds., WHO Classification of tumours of Haematolopoietic and lymphoid tissues, 2017, pp. 30–36.*

15. c. The characteristic immunophenotype of Sézary cells is CD3+, CD4+, CD8−, CD7±, CD26−, and CD25−. The demonstration of an elevated CD4/CD8 ratio, of increased proportions of CD4+/CD7− or CD4+/CD26− T-cells, and of clonal rearrangement of T-cell receptor genes are useful diagnostic criteria for Sézary syndrome. Adult T-cell leukemia/lymphoma (ATLL) is also CD3+, CD4+, and CD8−; however, CD25 is characteristically expressed in ATLL. Approximately 25% of T-cell prolymphocytic leukemia (T-PLL) cases are CD3+, CD4+, and CD8+, a feature unique to T-PLL. Cytotoxic T-cell lymphoma (e.g., subcutaneous panniculitis-like T-cell lymphoma, SPTCL) are CD3+, CD4−, and CD8+. *Swerdlow SH, et al, eds., WHO Classification of tumours of Haematolopoietic and lymphoid tissues, 2017, pp. 345–422.*

16. d. The lymphoblasts in precursor B-ALL are often positive for TdT, HLA-DR, CD10 (as known as common ALL antigen, CALLA), CD34, and B-cell marker CD19. In the most differentiated precursor-B ALL, so-called pre-B ALL, the blasts express cytoplasmic mu chains. Surface immunoglobulin is characteristically absent in precursor B-ALL. *Swerdlow SH, et al, eds., WHO Classification of tumours of Haematolopoietic and lymphoid tissues, 2017, pp. 199–214, McPherson RA et al., eds. Henry's Clinical Diagnosis and Management by Laboratory Methods, 23rd ed., 2011, pp. 665–671.*

17. d. In contrast to normal plasma cells which express CD19 and lack expression of CD56, neoplastic plasma cells lack CD19 and usually express adhesion molecule CD56. Myeloma cells are typically CD45 −/weak, CD79a+, CD38+, and CD138+. It is worthy to note that since myeloma cells lack surface immunoglobulin (Ig), to establish clonality (light chain restriction), membrane permeabilization is required to give antibodies access to intracellular Ig. *Swerdlow SH, et al, eds., WHO Classification of tumours of Haematolopoietic and lymphoid tissues, 2017, pp. 243–248; McPherson RA et al., eds. Henry's Clinical Diagnosis and Management by Laboratory Methods, 23rd ed., 2011, 2011, p. 671.*

18. a. NK cells share some markers with cytotoxic T-cells. They both can express CD2, CD7, CD8, CD56, and CD57. NK cells are often positive for the epsilon chain of CD3. However, unlike T-cells, NK-cells lack surface CD3 and are typically positive for CD16. CD16 and CD56 are the surface antigens most commonly used to identify NK cells. T-cell receptor genes in neoplastic NK cells are in germline configuration; clonality therefore has to be established by other methods such as cytogenetic analysis and pattern of X chromosome inactivation in female patients. *Swerdlow SH, et al, eds., WHO Classification of tumours of Haematolopoietic and lymphoid tissues, 2017, pp. 348–371, McPherson RA et al., eds. Henry's Clinical Diagnosis and Management by Laboratory Methods, 23rd ed., 2011, p. 669.*

19. b. Thymocytes from either hyperplastic thymus or thymoma are always CD4/CD8 positive, and therefore the presence of CD4/CD8 coexpression is not necessarily indicative of mediastinal T-lymphoblastic lymphoma (T-LBL). Likewise, thymocytes are often positive for CD10 and CD1a. However, thymocytes in thymoma or benign thymic hyperplasia display characteristic variable (smear-like) expression of surface CD3 antigen. This smeared pattern is never observed in surface CD3-positive T-LBL. *Swerdlow SH, et al, eds., WHO Classification of tumours of Haematolopoietic and lymphoid tissues, 2017, pp. 209–12, McPherson RA et al., eds. Henry's Clinical Diagnosis and Management by Laboratory Methods, 23rd ed., 2011, p. 669.*

20. c. Hematogones are lymphoid precursors, which are often increased in the regenerative marrow after chemotherapy. Hematogones may be difficult to distinguish immunophenotypically from residual leukemic B lymphoblasts. Both the hematogones and common B-ALL lymphoblasts are positive for TdT, CD10, CD34, and CD19, and lack cytoplasmic and surface immunoglobulin. Nevertheless, unlike leukemic lymphoblasts, the immunophenotype of hematogones is characteristic in that they lack aberrant antigen expression and that they show variable (smear-like) expression of markers associated with B-cell differentiation such as CD20. *Swerdlow SH, et al, eds., WHO Classification of tumours of Haematolopoietic and lymphoid tissues, 2017, pp. 199–208, McPherson RA et al., eds. Henry's Clinical Diagnosis and Management by Laboratory Methods, 23rd ed., 2011, p. 661.*

21. d. Acute myeloblastic leukemia, minimally differentiated (formerly known as FAB M0) is an acute leukemia with no evidence of myeloid differentiation by morphology and light microscopy cytochemistry; blasts are negative for myeloperoxidase. Immunophenotyping studies are essential to distinguish this disease from acute lymphoblastic leukemia. Blast cells express one and usually more pan-myeloid antigens including CD13, CD33, and CD117. *Swerdlow SH, et al, eds., WHO Classification of tumours of Haematolopoietic and lymphoid tissues, 2017, pp. 156–157.*

22. b. CD3 is a lineage-specific T-cell marker, and both helper and suppressor T-cells express CD3. The CD3 complex contains gamma, delta, and epsilon chains, which are associated with the T-cell receptor (TCR). CD3 itself does not bind antigen, but is involved in the transduction of signals into the T-cell after the TCR has bound antigen. NK cells do not have a CD3 complex, but usually express the epsilon chain in the cytoplasm. HIV selectively attacks CD4 T-cells, and therefore CD4/CD8 ratio and absolute CD4 count are decreased in HIV-infected individuals. *McPherson RA et al., eds. Henry's Clinical Diagnosis and Management by Laboratory Methods, 23rd ed., 2011, pp. 660–671.*

Chapter 30

Clinical chemistry of neurological and psychiatric diseases

Gyorgy Csako[a],* and Alan H.B. Wu[b]
[a]National Institute of Health, Washington, DC, United States, [b]University of California, San Francisco, CA, United States

1. Classical or principal neurotransmitters include all of the following except
   a. acetylcholine
   b. norepinephrine (NE)
   c. prostaglandin E_2 (PGE_2)
   d. gamma[γ]-aminobutyric acid (GABA)
   e. serotonin (5-hydroxytryptamine, 5-HT)
2. The number of unique proteins identified in the cerebrospinal fluid (CSF) proteome with a combination of recently available techniques is
   a. 10–50
   b. 51–100
   c. 101–1000
   d. 1001–2000
   e. >2000
3. According to protein electrophoresis coupled with immunostaining, beta [β]$_2$-transferrin, the asialo form of transferrin is present in all of the following body fluids except
   a. ear secretion
   b. saliva
   c. nose secretion
   d. blood
   e. tear

---

*Retired.

Self-assessment Q&A in Clinical Laboratory Science, III. https://doi.org/10.1016/B978-0-12-822093-1.00030-2
© 2021 Elsevier Inc. All rights reserved.

4. When blood contaminates a CSF specimen due to "spontaneous" bleeding within the central nervous system (e.g., subarachnoid hemorrhage) or a traumatic lumbar puncture, the CSF total protein (TP) still might be estimated by correcting for the amount of blood present using the following formula
   a. 1 mg/dL protein for every 100 RBC/μL
   b. 5 mg/dL protein for every 1000 RBC/μL
   c. 10 mg/dL protein for every 100 RBC/μL
   d. 1 mg/dL protein for every 1000 RBC/μL
   e. 50 mg/dL protein for every 1000 RBC/μL
5. Workup for patients who have ataxia without family history may include all of the following tests except
   a. C-reactive protein (CRP)
   b. α-fetoprotein
   c. vitamin B_{12}
   d. thyrotropin (thyroid stimulating hormone, TSH)
   e. urine organic acids
6. Blood-borne protein marker(s) found to be useful for the diagnosis of acute (ischemic or hemorrhagic) stroke within 3–6 h after onset include all of the following except
   a. S-100B (or S-100)
   b. von Willebrand factor (vWF)
   c. interleukin-2 (IL-2)
   d. matrix metalloproteinase-9 (MMP-9)
   e. B-type neurotrophic growth factor (BNGF)
7. Concussion (mild traumatic brain injury) can be reliably diagnosed in adults (>18 y/o) from blood sample by measuring a combination of
   a. S-100B (or S-100) and interleukin-2 (IL-2)
   b. creatine kinase (CK) and lactate dehydrogenase (LD)
   c. C-reactive protein (CRP) and amyloid A
   d. ubiquitin C-terminal hydrolase-L1 (UCH-L1) and glial fibrillary acidic protein (GFAP)
   e. brain natriuretic peptide (BNP) and S-100B (or S-100)
8. CSF lactate often is elevated in all of the following cases except
   a. aseptic (viral) meningitis
   b. cerebrovascular accidents (CVA)
   c. fungal or bacterial CNS infections
   d. CNS tumors
   e. head trauma
9. Monoclonal or oligoclonal bands are often or characteristically present in CSF specimens in association with all of the following cases except
   a. cerebral lymphoma
   b. xanthochromia
   c. multiple sclerosis (MS)

d. elevated IgG index

e. viral and postviral encephalitis (e.g., human immunodeficiency virus [HIV] encephalitis, post-polio syndrome)

10. The interpretation of CSF beta[β]-trace protein (prostaglandin D synthase) results for CSF leakage may be complicated by the presence of all of the following conditions except
    a. bacterial meningitis
    b. uncomplicated third trimester singleton pregnancy
    c. spinal canal stenosis
    d. uncomplicated first or second trimester singleton pregnancy
    e. hemodialysis

11. The diagnosis of multiple sclerosis (MS) is supported by finding
    1. myelin basic protein (MBP) in CSF
    2. elevated IgG index in CSF
    3. oligoclonal banding in CSF
    4. paraprotein in serum
       a. 1, 2, and 3
       b. 1 and 3
       c. 2 and 4
       d. 4 only
       e. all of the above

12. Brain disorder(s) that cannot be reliably diagnosed with currently available clinical laboratory test(s) include
    1. schizophrenia
    2. bipolar disorder
    3. epilepsy
    4. major depression disorder (MDD)
       a. 1, 2, and 3
       b. 1 and 3
       c. 2 and 4
       d. 4 only
       e. all of the above

13. Methods found to be particularly useful for improved identification and characterization of CSF proteins include
    1. high-sensitivity mass spectrometry (MS)
    2. gas chromatography
    3. two-dimensional liquid and gel electrophoresis
    4. atomic absorption
       a. 1, 2, and 3
       b. 1 and 3
       c. 2 and 4
       d. 4 only
       e. all of the above

14. In IgM paraproteinemic demyelinating neuropathies (IgM PDNs), the abnormal serum IgM may have reactivity with such antigens as
    1. myelin-associated glycoprotein (MAG)
    2. trisulfated heparin disaccharide
    3. gangliosides (e.g., GD1a, GD1b , GM1, GM2, etc.)
    4. sulfatide
        a. 1, 2, and 3
        b. 1 and 3
        c. 2 and 4
        d. 4 only
        e. all of the above

15. CSF tests found to be useful for differentiation of Alzheimer disease (AD) from other major forms of dementias include
    1. (hyper)phosphorylated tau_{181} protein (p-tau_{181})
    2. amyloid beta $protein_{1-42}$ ($A\beta_{1-42}$)
    3. total tau protein (t-tau)
    4. total protein
        a. 1, 2, and 3
        b. 1 and 3
        c. 2 and 4
        d. 4 only
        e. all of the above

16. Multiple system atrophy and pure autonomic failure, two autonomic failure syndromes of dysautonomias, can be best diagnosed/differentiated by measuring
    1. plasma serotonin
    2. plasma norepinephrine or norepinephrine metabolites
    3. urinary 5-hydroxyindoleacetic acid (5-HIAA)
    4. urinary norepinephrine or norepinephrine metabolites
        a. 1, 2, and 3
        b. 1 and 3
        c. 2 and 4
        d. 4 only
        e. all of the above

17. Markedly increased total CSF protein may occur in case(s) of
    1. encephalomyelitis
    2. xanthochromia
    3. meningitis
    4. Alzheimer disease (AD)
        a. 1, 2, and 3
        b. 1 and 3
        c. 2 and 4
        d. 4 only
        e. all of the above

18. Laboratory test(s) useful for (differential) diagnosis of peripheral (poly) neuropathies include
   1. detection of serum paraproteins by protein electrophoresis and immunofixation electrophoresis (IFE)
   2. measurement of fasting blood glucose level
   3. CSF examination
   4. measurement of urinary heavy metal levels
      a. 1, 2, and 3
      b. 1 and 3
      c. 2 and 4
      d. 4 only
      e. all of the above

19. CSF glucose usually is elevated when the patient has
   1. acute pyogenic meningitis
   2. CNS syphilis or epidemic encephalitis
   3. primary or metastatic tumors of the meninges
   4. high blood glucose
      a. 1, 2, and 3
      b. 1 and 3
      c. 2 and 4
      d. 4 only
      e. all of the above

20. "Single cause" inherited neurodegenerative diseases include
   1. Alzheimer disease (AD)
   2. amyotrophic lateral sclerosis (ALS)
   3. Parkinson disease (PD)
   4. Huntington disease (HD)
      a. 1, 2, and 3
      b. 1 and 3
      c. 2 and 4
      d. 4 only
      e. all of the above

Answers

1. c. In the human brain, about 10^{11} neurons have about 1000 synapses for communication with other neurons, with the Purkinje cells of the cerebellum receiving 100,000 contacts from input cells. Overall, the human brain may contain between 10^{14} and 10^{15} synaptic connections. The diverse chemical substances that carry information between neurons are neurotransmitters. Since the discovery of acetylcholine, >100 substances and many more receptors are implicated in synaptic transmission. Classical or principal neurotransmitters in the brain include about 10 small molecules, which in addition to acetylcholine, include norepinephrine, γ-aminobutyric acid, and serotonin (5-hydroxytryptamine). Neuropeptides refer to peptides found in neural tissue. There are >50 neuropeptides in

the mammalian brain (e.g., neuropeptide Y, somatostatin, and substance P). Nitrous oxide and carbon monoxide gas also act as neurotransmitters. Prostaglandin E_2 is a member of the prostaglandin group derived from unsaturated 20-carbon fatty acids, primarily arachidonic acid, via the cyclooxygenase pathway. All prostaglandins, including PGE_2, exert their biologic effects by binding to specific cell surface receptors causing an increase in the level of the intracellular second messenger cyclic AMP and GMP. Fever is a result of the action of PGE_2 on the brain and requires prostaglandin receptors. However, despite its documented effect on brain cells, PGE_2 is not a neurotransmitter. *Curr Biol 2005;15:R154–8; Rifai N, et al., eds. Tietz Textbook of Clinical Chemistry and Molecular Diagnostics, 6th ed., 2017, pp. 832–924.e14. Kaplan LA, et al. eds. Clinical Chemistry. Theory, Analysis, Correlation, 4th ed., 2003, pp. 787–808.*

2. e. Using multidimensional chromatography techniques, tandem mass spectrometry, and stringent proteomics criteria, as many as 2594 proteins were identified in well-characterized pooled human CSF samples in 2007. A large portion of the CSF proteome was directly related to central nervous system function or structure and only 405 proteins were identified in human plasma as well. Since the pooled CSF samples used in this study were obtained from a wide range of ages (21–85 y), the results were thought to be representative of the general CSF protein profile, i.e., the CSF proteome. An even larger CSF proteome was published from a pooled sample. Because high-abundance proteins can interfere with the detection of low-abundance proteins, the CSF sample first was depleted by immunoaffinity, then fractionated with off-gel electrophoresis and analyzed with LC-tandem MS. This approach allowed the identification of 20,689 peptides mapping on 3379 proteins. Among the CSF proteins identified, 34% corresponded to genes whose transcripts are highly expressed in brain according to the Human Protein Atlas. In another study, 12,344 peptides were mapped on 2281 proteins. In the context of the Chromosome-centric Human Proteome Project (C-HPP), seven "missing" proteins were found for validation. *Pan S. A combined dataset of human cerebrospinal fluid proteins identified by multidimensional chromatography and tandem mass spectrometry. Proteomics 2007;7:469–73. Larssen E. A rapid method for preparation of the cerebrospinal fluid proteome. Proteomics 2015;15:10–5. Macron C, et al. Deep dive on the proteome of human cerebrospinal fluid: A valuable data resource for biomarker discovery and missing protein identification. J Proteome Res 2018;17:4113–26. Macron C, et al. Identification of missing proteins in normal human cerebrospinal fluid. J Proteome Res 2018;17:4315–9.*

3. c. Detection of β_2-transferrin by electrophoretic methods is used for the differential diagnosis of CSF otorrhea and rhinorrhea. Similar to CSF, tear and ear secretions and plasma contain both β_1- and β_2-transferrin isoforms, but, in contrast to CSF, the concentration of the β_2-isoform relative to the β_1-isoform normally is much lower in these fluids. Of note is that severe

alcoholism can increase the relative β_2-transferrin concentration, changing the plasma transferrin isoform pattern closer to CSF-like. Saliva also contains both β_1- and β_2-transferrin isoforms but with slightly different mobilities. Only the nose secretion was found to be devoid of the β_2-isoform of transferrin. *Görögh T, et al. Separation of β_2-transferrin by denaturing gel electrophoresis to detect cerebrospinal fluid in ear and nasal fluids.Clin Chem 2005;51:1704–10.*

4. d. Assuming somewhat "extreme" combinations of RBC count and serum total protein (TP), the following relationships can be calculated: (1) If RBC count is abnormally high in the blood (e.g., $10 \times 10^6/\mu L$) and serum TP is low (e.g., 5 g/dL), then the presence of 1000 RBC/μL in the CSF would be associated with 0.5 mg TP/dL from the blood contamination. (2) If RBC count is relatively low (e.g., $5 \times 10^6/\mu L$) and serum TP is abnormally high (e.g., 10 g/dL), then the presence of 1000 RBC/μL in the CSF would be associated with 2.0 mg TP/dL from the blood contamination. (3) Combinations of the above low and low or high and high RBC and TP levels would result in exactly 1.0 mg/dL extra protein for 1000 RBC/μL in a blood-contaminated CSF. Because of the possibility of hemolysis, RBC counts need to be determined promptly after obtaining a CSF specimen. *Kaplan LA, et al. eds. Clinical Chemistry. Theory, Analysis, Correlation, 4th ed., 2003, pp. 787–808. Wu AHB, ed. Tietz Clinical Guide to Laboratory Tests, 4th ed., 2006, pp. 360–3.*

5. a. Ataxia, defined as incoordination or clumsiness of movement in the absence of muscle weakness, has many causes. With advent of molecular genetic testing, an increasing number of genetic abnormalities (>25) have been linked to ataxias. The most common acquired cerebellar ataxias are caused by perinatal trauma and hypercoagulable or hemorrhagic vascular events (ataxic cerebral palsy). Open or closed head injury, cerebellar stroke, infection of the CNS with a wide variety of agents (e.g., viral, bacterial and fungal pathogens, spirochetes, prions, and parasites such as Toxoplasma), primary or metastatic neoplasms, autoantibodies (e.g., in gluten enteropathy or celiac sprue), endocrine disorders (e.g., hypo- and hyperthyroidism), and toxic agents (e.g., alcohol, heavy metals, drugs such as cocaine, heroin, phencyclidine) all can produce ataxias. Lab workup in nonhereditary cases can be extensive and may be classified as first-, second-, and third-line blood studies, CSF studies, and paraneoplastic studies. Since the absence of family history may be in error, genetic workup may be justified. Despite these tests, the etiology of late-onset, sporadic cerebellar ataxia remains unknown in $>50\%$ of adults. While elevated CRP can occur in patients with acquired ataxia (e.g., due to infectious agents), it is neither useful for diagnosis. *Perlman SL. Ataxias. Clin Geriatr Med 2006;22:859–77.*

6. c. Since brain imaging during the hyperacute stage of stroke may not always be available and it is often normal in case of an ischemic stroke,

early blood biomarkers of cerebral tissue injury are desirable for diagnosis and timely therapy. While many biomarkers have been identified, none are sufficient in sensitivity and specificity for use in routine practice. In >50 plasma protein biomarkers, including both brain-specific markers (e.g., S-100B, BNGF), and nonspecific markers (e.g., inflammatory and hemostatic factors, acute-phase reactants, etc.) have been studied within 3–6 h after onset of acute cerebral ischemia. Current guidelines restrict thrombolytic therapy (rtPA) to within 3 h, and possibly 6 h after onset of an ischemic stroke. In one study, the best markers were S-100B, BNGF, vWF, MMP-9, and monocyte chemotactic protein-1. In other studies, vascular cell adhesion molecule, PARK7 (also called DJ-1) and nucleoside diphosphate kinase A has diagnostic utility. Further, measurement of S-100b may predict a malignant course of infarction, serve as a surrogate marker for successful clot lysis in hyperacute middle cerebral artery occlusion, and to indicate a higher risk of hemorrhagic transformation after thrombolytic therapy in acute stroke. Multimarker analysis can improve diagnostic specificity. IL-2, a product of activated T-cells, is not valuable in acute stroke diagnosis. *Makris K, et al. Blood biomarkers in ischemic stroke: potential role and challenges in clinical practice and research. Crit Rev Clin Lab Sci 2018;55:294–328. Reynolds MA, et al. Early biomarkers of stroke. Clin Chem 2003;49:1733–9. Allard L, et al. PARK7 and nucleoside diphosphate kinase A as plasma markers for the early diagnosis of stroke. Clin Chem 2005;51:2043–51.*

7. d. Traumatic brain injury is an insult to the brain caused by external physical force that disrupts normal brain function. This often results in severe cognitive and neurological impairment or physical functioning. The severity of a TBI may range from mild, accounting for 94.5% of cases to severe. The CDC estimated 2.5 million ED visits and 282,000 hospitalizations are due to TBI. The conventional care includes a head CT scan to rule out bleeding inside the skull but, unfortunately, >90% of patients with mild TBI have a negative scan. CT scan is costly, time consuming, often inconclusive, and exposes the patient to radiation. In 2018 the FDA approved the first blood test to assist in the detection of certain TBIs in adults. Banyan's Brain Trauma Indicator (BTI) measures two brain-specific proteins, ubiquitin C-terminal hydrolase-L1, reflecting neuronal cell body injury and glial fibrillary acidic protein reflecting astroglial injury, that appear in the blood within 12 h of a brain injury when bleeding has occurred. In the clinical trial of >1900 adults with suspected concussion, BTI and CT scan were compared. BTI predicted evidence of brain tissue damage on a CT scan 97.5% of the time and predicted no evidence of damage on a CT scan 99.6% of the time. *Korley F et al. Emergency department evaluation of traumatic brain injury in the United States, 2009–2010. J Head Trauma Rehabil 2016;31:379–87. Taylor CA et al. Traumatic Brain Injury-Related Emergency Department Visits, Hospitalizations, and*

Deaths—United States, 2007 and 2013. MMWR Surveill Summ 2017;66 (9):1–16; Keobeissy FH, ed. Brain neurotrauma: molecular, neurophysical, and rehabilitation aspects. 2015. J Neurotrauma 2019;36:2083–91.

8. a. Although a variety of diseases can increase lactate levels in the CSF (reference: 10–22 mg/dL or 1.1–2.4 mmol/L), measurement of CSF lactate may be most helpful in distinguishing bacterial or fungal CNS infections from "aseptic" (viral) meningitis. In bacterial and fungal meningitides, CSF lactate generally is elevated above 35 mg/dL (3.9 mmol/L), whereas it is almost invariably <35 mg/dL in viral meningitis. High CSF lactate levels represent poor prognosis after head injury. *Rifai N, et al., eds. Tietz Textbook of Clinical Chemistry and Molecular Diagnostics, 6th ed., 2017, pp. 631–3. Wu AHB, ed. Tietz Clinical Guide to Laboratory Tests, 4th ed., 2006, pp. 652–3.*

9. b. Increased intrathecal (local) production of immunoglobulins due to cerebral lymphoma, demyelinating diseases (e.g., MS), infection or other immune-mediated processes of the central nervous system (e.g., viral/post-viral encephalitis, neurosyphilis, Lyme disease) often is evaluated by an IgG index and is commonly associated with the appearance of monoclonal or oligoclonal bands, primarily of the IgG-type, in the CSF. Xanthochromia results from bleeding into the CSF and, therefore, the associated increase in CSF immunoglobulin levels reflects a usually polyclonal blood pattern. *Rifai N, et al., eds. Tietz Textbook of Clinical Chemistry and Molecular Diagnostics, 6th ed., 2017, pp. 400–2, Kaplan LA, et al. eds. Clinical Chemistry. Theory, Analysis, Correlation, 4th ed., 2003, pp. 787–808. Wu AHB, ed. Tietz Clinical Guide to Laboratory Tests, 4th ed., 2006, pp. 596–7, 926–7.*

10. d. Although both β2-transferrin and β-trace protein (prostaglandin D synthase) are useful immunological markers for the detection of CSF leaks, the electrophoretic methods for transferrin isoform determination are slow, cumbersome, and labor intensive. In contrast, rapid automated methods are available for quantitation of β-trace protein. β-trace protein is one of the most abundant proteins in the CSF. It is a small protein (23–29 kDa), produced in the epithelial cells of the plexus choroideus and leptomeninges. The ratio of CSF to serum β-trace protein is the highest (~35) among all CSF proteins, making it an ideal marker for CSF detection. For optimal performance, concomitant determination of β-trace protein in both CSF and serum is recommended. Combining a 0.68 mg/L cutoff in CSF secretion with a 4.9 CSF to serum ratio cutoff, 98% sensitivity and 100% specificity were obtained in the study of 176 "uncomplicated" patients. Due to alterations either in the CSF or serum β-trace protein concentrations, bacterial meningitis, hemodialysis, renal disease in general, spinal canal stenosis, and third trimester pregnancy may lead to falsely low or high results. Uncomplicated first or second trimester singleton pregnancies are not expected to interfere with the interpretation.

Reiber H, et al. Beta-trace protein as sensitive marker for CSF rhinorhea and CSF otorhea. Acta Neurol Scand 2003;108:359–62. Risch L, et al. Rapid, accurate and non-invasive detection of cerebrospinal fluid leakage using combined determination of β-trace protein in secretion and serum. Clin Chim Acta 2005;351:169–76. Kristensen K, et al. Temporal changes of the plasma levels of cystatin C, β-trace protein, $β_2$-microglobulin, urate and creatinine during pregnancy indicate continuous alterations in the renal filtration process. Scand J Clin Lab Invest 2007;67:612–8.

11. a. CSF analysis is useful if clinical or MRI criteria are incomplete for the diagnosis of MS or if an alternative diagnosis such as an infectious process is considered. Psychiatric diseases can be mistaken for MS. In addition to oligoclonal bands with a paired serum sample, IgG index, and MBP, suggested CSF examination includes cell counts, total protein, and glucose. Although the presence of MBP, oligoclonal bands, and elevated IgG index all support the diagnosis of MS, they neither alone nor in combination are specific enough for a definitive diagnosis. MBP can be elevated secondary to any disruption of CNS tissue. An elevated IgG index indicates intrathecal production of immunoglobulins and oligoclonal bands are common in MS but both can occur with infections or other immune-mediated processes. According to the McDonald criteria, in patients with a typical clinically isolated syndrome and MRI showing dissemination in space, the presence of CSF-specific oligoclonal bands (isoelectric focusing preferably coupled with immunofixation, IFE) allows a diagnosis of MS. Serum paraproteins are not associated with MS, and IFE of a paired serum specimen is needed to prove that the serum is either free of paraprotein(s) or that the serum bands are different from those in the CSF. In >80% of cases of MS, as another supporting test, the CSF IgG index >0.70. The rate of intrathecal IgG synthesis in mg/day is estimated by Tourtellotte's formula. Values >8 mg/day are found in most cases of MS, but this calculation provides no more clinical information than the IgG index. *Rifai N, et al., eds. Tietz Textbook of Clinical Chemistry and Molecular Diagnostics, 6th ed., 2017, pp. 373–403.e5. Thompson AJ, et al. Diagnosis of multiple sclerosis: 2017 revisions of the McDonald criteria. Lancet Neurol 2018;17:162–73. Solomon AJ, et al. Misdiagnosis of multiple sclerosis: Impact of the 2017 McDonald criteria on clinical practice. Neurol 2019;92:26–33.*

12. e. Although a variety of blood and CSF markers have been tested and many abnormalities have been discovered (e.g., low CSF levels of norepinephrine and 5-HIAA in depressed patients), so far no clinical laboratory test alone or in combination with others has been found reliable for a definitive diagnosis of these major psychiatric disorders. *Kaplan LA, et al. eds. Clinical Chemistry. Theory, Analysis, Correlation, 4th ed., 2003, pp. 787–808.*

13. b. For state-of-the-art CSF protein mapping, proteins first are separated by two-dimensional liquid or gel electrophoresis then are identified by high-sensitivity mass spectrometric techniques. Based on the mode of ionization and detection, commonly used approaches for CSF proteomics include matrix-assisted laser desorption/ionization time-of-flight mass spectrometry, SELDI-TOF-MS, LC/MS/MS, LC-electrospray ionization-quadrupole-ion trap-MS, and LC-ESI-Q-TOF-MS. Differentially expressed CSF proteins identified by proteomic techniques are promising for improved and/or reliable diagnosis of various CNS disorders such as Alzheimer disease and multiple sclerosis. Gas chromatography and atomic absorption obviously are not proteomic techniques. *De Jong D, et al. Current state and future directions of neurochemical biomarkers for Alzheimer's disease. Clin Chem Lab Med 2007;45:1421–34. Lovestone S, et al. Proteomics of Alzheimer's disease: understanding mechanisms and seeking biomarkers. Expert Rev Proteom 2007;4:227–38. Lehmensiek V, et al. Cerebrospinal fluid proteome profile in multiple sclerosis. Mult Scler 2007;13:840–9.*

14. e. In approximately two-thirds of the patients with IgM PDN, the monoclonal (M)-protein reacts with a variety of neural antigens such as MAG (∼50%), sulfatide (∼6%), chondroitin sulfate C (<2%), several gangliosides (all combined: up to ∼10%), and cytoskeletal proteins. In patients with a distinct clinical presentation of PDN (∼8%), the M-protein reacted with trisulfated heparin disaccharide that is the most abundant disaccharide component of heparin oligosaccharides and is also present in heparan-sulfate glycosaminoglycan *Nobile-Orazio E. IgM paraproteinaemic neuropathies. J Peripher Nerv Syst 2006;11:9–19.*

15. a. Although no CSF neurochemical biomarker has been yet identified for reliable diagnosis of AD, $A\beta_{1-42}$, t-tau, and p-tau$_{181}$ (and the $A\beta_{1-42}$/p-tau$_{181}$ or p-tau$_{181}$/t- tau ratios) in the CSF were all found to be useful for differentiating AD from other major forms of dementias (vascular dementia, dementia with Lewy bodies, frontotemporal lobar degeneration, and Creutzfeldt-Jakob disease). In recent studies, multiplex analysis of $A\beta_{1-42}$, t-tau, and p-tau$_{181}$ with ELISA or Luminex xMAP technology has increased their discriminative power for diagnosing AD vs either vascular dementia or mild cognitive impairment. Other CSF and circulating (blood-borne) biomarkers are also being evaluated for the diagnosis/differentiation of cognitive disorders and dementias. CSF total protein is normal or only mildly elevated in AD and is not a useful test for diagnosing or differentiating various dementias. *De Jong D, et al. Current state and future directions of neurochemical biomarkers for Alzheimer's disease Clin Chem Lab Med 2007;45:1421–34. Sjögren M, et al. Advances in the detection of Alzheimer's disease—use of cerebrospinal fluid biomarkers. Clin Chim Acta 2003;332:1–10. Solfrizzi V, et al. Circulating biomarkers of cognitive decline and dementia. 006;364:91–112.*

16. c. Abnormalities of blood pressure control and dominate the clinical manifestations of dysautonomias and orthostatic hypotension is the abnormality in pure autonomic failure and multiple system atrophy. Since this hypotension occurs due to the failure of neurogenic vasoconstrictor responses secondary to defective sympathoneural release of norepineph-rine, measurement of norepinephrine and norepinephrine derivatives in plasma, urine, or both is diagnostic. Urinary 5-HIAA and, less commonly, plasma serotonin primarily are measured for the diagnosis of carcinoid tumors and, along with CSF 5-HIAA levels, may be used for the diagnosis of some other dysautonomias. *Jameson JL, et al. eds. Harrison's Principles of Internal Medicine, 20th ed., 2018, pp. 1421.o26–o75. Wu AHB, ed. Tietz Clinical Guide to Laboratory Tests, 4th ed., 2006 pp. 584–7, 982–3.*

17. a. Under physiologic conditions, only ~20% of the CSF protein originates from intrathecal synthesis, and the remainder is produced from plasma by ultrafiltration. Breakdown of the blood-central nervous system barrier due to infection such as bacterial meningitis or other inflammatory conditions leads to the "influx syndrome" with increasing protein levels in the CSF. Alternatively, contamination of the CSF with blood due to "spontaneous hemorrhage" or traumatic lumbar puncture produces abnormally high CSF protein levels. Presence of xanthochromia indicates that the bleeding occurred at least 2–4 h earlier. CSF protein concentrations over 130 mg/dL may be seen occasionally in premature infants. Patients with AD have normal or only mildly elevated levels of CSF total protein. *Kaplan LA, et al. eds. Clinical Chemistry. Theory, Analysis, Correlation, 4th ed., 2003, pp. 787–808. Wu AHB, ed. Tietz Clinical Guide to Laboratory Tests, 4th ed., 2006, pp. 918–9.*

18. e. Detection of serum paraproteins supports the diagnosis of a paraproteine-mic demyelinating neuropathy (e.g., IgM PDN). Abnormally high fasting blood glucose suggests diabetes mellitus-associated neuropathy occurring in ~60% of these patients. Examination of the CSF would reveal, for instance, a high protein concentration with normal cell count in case of acute idiopathic polyneuropathy (Guillain–Barre syndrome). Finding of elevated levels of arsenic, lead, or mercury in the urine would be consistent with a toxic polyneuropathy. As guided by history and clinical presentation, these lab tests along with complete blood cell count, kidney, liver and thyroid function tests, serum electrolytes, vitamins B_{12} and B_1, tests for rheumatoid factor, antinuclear antibody, hepatitis B surface antigen, and syphilis, etc., should be ordered selectively. In about half of all peripheral (poly)neurop-athy cases, no specific cause can be established by routine laboratory testing and only less than half of these cases can be identified as hereditary. *Nobile-Orazio E. IgM paraproteinaemic neuropathies. Curr Opin Neurol 2004;17:599–605. Kaplan LA, et al. eds. Clinical Chemistry. Theory, Analysis, Correlation, 4th ed., 2003, pp. 582, 593.*

19. c. Most nonviral infections (e.g., acute pyogenic meningitis, tuberculous meningitis) and primary or metastatic tumors of the meninges reduce, whereas CNS syphilis and high blood glucose levels (e.g., in diabetic hyperglycemia) increase the CSF glucose levels. Normally, CSF glucose levels approximate 60%–80% of the plasma glucose level, but the percentage is decreasing with increasing blood glucose levels. For proper interpretation, CSF glucose levels must be compared to concomitantly measured blood glucose levels. Because it takes $\sim$4 h to reach equilibrium with the blood glucose, the CSF glucose at any given time reflects the blood glucose during the past 4 h. *Kaplan LA, et al. eds. Clinical Chemistry. Theory, Analysis, Correlation, 4th ed., 2003, pp. 791–2. Wu AHB, ed. Tietz Clinical Guide to Laboratory Tests, 4th ed., 2006, pp. 450–1.*

20. d. Neurodegenerative diseases afflict $\sim$2% of the population in the developed world at any time and many of these cases involve genetic abnormalities. Among these diseases, HD is unique because it has a single cause in all patients: inheritance of an expanded CAG trinucleotide repeat that normally consists of 6–34 CAG units. The length of the expanded trinucleotide repeat is the primary determinant of the age when the patients becomes clinically symptomatic, accounting for $\sim$70% of the variance observed for age at onset. The expanded CAG-repeat of the HD gene leads to an elongated polyglutamine tract in the large huntingtin protein ($>$3100 amino acids) and the expanded polyglutamine confers an altered physical property on the protein with neuronal toxicity. In contrast to the homogeneous genetic nature of HD, AD, ALS, PD, and most other well-known inherited dementias, ataxias, and motor-neuron diseases in all four classes of the neurodegenerative diseases ("tautopathies," "synucleinopathies," ubiquitin inclusion diseases, and polyglutamine diseases) have heterogeneous etiologies with varying genetic, cellular, and biochemical abnormalities. *Hardy J, et al. Genetics of Parkinson's disease and parkinsonism. Ann Neurol 2006;60:389–98. Hardy J, et al. The genetics of neurodegenerative diseases. J Neurochem 2006;97:1690–9. Gusella JF, et al. Huntington's disease: seeing the pathogenic process through a genetic lens. Trends Biochem Sci 2006;31:533–40.*

Chapter 31

Newborn screening

Rasoul A. Koupaei
California Department of Public Health, Richmond, CA, United States

1. Which of the following sentence is correct with respect to "clinical biochemical genetics"?
   a. The clinical biochemical genetics is a discipline that covers the biochemical diagnosis of IEM by metabolite and enzymatic analysis of physiological fluids and tissue.
   b. It is concerned with the evaluation and diagnosis of patients and families with inherited metabolic disease.
   c. It is concerned with monitoring of treatment and distinguishing heterozygous carriers from noncarriers.
   d. all of above
2. What are the main pattern(s) of acute clinical presentation in inborn error metabolism (IEM) diseases?
   a. intoxication
   b. energy deficiency
   c. intoxication and energy deficiency
   d. none the above
3. Which of the following sentence is incorrect?
   a. Lactic acidemia is rarely seen in organic aciduria and fatty acid oxidation.
   b. Fasting hypoglycemia mostly is seen in organic aciduria and fatty acid oxidation.
   c. Metabolic acidosis is a typical finding in organic aciduria.
   d. Ketonuria (ketone bodies) is seen in most of organic aciduria.
4. Which the following statement is the most accurate statement with respect to prenatal diagnostic of IEM disease?
   a. These disorders typically follow an autosomal recessive mode of inheritance and the recurrence risk in subsequent pregnancies is 25%.
   b. Prenatal diagnosis of IEM may include CVS sampling and/or amniocentesis.

c. CVS sampling used in prenatal diagnosis of IEM is performed at 10–12 weeks of gestational age, and has a higher risk of fetal loss, and entails the occurrence of artifactual results caused by contamination with maternal tissue.

d. all of the above

5. Which of the following statement is the accurate statement with respect to newborn screening programs?

   a. Newborn screening has become an important and effective component of preventive medicine, allowing detection of treatable disorders before irreversible clinical symptoms manifest themselves.

   b. Originally newborn screening instituted in the 1960s for the early detection of phenylketonuria (PKU).

   c. Most newborn screenings are done using neonatal blood spots collected at early hours of life (24–72 h of life).

   d. Some inborn errors of metabolism are undetectable by MS/MS analysis of amino acids and acylcarnitines; a negative newborn screening outcome should not be regarded as a blanket exclusion of metabolic disorders in patients presenting with unexplained clinical manifestations at any age.

   e. all of the above

6. Which of the following clinical laboratory testing is the first line laboratory testing when a patient with acute metabolic decompensation and possible inborn error of metabolism is brought to medical attention?

   a. general chemistry and general hematology

   b. quantitative profiling of amino acids, carnitine, acylcarnitines, and fatty acids in plasma

   c. quantitative profiling organic acids, and acylglycines in urine

   d. molecular testing

7. Which of the following factors could increase the risk of an underlying metabolic disorder in the case of sudden unexplained death?

   a. family history of sudden infant death syndrome (SIDS) or other sudden, unexplained deaths at any age

   b. family history of Reye's syndrome

   c. maternal complications of pregnancy (acute fatty liver of pregnancy, HELLP syndrome, others)

   d. macroscopic findings at autopsy

   e. all of above

8. The concentrations of individual amino acids in physiological fluids reflect a balance between following factors except:

   a. intestinal uptake

   b. body mass index

   c. anabolic use by the liver

   d. the synthesis and turnover of the body's structural proteins

   e. the integrity of renal functions (filtration and tubular reabsorption)

9. Which of the following statement is incorrect about phenylalanine?
   a. Phenylalanine is an essential amino acid.
   b. Dietary intake in excess of anabolic needs is converted to tyrosine by phenylalanine hydroxylase and further degraded via a ketogenic pathway.
   c. A primary or cofactor-related defect of phenylalanine hydroxylase activity causes accumulation of phenylalanine, phenylketones, and phenylamine.
   d. Dietary intake in excess of anabolic needs is converted to tyrosine by phenylalanine oxidase and further degraded via a ketogenic pathway.
10. Which of the following statement is incorrect about phenylketonuria (PKU)?
   a. Patients with classic PKU are clinically silent at birth and neurological manifestations typically do not become evident until a few months of age.
   b. To treat a PKU cases, infants should be screened within first couple of days of life.
   c. PKU infants may lose about 50 points in their adult IQ if left untreated until the end of the first year of life.
   d. PKU is inherited as autosomal dominant traits.
11. Which of the following observation is correct about the criteria for a biochemical diagnosis of untreated classical Phenylketonuria (PKU)?
   a. A plasma phenylalanine level above 20 mg/dL (1.2 mmol/L).
   b. A phenylalanine/tyrosine ratio >3.
   c. Increased urinary levels of metabolites of phenylalanine (i.e., phenylpyruvic and 2-hydroxyphenylacetic acids.
   d. A normal concentration of the cofactor tetrahydrobiopterin (BH4).
   e. all of the above
12. Which of the following laboratory testing is not the common practice for PKU diagnosis?
   a. in vitro assay of enzyme catalytic activity and genotyping
   b. plasma and urine amino acid analysis
   c. urinary amino acid analysis
   d. urinary organic acid profile
13. Which of the following statement is incorrect with respect to the best PKU treatment practice?
   a. Dietary treatment should be started as soon as possible.
   b. Although a less strict dietary regimen is possible after 6 years of age, some forms of restriction are necessary indefinitely.
   c. Pregnant women with PKU who are not on a low phenylalanine diet and have phenylalanine levels > 360 μmol/L (6 mg/dL) are at higher risks.
   d. Pregnant women with PKU who are not on a low phenylalanine diet and have phenylalanine levels < 360 μmol/L (6 mg/dL) are at higher risks.

14. Which of the following statement is correct about BH4 deficiency?

   a. In about 2% of cases of hyperphenylalaninemia is due to a deficiency of either biosynthesis or recycling of BH4.

   b. BH4 is the phenylalanine hydroxylase enzyme cofactor.

   c. BH4 deficiency cases are clinically indistinguishable from classical PKU when detected by newborn screening at birth.

   d. Measurement of urinary neopterin and biopterin provides a tool for the differential diagnosis of BH4 deficiency cases.

   e. all of the above

15. Which of the following statement is incorrect about hepatorenal tyrosine-mia/tyrosinemia type I deficiency?

   a. Hepatorenal tyrosinemia is an autosomal recessive disease.

   b. Tyrosinemia type I (TYR-I) is an autosomal dominant disease.

   c. Tyrosinemia type I is caused by a deficiency of the enzyme fumaryla-cetoacetate hydrolase (FAH).

   d. The incidence of TYR-I is approximately I in 100,000, with well-known clustering of cases in Scandinavia.

   e. Tyrosinemia type I symptoms include acute liver failure, cirrhosis, hepatocellular carcinoma, renal Fanconi syndrome, glomerulosclero-sis, and crises of peripheral neuropathy.

16. Which of the following statement is incorrect about hepatorenal tyrosine-mia/tyrosinemia type I deficiency?

   a. An elevated concentration of tyrosine in plasma is expected with ade-quate specificity in TYR-I cases.

   b. Biochemical diagnosis of Tyr-1 is based on the detection of succinyla-cetone (4,6-dioxaneheptanoic acid) either in urine or plasma.

   c. Pharmacological doses of NTBC(2-(2-nitro-4-trifluoro-methylben-zoyl)-1,3-cyclo-Hexanedione) virtually prevent the synthesis of succinylacetone.

   d. NTBC have become the treatment of choice for TYR-I in combination with dietary restriction and liver transplantation.

17. Which of the following statement is incorrect about homocystinuria?

   a. Homocystinuria is caused by a defect of cystathionine β-synthase enzyme.

   b. Defect of cystathionine β-synthase enzyme is an autosomal recessive disorder.

   c. The biochemical phenotype of homocystinuria is mostly characterized by decreased plasma concentrations of methionine.

   d. Patients with cblC, cblD, and cblF defects have methylmalonic aciduria in addition to homocystinuria.

18. Which of the following statement is correct about maple syrup urine dis-ease (MSUD)?

   a. Leucine, isoleucine, and valine are the three essential branched-chain amino acids involved in MSUD.

b. A defect on any component of the branched-chain α-keto acid dehydro-genase complex causes maple syrup urine disease (MSUD).

c. Maple syrup urine disease (MSUD) is autosomal recessive.

d. At the bedside, mixing urine with a solution of 2,4-dinitrophenylhy-drazine(DNPH) is a simple and rapid test to confirm a clinical suspicion.

e. all of the above

19. Which of the following statement is incorrect about maple syrup urine dis-ease (MSUD)?

a. Cornerstones of treatment are dietary restriction of branched-chain amino acids and high dose thiamine.

b. Leucine, isoleucine, and valine are the three essential branched-chain amino acids involved in MSUD.

c. Main laboratory presentations of MSUD include marked elevation of leucine, isoleucine, and valine and decreased level of plasma L-alloisoleucine.

d. Marked elevation of leucine, isoleucine, and valine plus the patho-gnomonic presence of L-alloisoleucine is seen in most MSUD cases.

20. Which of the following statement is incorrect about urea cycle defects?

a. The dominant laboratory findings in urea cycle defects are hyperammo-nemia and respiratory alkalosis.

b. Plasma amino acids and urine orotic acid levels are necessary for a dif-ferential diagnosis of urea cycle defects.

c. Lower glutamine and alanine concentrations are common for these disorders.

d. Low or undetectable citrulline profile is consistent with either OTC deficiency or CPS deficiency.

21. Which of the following statement is incorrect about organic acidemias?

a. The biochemical diagnosis of individual organic acidemias relies on urine organic acid analysis by gas chromatography-mass spectrometry (GC-MS).

b. The incidence of individual inborn errors of organic acid metabolism varies from 1: 10,000 to >1: 1,000,000 live births.

c. Organic acids are water-soluble compounds containing one or more carboxyl groups.

d. The biochemical diagnosis of individual organic acidemias relies on urine organic acid analysis by ion exchange chromatography.

22. Which of the following statement is incorrect about propionic acidemia (PA)?

a. Patients with PA typically are born at term with no pregnancy complications.

b. Severe metabolic acidosis, lactic acidemia, hypoglycemia, and hyper-ammonemia are often present.

c. Plasma amino acids analysis presents with hyperglycinemia.

    d. The urine organic acid profile from a sample from an acutely affected individual could reveal lactic aciduria, ketonuria, and a very low level of propionate metabolites.

23. Which of the following statement is incorrect about isovaleric acidemia?

    a. Isovaleric acidemia (IVA) is a disorder of isoleucine catabolism.

    b. Isovaleric acidemia (IVA) is a disorder of leucine catabolism.

    c. Isovaleric acidemia (IVA) is caused by a deficiency of isovaleryl-CoA dehydrogenase.

    d. Isovaleric acidemia (IVA) presents with the characteristic odor of "sweaty feet."

    e. In practice, the diagnosis is confirmed by excretion of organic acids isovalerylglycine accumulating as the major by-product followed by 3-hydroxyisovaleriac acid.

24. Which of the following statement is correct about glutaric acidemia type I (GA-I)?

    a. Glutaric acidemia type I (GA-I) results from an inherited defect in glutaryl-CoA dehydrogenase.

    b. Glutaryl-CoA dehydrogenase is an enzyme involved in the degradation of lysine and tryptophan.

    c. Characteristic clinical findings of GA-I are macrocephaly and frontal bossing.

    d. The classic biochemical phenotype consists of massive glutaric aciduria with increased concentrations of glutarylcarnitine in plasma.

    e. all of the above

25. Which of the following statement is incorrect about fatty acid oxidation (FAO) disorder?

    a. Medium- and short-chain fatty acids enter the mitochondria independently of the carnitine cycle.

    b. Fatty acids are oxidized to acetyl-CoA as their final product.

    c. Typical manifestations of FAO disorders are hypoketotic hypoglycemia.

    d. Typical manifestations of FAO disorders are hyperketotic hypoglycemia.

26. Which of the following statement is incorrect about very-long-chain acyl-CoA dehydrogenase (VLCAD) deficiency?

    a. Very-long-chain acyl-CoA dehydrogenase (VLCAD) deficiency is a fatty acid oxidation (FAO) disorder.

    b. Very-long-chain acyl-CoA dehydrogenase (VLCAD) deficiency is an aminoacidopathy.

    c. Very-long-chain acyl-CoA dehydrogenase (VLCAD) deficiency is an autosomal recessive condition.

    d. Laboratory indicators during acute episodes include hypoketotic hypoglycemia and creatine kinase elevations.

27. Which of the following statement is incorrect about medium-chain acyl-CoA dehydrogenase (MCAD) deficiency?

a. MCAD deficiency is responsible for the initial dehydrogenation of acyl-CoAs with a chain length between 4 and 12 carbon atoms.

b. MCAD deficiency present with hypoketotic hypoglycemia, vomiting, and lethargy.

c. MCAD deficiency is a potentially lethal disease and early diagnosis is crucial.

d. In cases with MCAD deficiency, analysis of plasma acylcarnitines by MS/MS reveals accumulation of C6–C10 acylcarnitine species with prominent concentration of C8.

e. In cases with MCAD deficiency, analysis of plasma acylcarnitines by MS/MS reveals accumulation of C10–C16 acylcarnitine species with prominent concentration of C14.

28. Which of the following statement is incorrect about medium-chain acyl-CoA dehydrogenase (MCAD) deficiency?

a. Patients with secondary carnitine deficiency may not have a significant elevation of C6–C10 acylcarnitine concentration.

b. MCAD deficiency is responsible for the initial dehydrogenation of acyl-CoAs with a chain length between 4 and 12 carbon atoms.

c. Hexanoylglycine, phenylpropionylglycine, and suberylglycine are only detected in urine of symptomatic patients.

d. Detection of C6–C10 acylcarnitine neonatal blood spots by MS/MS is the basis for newborn screening of MCAD deficiency.

29. Which of the following statement is incorrect about amino acid analysis?

a. Quantitative analysis of amino acids in plasma, urine, and CSF is a very common clinical investigation of most of IEM disorders.

b. Currently, ion exchange chromatography and MS/MS analysis of amino acids are the most widely used methods to identify defects of amino acid metabolism.

c. Screening methods, such as thin layer chromatography, should be considered as gold standard methodology.

d. Fasting plasma (heparin) is the preferred specimen type for amino acid analysis.

30. Which of the following statement is incorrect about amino acid analysis?

a. In newborns and infants, blood should be collected immediately before the next scheduled feeding.

b. Hemolyzed specimens show lower concentrations of amino acids that are found in higher concentration in blood cells (taurine, aspartic acid, and glutamic acid).

c. Urine collection should avoid fecal contamination and the addition of preservatives.

d. Cerebrospinal fluid (CSF) must be collected free of blood contamination.

31. Which of the following statement is correct about organic acid analysis?
   a. Organic acids (OA) analysis is mostly done by organic acids (OA) extraction from a urine sample by liquid-liquid extraction.
   b. Organic acids (OA) analysis is mostly done by organic acids (OA) extraction from a Plasma sample by liquid-liquid extraction.
   c. Organic acids (OA) analysis is mostly done by organic acids (OA) extraction from a serum sample by liquid-liquid extraction.
   d. Organic acids (OA) analysis is mostly done by organic acids (OA) extraction from a CSF sample by liquid-liquid extraction.
32. Which of the following statement is incorrect about acylcarnitine analysis?
   a. Plasma/serum are the preferred specimen types in diagnostic setting.
   b. Acylcarnitine analysis of dried blood on filter paper for newborn screening is the preferred specimen.
   c. The acylcarnitine analysis is mainly done through electrospray tandem mass spectrometry.
   d. The acylcarnitine analysis is mainly done through an ion exchange chromatography.

Answers

1. d. Newborn screening tests consist of a variety of different analytes, specimens, and analytical techniques. *Burtis CA, et al. Tietz Textbook of Clinical Chemistry, 4th ed., 2006, p. 2207.*
2. c. Some classifications of inborn errors (Saudubray) identified intoxication as acute or progressive accumulation of toxic compounds and energy deficiency as deficits in production or utilization. *Burtis CA, et al. Tietz Textbook of Clinical Chemistry, 4th ed., 2006, p. 2207.*
3. a. Lactate acid is not a typical end product of organic or fatty acid acidemia. *Burtis CA, et al. Tietz Textbook of Clinical Chemistry, 4th ed., 2006, p. 2208.*
4. d. Prenatal testing is an essential part of screening for inborn errors of metabolism and involves both genetic and biochemical testing of maternal blood and amniotic fluid. *Burtis CA, et al. Tietz Textbook of Clinical Chemistry, 4th ed., 2006, p. 2208.*
5. e. Newborn screening is performed in all states of the United States and rely on dried blood spots using the Guthrie card. *Burtis CA, et al. Tietz Textbook of Clinical Chemistry, 4th ed., 2006, p. 2209.*
6. d. Disorders of intermediate metabolism can produce a life-threatening decompensation which includes encephalopathy, neurologic injury, and can lead to death. They can be caused by defects in organic and amino acid metabolism, and all are important tests, although molecular analysis can pinpoint the specific defect. *Burtis CA, et al. Tietz Textbook of Clinical Chemistry, 4th ed., 2006, p. 2210.*
7. e. As many metabolic disorders are inherited, family history is essential. However, acquired diseases during pregnancy are also important factors in a death investigation. *Burtis CA, et al. Tietz Textbook of Clinical Chemistry, 4th ed., 2006, p. 2211.*

8. b. The blood concentration is dependent on dietary intake, hepatic synthesis, metabolism, and renal clearance but not body mass. *Burtis CA, et al. Tietz Textbook of Clinical Chemistry, 4th ed., 2006, p. 2211.*

9. d. The principal enzyme deficiency in phenylalanine deficiency is the phenylalanine hydrolase enzyme, not the oxidase. Neonates with this deficiency should be on phenylalanine dietary restrictions. *Burtis CA, et al. Tietz Textbook of Clinical Chemistry, 4th ed., 2006, p. 2211.*

10. d. Early phenylalanine testing for PKU, an autosomal recessive defect, is essential because it is silent and measures can be taken to avoid impairment of mental development. *Burtis CA, et al. Tietz Textbook of Clinical Chemistry, 4th ed., 2006, p. 2211.*

11. e. Increased phenylalanine, its ratio to tyrosine, and increased presence of urinary metabolites are hallmarks of untreated patients. BH4 concentrations are unaffected. *Burtis CA, et al. Tietz Textbook of Clinical Chemistry, 4th ed., 2006, p. 2211.*

12. a. It is sufficient to detect the presence of the amino acids, organic acids, and metabolites in PKU diagnosis. Enzyme activity for phenylalanine hydroxylase is not required. *Burtis CA, et al. Tietz Textbook of Clinical Chemistry, 4th ed., 2006, p. 2217.*

13. d. Individuals with PKU should be on some form of lifelong dietary restriction of phenylalanine. *Burtis CA, et al. Tietz Textbook of Clinical Chemistry, 4th ed., 2006, p. 2217.*

14. e. Neopterin and biopterin testing is part of many newborn screening programs but not all, since BH4 is not the major cause of PKU. *Burtis CA, et al. Tietz Textbook of Clinical Chemistry, 4th ed., 2006, p. 2217.*

15. b. Like PKU, tyrosinemia is an autosomal recessive disease, with a prevalence about 10 times lower. The major complications are hepatic, renal, and neurological. *Burtis CA, et al. Tietz Textbook of Clinical Chemistry, 4th ed., 2006, p. 2218.*

16. a. Many children with tyrosine type I deficiency, unlike type II deficiency, do not have increased blood tyrosine during the first 48 h, thus the reliance on testing for succinylacetone. *Burtis CA, et al. Tietz Textbook of Clinical Chemistry, 4th ed., 2006, p. 2218.*

17. c. Increased methionine is the precursor molecule to β-synthase and levels can increase 10-fold over normal causing significant pathology. *Burtis CA, et al. Tietz Textbook of Clinical Chemistry, 4th ed., 2006, p. 2220.*

18. e. Maple syrup urine disease, so named because of the distinctive odor produced by afflicted children, has an incidence of about 1 in 200,000. *Burtis CA, et al. Tietz Textbook of Clinical Chemistry, 4th ed., 2006, p. 2220.*

19. c. The highest concentrations relative to normal are seen for ʟ-alloisoleucine. *Burtis CA, et al. Tietz Textbook of Clinical Chemistry, 4th ed., 2006, p. 2220.*

20. c. Defects in the enzymes within the urea cycle inhibit the metabolism of amino acids and lead to increase of these amino acids and decreased removal of ammonia concentrations. *Burtis CA, et al. Tietz Textbook of Clinical Chemistry, 4th ed., 2006, p. 2220.*

21. d. Most methods for organic acid analysis rely on mass spectrometry for detection, with gas or more recently, liquid chromatography. *Burtis CA, et al. Tietz Textbook of Clinical Chemistry, 4th ed., 2006, p. 2221.*

22. d. Propionic acidemia is characterized by a deficiency of propionyl-COA carboxylase and leads to an accumulation of 3-OH-propionic acid, 30methylcitrate, and propionylcarnitine. *Burtis CA, et al. Tietz Textbook of Clinical Chemistry, 4th ed., 2006, p. 2222.*

23. a. Leucine but not isoleucine is a precursor to isovaleryl-CoA dehydrogenase. *Burtis CA, et al. Tietz Textbook of Clinical Chemistry, 4th ed., 2006, p. 2221.*

24. e. The incidence of glutaric acidemia type I is 1 in 100,000. *Burtis CA, et al. Tietz Textbook of Clinical Chemistry, 4th ed., 2006, p. 2230.*

25. d. Complications of hypoketotic hypoglycemia include convulsion, coma, and brain damage. *Burtis CA, et al. Tietz Textbook of Clinical Chemistry, 4th ed., 2006, p. 2232.*

26. b. Short-chain, medium-chain, long-chain, and very long-chain fatty acid disorders refer to the length of the fatty acid involved in the disease and does not involve amino acids. *Burtis CA, et al. Tietz Textbook of Clinical Chemistry, 4th ed., 2006, p. 2232.*

27. e. Accumulation of C10–C16 acylcarnitine species is characteristic of long-chain fatty acid disorders. *Burtis CA, et al. Tietz Textbook of Clinical Chemistry, 4th ed., 2006, p. 2236.*

28. c. Detection of hexanoylglycine, phenylprionylglycine, and suberylglycine in urine is observed in asymptomatic patients. *Burtis CA, et al. Tietz Textbook of Clinical Chemistry, 4th ed., 2006, p. 2236.*

29. c. Thin-layer chromatography is an antiquated technique for amino acid analysis and should be discarded in favor of mass spectrometric analysis, except, perhaps in resource-limited laboratories. *Burtis CA, et al. Tietz Textbook of Clinical Chemistry, 4th ed., 2006, p. 2237.*

30. b. Contamination of the sample can cause inaccuracies in amino acid analysis. *Burtis CA, et al. Tietz Textbook of Clinical Chemistry, 4th ed., 2006, p. 2237.*

31. a. Urine is the specimen of choice for organic acidurias. *Burtis CA, et al. Tietz Textbook of Clinical Chemistry, 4th ed., 2006, p. 2237.*

32. d. Ion-exchange chromatography has been largely replaced with liquid chromatography/mass spectrometry. *Burtis CA, et al. Tietz Textbook of Clinical Chemistry, 4th ed., 2006, p. 2241.*

Section E

General laboratory topics

Chapter 32

Preanalytical factors and test interferences

Alan H.B. Wu
University of California, San Francisco, CA, United States

1. Which of the following is true for capillary blood gases?
   a. All results are similar to venous
   b. All results are similar to arterial
   c. Almost exactly in between arterial and venous
   d. Largely equivalent to arterial except slightly lower oxygen
   e. Largely equivalent to arterial except slightly lower carbon dioxide
2. Which protein has the highest differential between CSF interstitial fluid and blood?
   a. β2-transferrin
   b. α1-antitrypsin
   c. IgG
   d. ceruloplasmin
   e. β-trace protein
3. What is true regarding capillary measurements?
   a. potassium values are higher due to tissue breakdown
   b. potassium values are lower due to dilution
   c. sodium values are higher due to tissue breakdown
   d. glucose is higher due to tissue utilization
   e. all are the same
4. While of the following CBC parameters exhibit lower counts in capillary blood vs venous?
   a. hemoglobin
   b. hematocrit
   c. RBC count
   d. WBC count
   e. platelet count

5. Which phase of testing accounts for the highest rate of laboratory errors?
   a. preanalytical
   b. analytical
   c. postanalytical
   d. roughly the same (within 5%)
   e. dependent on the laboratory situation

6. Which of the following is not an example of a postanalytical error?
   a. invalid reference range
   b. delay in reporting
   c. misunderstanding of the result by caregivers
   d. accuracy of reporting results
   e. transcription error

7. Which analyte is not stable for more than 24 h at room temperatures prior to separation?
   a. cortisol
   b. total calcium
   c. parathyroid hormone
   d. creatinine
   e. NT-proBNP

8. Which of the following is the most common preanalytical interference?
   a. hemolysis
   b. lipidemia
   c. hyperbilirubinemia
   d. therapeutic drugs
   e. antioxidant use

9. Which of the following is not a common cause of hemolysis?
   a. rapid draw from a syringe
   b. use of a straight needle venipuncture vs collection from an IV
   c. use of small gauge needles
   d. use of medical staff instead of phlebotomists
   e. number of attempts at venipuncture

10. Which of the following is the mechanism for biotin interference?
    a. Endogenous biotin binds directly to solid-phase particles.
    b. Dietary biotin binds to both the capture and signal antibodies producing a signal in the absence of the analyte.
    c. Biotin binds to solid-phase particles linked to streptavidin interfering with antibodies linked to biotin.
    d. Biotin quenches both fluorescence and chemiluminescence signals.
    e. Interference occurs in the absence of dietary biotin.

11. An abnormal serum chloride (135 mmol/L) with a normal sodium is reported on a patient who has epilepsy and a normal calcium.
    a. The patient is dehydrated.
    b. Bromide interferes with ion-selective electrodes for chloride.
    c. The sample is contaminated with intravenous normal saline.
    d. The patient has primary hyperparathyroidism.
    e. The patient has pseudohyponatremia due to multiple myeloma.

12. Which of the following analytes has been reported to absorb into the gel of SST tubes thereby reducing its concentration?
    a. cholesterol
    b. phenytoin
    c. sodium
    d. thyroid stimulating hormone
    e. nucleic acids

13. A patient is admitted with ethylene glycol exposure. Which of the following test may be adversely affected?
    a. serum ethanol with the alcohol dehydrogenase method
    b. methanol with the head space gas chromatography method
    c. lactate with the lactate oxidase enzymatic assay
    d. lactate dehydrogenase utilizing the pyruvate to lactate direction
    e. lactate dehydrogenase utilizing the lactate to pyruvate direction

14. Estimate the glucose concentration, starting from 100 mg/dL after a tube of blood is left uncentrifuged at room temperature for 8 h.
    a. no change
    b. 90 mg/dL
    c. 75 mg/dL
    d. 50 mg/dL
    e. 20 mg/dL

15. A 55-year-old female has a serum hCG concentration of 10 U/L (reference range 0–5 IU/mL) in a perimenopausal female who states that she is not sexually active. Repeat testing on dilution produces the appropriately expected result. Her follicle stimulating hormone was increased. What is the best explanation for this result?
    a. She has a gestational tumor.
    b. She is not truthful about her sex history and is actually pregnant.
    c. She has an ectopic pregnancy.
    d. She has human antimouse antibodies.
    e. The hCG was produced from the anterior pituitary.

16. A patient exposed to sodium nitrite has oxygen saturation measurements by cooximetry and pulse oximetry. Which of the following is likely?
    a. Results will be the same within precision limits.
    b. The apparent oxygen saturation will be falsely high in the pulse oximetry measurement.

    c. The carboxyhemoglobin concentration will be falsely high in the pulse oximetry measurement.

    d. The pulse oximetry measurement is the most accurate overall.

    e. Accuracy is dependent on the total hemoglobin concentration.

17. Which of the following has been identified as a preanalytical cause of false positive interference with the troponin assay with plasma?

    a. presence of heterophile antibodies

    b. presence of autoantibodies

    c. presence of microclots

    d. presence of lipids

    e. underfilling Vacutainer tubes

18. Which of the following is not a cause of pseudohyponatremia with use of indirect reading ion-selective electrodes?

    a. high lipid concentrations such as seen in acute pancreatitis

    b. high proteins due to monoclonal gammopathy

    c. increased concentration of lipoprotein-x as seen in obstructive jaundice

    d. polycythemia vera

    e. all of the above

19. Which is the correct order of blood draw when multiple tubes are needed at the same time?

    a. citrate, serum, heparin, EDTA, sodium fluoride, blood cultures

    b. blood cultures, serum, heparin, EDTA, sodium fluoride, citrate

    c. blood cultures, citrate, serum, heparin, EDTA, sodium fluoride

    d. sodium fluoride, blood cultures, citrate, serum, heparin, EDTA

    e. EDTA, citrate, sodium fluoride, blood cultures, serum, heparin

20. The following results were obtained on a patient two days apart. Assuming that the same patient was collected and there were no pathologic changes, which of the following is a possible explanation?

	Na	K	Cl	CO$_2$	BUN	Creat	Gluc	Ca
Day 1	137	4.1	102	25	20	1.35	101	9.6
Day 2	142	2.7	116	18	14	0.77	75	6.4

    a. blood drawn downstream of an IV of D5W

    b. blood drawn downstream of an IV of normal saline

    c. patient has psychogenic polydipsia and drank copious amount of water before day 2 labs

    d. used wrong blood collection tube, under-filled a citrate tube

    e. use of sodium heparin blood collection tube

21. The potassium content of per unit of hemoglobin from red cells has been determined. Based on this, calculations of potassium concentrations can be made from the plasma hemoglobin concentrations and the measured potassium of the hemolytic sample. Which of the following is the best reason for failure of this approach?

    a. Analytical assays for free plasma hemoglobin are insufficiently accurate at low concentrations.

    b. There is significant variability in the potassium content leading to significant range of calculated values.

    c. Physicians are not interested in calculated estimates.

    d. Hemoglobin produces an analytical interference with ion-selective electrode assays for potassium.

    e. CLIA regulations prohibit this calculation.

22. What is the effect of oxygen exposure from the air to whole blood ionized calcium measurements?

    a. decreases

    b. increases

    c. magnitude of change depends on the albumin concentration

    d. magnitude of change depends on the oxygen content

    e. no change

23. Which of the following does not require a fasting specimen for accuracy?

    a. glucose

    b. triglycerides

    c. direct LDL cholesterol

    d. lipid panel

    e. insulin and C-peptide

24. How much does use of a full lithium heparin blood collection tube add to the plasma measurement and does it cause an error in therapeutic drug monitoring interpretation?

    a. no change, does not affect TDM interpretation

    b. 0.5 mmol/L, does not affect TDM interpretation

    c. mmol/L, does not affect TDM interpretation

    d. mmol/L, does affect TDM interpretation

    e. 2.0 mmol/L, does affect TDM interpretation

Answers

1. d. The capillary bed is useful for blood gas analysis particularly in neonates. *Heidari K, et al. Correlation between capillary and arterial blood gas parameters in an ED. Am J Emerg Med 2013;31:326–9.*

2. a. Sometimes it is necessary to determine if a fluid is of cerebrospinal origin. *Nandapalan V, et al. Beta-2-transferrin and cerebrospinal fluid rhinorrhopea. Clin Otolaryngol Allied Sci. 1996;21:259–64.*

3. e. Capillary measurements are not as accurate as measurements in serum or plasma. *Karon B. Phlebotomy top gun: Measuring potassium in capillary blood samples. https://cdn.prod-carehubs.net/n1/96e99366cea7b0de/uploads/2018/01/2018-01-08-phlebotom-top-gun-handout.pdf.*

4. e. While most CBC parameters exhibit similar values in capillary than venous, platelets are lower, due to local aggregation, especially at higher counts. *Chavan P, et al. Comparison of complete blood count parameters*

between venous and capillary blood in oncology patients. J Lab Phys 2016;8:65–6.

5. a. Preanalytical is defined as errors prior to testing. Estimates are 46%– 68% of all errors. *Plebani M, et al. Errors in clinical laboratories or errors in laboratory medicine? Clin Chem Lab Med 2006;44:750–9.*

6. d. Use of an invalid reference range is a postanalytical error, even though the erroneously published reference range may predate testing. Accuracy in reporting could be considered an analytical error. *Villaester HP. Analytical and post analytical errors in laboratory. https://www.slideshare.net/ chewmeyellow/analytical-and-post-analytical-errors-in-laboratory.*

7. c. PTH concentrations decline when stored at room temperature in whole blood. NT-proBNP but not BNP is stable. *Oddoze C, et al. Stability study of 81 analytes in human whole blood, in serum and in plasma. Clin Biochem 2012;45:464–9.*

8. a. Hemolysis is a common preanalytical error, especially from the emergency room. The mean rate of hemolysis according to a national survey was 18%. *Phelan MP, et al. Estimated national volume of laboratory results affected by hemolyzed specimens from emergency departments. Arch Pathol Lab Med 2016;140:621.*

9. b. There are many causes of hemolysis; however, use of straight needle venipuncture has been shown to reduce the hemolysis rate. *Heyer NJ, et al. effectiveness of practices to reduce blood sample hemolysis in EDs: A laboratory medicine best practices systematic review and meta-analysis. Clin Biochem 2012;45:1012–32.*

10. c. Some immunoassays use solid-phase particles that are linked to streptavidin. When biotin-labeled antibodies are added, they form a complex with streptavidin and the solid-phase reagent to capture the analyte. The addition of a second antibody labeled with a fluorophore or chemiluminescent tag is added to complete the sandwich and enable detection of the complex. Paramagnetic particles linked to streptavidin are not specific to any analyte. *Colon PJ, et al. Biotin interference in clinical immunoassays. J Appl Lab Med 2018;2:941–51.*

11. c. Potassium bromide is used to treat certain types of epileptic seizures especially children. The chloride ion selective electrode can cause false positive results in the presence of bromide ions. *Wang T, et al. Variable selectivity of the Hitachi chemistry analyser chloride ion-selective electrode toward interfering ions. Clin Biochem 1994;27:37 41.*

12. b. Many therapeutic drugs have been reported to bind to the gels used in SST tubes, although the effect varies from manufacturer to manufacturer. *Dasgupta A, et al. Absorption of therapeutic drug by barrier gels in serum separator blood collection tubes. Volume- and time-dependent reduction in total and free drug concentrations. Am J Clin Pathol 1994;101:456–61.*

13. c. The presence of glycolate, an ethylene glycol metabolite and similar in structure to lactate, causes interferences with many lactate assays. *Tintu A, et al. Interference of ethylene glycol with L-lactate measurement is assay-dependent. Ann Clin Biochem 2013;50:70–2.*

14. d. Glucose declines at a rate of 5%–7% per hour at room temperature. *Turchiano M, et al. Impact of sample collection and processing methods on glucose levels in community outreach studies. J Environ Pub Health 2013; Article ID 256151.* https://doi.org/10.1155/2013/256151.

15. e. Perimenopausal and postmenopausal nonpregnant women can have mildly increased hCG that is accompanied by a high FSH. Pregnant women have low FSH and much higher hCG. *Wu AHB, et al. Mild positive hCG in a perimenopausal female: normal, malignancy, or phantom? Lab Med 2009;40:463–6.*

16. b. Pulse oximetry only measures oxygenated and reduced hemoglobin while cooximetry will also measure methemoglobin, which would be expected to be high in this case, and carboxyhemoglobin, expected to the normal in this case. Pulse oximetry will produce a falsely high value methemoglobin level. *Jurban A. Pulse Oximetry. Crit Care 2015;19:272. doi:10.1186/s13054-015-0984-8.*

17. c. Insufficient or inadequate sample processing of plasma can lead to microclots producing falsely high troponin results. Lipids can produce low values. Presence of atypical antibodies is an analytical not preanalytical interference. *Hermann DS, et al. Variability and error in cardiac troponin testing. An ACLPS critical review. Am J Clin Pathol 2017;148:281–95.*

18. d. High hematocrit as seen in polycythemia vera is not a cause of pseudo-hyponatremia unless an indirect ion-selective sodium measurement is made on whole blood. *Fortgens P, et al. Pseudohyponatremia revisited. A modern-day pitfall. Arch Pathol Lab Med.2011;135:516–9.*

19. c. Blood cultures are always first, when ordered, in order to minimize bacterial skin contamination (e.g., *Staphylococcus* sp.). This is followed by citrate, because there are additives added to plastic tubes that can interfere with clotting. EDTA and fluoride are last as they can contaminate the previous tubes affecting lab results, e.g., potassium and sodium measurements. *Clinical & Laboratory Standards Institute. Order of Blood Draw Tubes and Additives. https://clsi.org/about/blog/order-of-blood-draw-tubes-and-additives/.*

20. b. Collection downstream from a line can contaminate the blood. D5W contains a high glucose concentration (5 g/dL), while saline contains slightly high sodium (154 mmol/L) and a high chloride (154 mmol/L) content. *Patel DK, et al. Method to identify saline-contaminated electrolyte profiles. Clin Chem Lab Med 2015;53:1515–91.*

21. b. In the literature, there is a significant range of estimates of potassium increases per 100 mg of hemoglobin (mean, about 0.3 mmol/l K^+ per 1 g/dL hemoglobin), depending on how the study is conducted. Given the tight reference range for potassium, there is little support for estimating potassium concentrations in a hemolyzed specimen, in lieu of recollection. *Mansour MMH, et al. Correction factors for estimating potassium concentrations in samples with in vitro hemolysis. A detriment to patient safety. Arch Pathol Lab Med 2009;133:960–6.*

22. a. The pH of blood increases with exposure to air due to the shift from bicarbonate to carbonic acid. Ionized concentrations competes with $[H^+]$ with binding to albumin and decreases as $[H^+]$ concentrations decline. *Wang S, et al. pH effects on measurements of ionized calcium and ionized magnesium in blood. Arch Pathol Lab Med 2002;126:947–50.*

23. c. The direct LDL assay does not require a fasting specimen, unlike the calculated LDL from the Friedewald equation, that relies on an accurate triglyceride measurement, which is inaccurate under nonfasting conditions. *Sathiyakumar V, et al. Fasting vs non-fasting and low-density lipoprotein-cholesterol accuracy. Circulation 2017; https://www.acc. org/latest-in-cardiology/journal-scans/2017/10/27/11/02/fasting-vs-nonfasting-and-ldlc-accuracy.*

24. d. Lithium is used as a therapeutic agent for depression. The therapeutic range is 0.5–0.8 mmol/L. *Wills BK, et al. Factitious lithium toxicity secondary to lithium heparin-containing blood tubes. J Med Toxicol 2006;2:61–3.*

Chapter 33

Reference ranges

Dina N. Greene[a] and Alan H.B. Wu[b]
[a]Kaiser Permanente, Seattle, WA, United States, [b]University of California, San Francisco, CA, United States

1. Which of the following is true when comparing reference intervals to clinical decision points?
   a. Only clinical decision points are based on clinical outcome studies.
   b. Both are considered the gold standard/highest quality for interpreting laboratory tests.
   c. Only clinical decision points are intended to distinguish individuals with disease from individuals without.
   d. Only reference intervals require partitioning by age, sex, and other demographics.
   e. Clinical decision points are only used when reference intervals are not preestablished.
2. Which of the following is/are true about the term "reference range"?
   a. This is the range used to evaluate if a lab result is outside of what is considered normal or healthy.
   b. This is the complete range of values observed in a particular population and does little to differentiate normal or healthy.
   c. This range is often defined as the central 95th percentile of a distribution.
   d. This range is utilized when clinical decision points are not available.
   e. Using this range as a comparator for laboratory results will lead to over-flagging results.
3. Which of the following tests are usually compared to a reference interval and not a clinical decision point?
   a. digoxin
   b. HbA1c
   c. calcium
   d. eGFR
   e. neonatal bilirubin
4. Most accrediting agencies
   a. require reference intervals to be appended to every test.
   b. recommend reference intervals be appended to every tests.

    c. have no opinion on reference intervals.

    d. provide laboratories with appropriate reference intervals.

    e. provide laboratories with samples from healthy subjects to validate or establish reference intervals.

5. Which organization(s) provides detailed guidance for establishing and verifying reference intervals?

    a. CAP

    b. CLIA

    c. AACC

    d. CLSI

    e. IFCC

6. Which of the following are not required when establishing reference intervals?

    a. specimens from healthy subjects

    b. awareness of analytical and preanalytical issues that can alter results

    c. preliminary approximation of the expected concentration

    d. prospective knowledge of statistical methods to apply to the results

    e. prospective or retrospective analysis of potential covariates

7. When establishing population-based reference intervals, increasing the sample size

    a. dismisses the need to derive confidence intervals for the reference limit.

    b. minimizes the likelihood that the data set will contain outliers.

    c. has no effect on the precision of the reference limits.

    d. decreases the precision of the reference limit.

    e. increases the precision of the reference limit.

8. A disadvantage of direct sampling over indirect sampling is

    a. direct sampling is more expensive and does not improve clinical sensitivity.

    b. there are no disadvantages; direct is always the best way.

    c. accessing patients for direct sampling is difficult.

    d. direct sampling uses results directly available from the LIS using previous results, so it skews results.

    e. regulatory agencies do not recommend direct sampling.

9. Which of the following best describes Dixon's test for outlier exclusion ($n = 16$ observations)?

    a. Outlier removal is an iterative process; once the first outlier is removed the data should be inspected for additional outliers and the test reapplied, as needed.

    b. A value can be considered an outlier if the gap between the suspected result and the range of results is greater than 33%.

    c. A value can be considered an outlier if it exceeds the 25% or 75% boundaries by 1.5 times the interquartile range.

    d. A value can be considered an outlier if it exceeds the 25% or 75% boundaries by two times the interquartile range.

e. In order to apply this test, data must be normally distributed and transforming the data to make it Gaussian is not appropriate.

10. Which of the following best describes Tukey test for outlier exclusion?

a. A value can be considered an outlier if it exceeds the 25% or 75% boundaries by 1.5 times the interquartile range.

b. A Tukey test is very powerful because it does not require data to be normally distributed or transformed.

c. A value can be considered an outlier if the gap between the suspected result and the range of results is greater than 33%.

d. Tukey method is best used on values that are high values, which is due to the nature of the equation. For low values, the data must be transformed before analysis.

e. Outlier removal cannot be iterative using a Tukey test. Once an outlier has been removed the test cannot be reapplied to additional points that may be skewing results.

11. Parametric distributions

a. are analyzed identically to nonparametric distributions.

b. indicate that 95% of values lie within ± 1.96 standard deviations of the mean.

c. must be how the data are distributed in order to calculate a clinical decision point.

d. usually require transformation before the reference interval can be calculated.

e. indicate that the analyte of interest has a small physiological concentration range.

12. Nonparametric distributions

a. indicate that 99% of values lie within ± 1.96 standard deviations of the mean.

b. are skewed, and therefore can only be used to determine one-sided reference intervals.

c. can be mathematically transformed to adopt a Gaussian distribution.

d. use simple percentages to calculate the central 95 percentile.

e. usually indicate that the reference population was not screened properly and includes several samples from participants with subclinical pathology.

13. Which of the following may need reference intervals specific to their population?

a. pediatrics

b. transgender people

c. geriatrics

d. professional athletes

e. cisgender people

f. all of the above

14. Which of the following would not be an acceptable way to implement reference intervals via transference?

    a. Adopt the reference interval from a previous method after method comparison between the old and new methods with patient samples spanning the AMR shows <5% systematic bias between the two methods.

    b. Adopting a manufacturer reference interval based on the published package insert of the assay.

    c. Adopting a reference interval established by a third party based on the published scientific literature.

    d. Evaluating the distribution of results from a population at random and using a parametric or nonparametric approach to derive the reference interval.

    e. Use a correction factor to adjust the reference interval from a previous method after method comparison between the old and new methods with patient samples spanning the AMR shows 35% systematic bias between the two methods.

15. Harmonization of reference intervals

    a. has been universally implemented since 2016.

    b. can only be accomplished for pediatric cohorts.

    c. are easiest to accomplish for immunoassays.

    d. would simplify laboratory result interpretation significantly.

    e. are not recommended by national and international organizations.

16. Which of the following reference intervals have been harmonized?

    a. testosterone

    b. sodium

    c. albumin

    d. hemoglobin

    e. none of these

17. The 99th percentile is used for analytes like cardiac troponin rather than the 97.5th percentile because

    a. it is not; we always use the central 95th percentile or upper 97.5th percentile when establishing population-based reference intervals.

    b. using the 99th percentile reduces the number of healthy patients who would be further worked-up relative to the 97.5th percentile.

    c. using the 99th percentile increases the number of patients who would be further worked-up relative to the 97.5th percentile and makes sure no one is missed.

    d. the 99th percentile was derived from several clinical trials and is therefore the best upper reference limit for detecting acute myocardial infarction.

    e. in conventional cardiac troponin assays, only 1 person out of 100 had detectable cTn and therefore the 99th percentile was a practical decision point to use.

18. If outliers are not identified and removed before establishing a reference interval the resulting interval will be

a. wider than would be seen in a completely healthy population.

b. narrower than what would be seen in a healthy population.

c. appropriate to implement—outliers are expected to be within the data and that is why the central 95 percentile is calculated.

d. unsuitable for clinical use due to the excess of false positives that will occur.

e. the distribution of data will be right skewed.

19. Which of the following is false about reference intervals for nonstandard body fluids?

a. Because they are only found (or collected) in pathological states establishing reference intervals that define healthy is impossible.

b. Regulatory bodies understand that it is impossible to derive reference intervals and therefore no result interpretation is required.

c. Clinical trials or retrospective data mining can be used to establish medical decision points.

d. Light's criteria is one of the most well-known methods for determining if pleural fluid is exudative based on laboratory findings.

e. Any amount of detectable bilirubin in CSF, particularly in the presence of oxyhemoglobin, should cause concern for a traumatic brain injury.

20. The following data was obtained to compute a normal range study. Which of the following is the most appropriate range to implement?

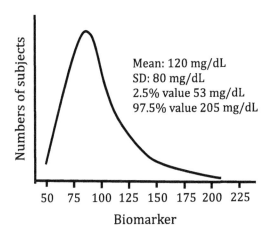

Mean: 120 mg/dL
SD: 80 mg/dL
2.5% value 53 mg/dL
97.5% value 205 mg/dL

a. 40–200 mg/dL

b. -30–270 mg/dL

c. 53–205 mg/dL

    d. 50–225 mg/dL

    e. cannot determine without the actual data points

21. Which attribute of a biological variation parameter determines if reference ranges are appropriate?

    a. reference change value

    b. preanalytical variation

    c. index of individuality

    d. homeostatic set point

    e. reference values needed for all tests

22. Without looking them up values in a reference text, which of the following reference ranges is incorrect?

    a. potassium 3.5–5.0 meq/L

    b. total CO_2 35–45 mmHg

    c. ionized calcium 1.15–1.32 mmol/L

    d. creatinine males 0.9–1.3

    e. pH 7.35–7.45

23. Some expert groups have advocated reporting hemoglobin A1c values into average glucose (AG). What is the AG for a A1c of 6%?

    a. 115 mg/dL

    b. 120 mg/dL

    c. 126 mg/dL

    d. 130 mg/dL

    e. 150 mg/dL

24. Which of the following results is physiologically inconsistent with its reference interval?

    TSH: 0.4–4.2 μ/mL

        Free T4: 0.8–2.7 ng/dL

        Free T3: 260–480 pg/dL

    a. TSH: 3.0 (N), Free T4: 1.0 (N), Free T3: 300 (N)

    b. TSH: 8.0 (H), Free T4: 0.3 (L), Free T3: 150 (L)

    c. TSH: 0.2 (H), Free T4: 0.3 (L), Free T3: 600 (H)

    d. TSH: 0.2 (L), Free T4: 2.5 (N), Free T3: 600 (H)

    e. c. and d.

25. For a transgender woman, which of the following laboratory tests should the female reference range be used?

    1. creatinine

    2. gonadotropins

    3. creatinine kinase

    4. CBC parameters

        a. 1, 2, 3

        b. 1, 3

        c. 2, 4

        d. 4 only

        e. all of the above

26. Which of the following produces significantly different reference ranges relative to arterial when a venous sample is tested for blood gases?

   a. pH

   b. PO_2

   c. PCO_2

   d. bicarbonate

   e. all of the above tests are interchangeable

27. The International System of Units has recommended that laboratory tests be converted from mass units (e.g., mg/dL) to molar units (e.g., μmol/L). Which of the following is not a reason why some tests have not been converted?

   a. Most laboratories in the United States have retained mass units in routine clinical practice.

   b. Some analytes, e.g., total protein, are heterogenous and a molar concentration cannot be determined.

   c. The molecular weight of the tested compound is unknown.

   d. Some analytes, e.g., troponin, undergo degradation.

   e. Some analytes, e.g., tumor markers undergo posttranslational modifications.

Answers

  1. a. Clinical decision points are established using data derived from large clinical trials and generally can only universally apply to analytes for which there is some amount of standardization or harmonization. *E.g., Expert Panel on Detection, Evaluation, and Treatment of High Blood Cholesterol in Adults. Executive summary of the third report of the National Cholesterol Education Program (NCEP) expert panel on detection, evaluation, and treatment of high blood cholesterol in adults (adult treatment panel III). JAMA 2001;285:2486–97.*

  2. b. The term reference interval is the proper way to describe derived reference limits within a range of observed values in a defined population. *Katayev A, et al. Establishing reference intervals for clinical laboratory test results: Is there a better way? Am J Clin Pathol 2010;133:180.6.*

  3. c. All other analytes listed either use therapeutic ranges or clinical decision point/clinical decision nomograms (neonatal bilirubin). *Serum calcium. Medscape. https://emedicine.medscape.com/article/2087447-overview.*

  4. a. Regulatory agencies recognize that laboratory tests cannot be interpreted independent of reference values and therefore reference intervals are required by most regulatory agencies. *Katayev A, et al. Establishing reference intervals for clinical laboratory test results: Is there a better way? Am J Clin Pathol 2010;133:180.6.*

  5. d. and e. The IFCC and CLSI have published guidelines for reference intervals. Soldberg HE. The IFCC recommendation on estimation of reference intervals. *The RefVal program. Clin Chem Lab Med. 2004;42(7):710–4. CLSI. Defining, establishing, and verifying reference intervals in the* *clinical laboratory; approved guideline 3rd edition CLSI document C28-A.3;2008.*

6. c. knowledge of the necessary measuring range is necessary before an assay is developed, but is not a requirement for establishing a reference interval. *Katayev A, et al. Establishing reference intervals for clinical laboratory test results: Is there a better way? Am J Clin Pathol 2010;133:180.6.*

7. e. increasing a sample size will make the reference limits more precise, which is apparent by the decreased confidence interval of the limit that occurs. *Burtis CA, et al. Tietz Textbook of Clinical Chemistry and Molecular Diagnostics. Saunders, 2006, p. 435.*

8. c. Direct sampling improves sensitivity and is the gold standard, but is difficult to perform because collecting samples from different cohorts of populations (particularly special populations) can be difficult. *Ozarda Y. Reference intervals: current status, recent developments and future considerations. Biochemia medica 2016;26:5–16. doi:10.11613/BM.2016.001.*

9. b. The Dixon criteria is calculated as $Q = $ gap/range. If Q exceeds a certain range, it is considered an outlier. *Critical values for expanded Dixon outlier test. http://webspace.ship.edu/pgmarr/Geo441/Tables/Dixon%20Table,%20Expanded.pdf.*

10. a. Data is divided into quartiles with $q1$ being the 25th percentile and $q3$ the 75 percentile. The inter-quartile range (IQR) is defined as the interval between $q1$ and $q3$. The range of acceptable values is between $q1 - (1.5 * iqr)$ and $q3 + (1.5 * iqr)$. *Seo S. A review and comparison of methods for detecting outliers in univariate data sets. http://d-scholarship.pitt.edu/7948/1/Seo.pdf.*

11. a. If 95% values lie outside ± 1.96 standard deviations, it is considered nonparametric. *The relationship Between Confidence Intervals and p-values. https://www.ucl.ac.uk/child-health/short-courses-events/about-statistical-courses/research-methods-and-statistics/chapter-6-content-2.*

12. c. and d. Nonparametric data can be transformed to parametric data or use of the central 95% percentile for normal range calculations. *Demystifying statistical analysis 7: Data transformations and nonparametric tests. https://learncuriously.wordpress.com/2018/10/06/data-transformations-and-nonparametric-tests/.*

13. f. All of these populations might need specialized reference intervals; even partitioning between cisgender men and women can be necessary for some laboratory tests. *Greene DN, et al. Hematology reference intervals for transgender adults on stable hormone therapy. Clin Chim Acta 2019;492:84–90.*

14. d. This would be an imperfect way of establishing a reference interval (a population selected for healthy individuals is preferred). Transference of reference intervals begins with a previously established interval. *https://academic.oup.com/clinchem/article/61/8/1012/5611529.*

15. d. Harmonization is a step closer to standardization, by producing lab results that are in agreement with another despite use of different methods or even units of measure. *Tate JR, et al. Harmonization of clinical laboratory test results. eJIFCC 2016;27(1):5–14. Published 2016 Feb 9.*

16. e. While there is very little bias between instruments for sodium, albumin, and hemoglobin, reference interval harmonization is still currently uncommon. *Producing lab results that Harmonization of Clinical Laboratory Test Results. eJIFCC. 2016;27(1):5–14. Published 2016 Feb 9.*

17. b. Clinical trials have validated the 99th percentile, but the value was not selected because a clinical trial defined this as the best upper reference limit. The 99th percentile is used because it provides a clinically acceptable balance between sensitivity and specificity. *Thygesen K, et al. Fourth universal definition of myocardial infarction (2018). J Am Coll Cardiol 2018;2231–64.*

18. a. Outliers will produce wider reference intervals, reducing the ability of the test to detect an abnormality. *https://pubmed.ncbi.nlm.nih.gov/11719478/.*

19. b. Although classical reference intervals are likely unable to be established for body fluids, some regulatory agencies still require the laboratory to provide a method for result interpretation. *Akgul M, et al. Body fluid testing. What you should know before you hit go. Clin Lab News 2015; https://www.aacc.org/publications/cln/articles/2015/october/body-fluid-testing-what-you-should-know-before-you-hit-go.*

20. c. For parametric distribution, the mean ±2SD determines the reference range. The distribution of this data is nonparametric. Therefore data points are placed into rank order and the lowest 2.5% and highest 97.5% value determines the reference range. *Clin Lab Navigator. http://www.clinlabnavigator.com/reference-ranges.html.*

21. c. The index of individuality is the ratio of the intra-individual variation divided by the inter-individual variation. Values below 0.6 indicate that reference intervals will be of little value. *Fraser CS. Biological Variation, AACC Press, 2001.*

22. b. The reference ranges for some lab tests should be known by practicing clinical laboratorians without lookup. The range for CO^2 is for pCO_2 from a blood gas, not total CO_2. *Wu AHB, ed. Tietz Clinical Guide to Laboratory Tests, 4th ed., Saunders, 2006.*

23. c. The upper reference limit for hemoglobin A1c is <5.7%. The American Diabetes Association has a conversion of average glucose $= 28.7 \times$ A1c $- 46.7$. A 6% A1c is an average glucose of 126 mg/dL. *Nathan DM, et al. Translating the A1c assay into estimated average glucose values. Diab Care 2008;31:1–6. https://professional.diabetes.org/diapro/glucose_calc.*

24. c. Certain combinations of lab tests are linked to physiology and fall into specific patterns. Low thyroid stimulating hormone is usually accompanied by high free T3 and free T4 and vice versa. Patients with T3 thyrotoxicosis have high T3, low TSH, and normal T4. *Sterling K, et al. T3 thyrotoxicosis: Thyrotoxicosis due to elevated serum triiodothyronine levels.* JAMA *1970;213(4):571–5.*

25. c. Transgender subjects treated with sex hormones will require the transgender reference range for these hormones. Other tests such as hemoglobin and hematocrit testing should use the nonbirth gender ranges. *Greene DN, et al. Hematology reference intervals for transgender adults on stable hormone therapy.* Clin Chim Acta *2019;492:84–90.*

26. d. Oxygen content is reduced in venous blood because it is utilized by tissues and organs. *Malatesha G, et al. Comparison of arterial and venous pH, bicarbonate, PCO₂ and PO₂ in initial emergency department assessment.* Emerg Med J *2007;24:569–71.*

27. c. The molecular weight of tested analytes, including proteins is known, and is not a reason why SI units cannot be used. *Young DS. Implementation of SI units for clinical laboratory data: Style specifications and conversion tables.* Am J Clin Nutr *1993;57:98–113.*

Chapter 34

Laboratory statistics

Alan H.B. Wu

University of California, San Francisco, CA, United States

1. Which of the following is true regarding biological variation of a lab test?
   a. A high index of individuality indicates that reference values will be of little utility.
   b. A minimum of 120 subjects are required for biological variation studies.
   c. Biological variation studies are usually conducted on healthy subjects.
   d. Reference change values refer to absolute changes in serial biomarker results.
   e. The number of samples needed to maintain a homeostatic set point refers to the sensitivity of the test.

2. What is a meta-analysis in the context of reviews?
   1. The statistical combination of several similar clinical trials data in order to achieve statistical significance
   2. Equivalent to the STARD (STAndards for the Reporting of Diagnostic accuracy studies) checklist
   3. A systematic review of clinical trial study findings
   4. Uses unique statistical calculations for hazards ratio
      a. 1, 3
      b. 2, 4
      c. 1, 2, 3
      d. all of the above
      e. none of the above

3. How can knowledge of the biological variation for an analyte improve quality control testing?
   a. Reduce the number of controls from 3 to 2.
   b. Enables a relaxation of QC rules when BV is low.
   c. Eliminates the need for proficiency testing.
   d. Can accept some QC results outside the 3 SD limit.
   e. Eliminate the need for preassayed control values.

Self-assessment Q&A in Clinical Laboratory Science, III. https://doi.org/10.1016/B978-0-12-822093-1.00034-X

4. Which of the tests listed has the lowest index of individuality of those tests listed?
   a. creatinine
   b. troponin
   c. AST
   d. cystatin C
   e. serum iron

5. Which of the following is not part of the STARD (STAndards for the Reporting of Diagnostic accuracy) checklist
   a. flow diagram for participant enrollment
   b. subject eligibility criteria
   c. rationale for choosing the reference test
   d. how missing data were handled
   e. concentrations of the quality controls used

6. In a clinical trial, 500 subjects with high cholesterol (>240 mg/dL) and 225 with normal cholesterol (<200). Of those with high cholesterol, 100 suffered an adverse cardiac event, while 25 suffered a cardiac event with a normal cholesterol. What is the odds ratio (OR) for the following data?
   a. 1.0
   b. 2.0
   c. 0.5
   d. 0.75
   e. 4.0

7. Which plot is often used to compare the results to two analytical assays whereby the differences are plotted on the y-axis?
   a. Bland-Altman
   b. Levy Jennings
   c. Receiver operating characteristic curve
   d. Forest plot
   e. Youden plot

8. Which plot is often used for meta-analyses?
   a. Bland-Altman
   b. Levy Jennings
   c. Receiver operating characteristic curve
   d. Forest plot
   e. Youden plot

9. Which of the following best describes machine learning with regards to improving laboratory diagnostics?
   a. An algorithm that branches based on lab decision cutoffs (classification and regression tree analysis).
   b. Combining digital imaging with lab diagnostics.
   c. Mathematical models that automatically maps input to desired outputs.
   d. Not subjected to FDA regulations.
   e. Will be transferrable from institution to institution.

10. The process of ensuring that the results of different laboratories using different clinical laboratory tests to measure the same substance are equivalent within clinically meaningful limits is termed
    a. harmonization.
    b. standardization.
    c. calibration.
    d. commutability.
    e. traceability.
11. Which is the correct definition of commutability?
    a. The ability to link the calibration of a laboratory test result back to a reference measurement procedure.
    b. A reference material's ability to react in the same way as patient specimens.
    c. A reference method's ability to produce the same result as a clinical assay.
    d. The manner by which a reference material is certified as a reference material.
    e. The assignment of the value of a reference material.
12. What is the equation for total analytic error?
    a. (standard deviation/mean) $\times 100$
    b. mean of blank plus 3 SD
    c. mean of blank plus 10 SD
    d. bias $+ 2 \times$ (allowable imprecision)
    e. false positives/(false positives $+$ true negatives)
13. Which statistical technique is most useful for variables classified as categorical (i.e., disease present or absent, biomarker positive or negative)?
    a. receiver operating characteristic curve analysis
    b. odds ratio
    c. Deming regression
    d. Passing Bablok analysis
    e. Westgard rules
14. Formulating questions is an important step for performing a systematic literature review. Which of the following does "PICO" represent?
    a. parentage, inspection, collaboration, outcome
    b. prognosis, intervention, cost, overview
    c. patient, intervention, comparator, outcome
    d. procedures, intervention, controls, observation
    e. practice, interval, Cochran, odds

15. The following pretest nomogram was constructed for a lab test. What is the posttest probability of a test at a concentration of 1 unit and a pretest probability of 70%

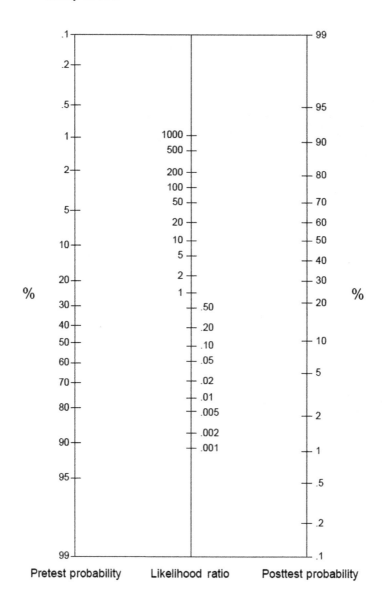

Pretest probability   Likelihood ratio   Posttest probability

a. 90%
b. 70%
c. 50%
d. 10%
e. cannot determine

16. Which of the following is an acceptable strategy for establishing a cutoff for a tumor marker?
    a. use of the 95th percentile of a healthy population
    b. use of the 99th percentile of a healthy population
    c. use of a cutoff concentration derived from ROC curve analysis plotting disease vs relevant benign conditions
    d. guidelines established by the National Cancer Institute
    e. values that predict cancer mortality

17. Which of the following statistical approaches is used for data where the distribution of results is nonparametric and skewed to the left?
    a. calculation of the mean $\pm 2SD$
    b. calculation of the mean $\pm 3SD$
    c. calculation of the median $\pm 2SD$
    d. determination of the central 95th percentile of results
    e. only an upper reference limit can be determined as the lower limit produces a negative result

18. Which of the Westgard rules account for random errors?
    a. $10 \times$
    b. R_{4s}
    c. 1_{3s}
    d. 2_{2s}
    e. 4_{1s}

19. Which of the Westgard rules account for long-term trend?
    a. $10 \times$
    b. R_{4s}
    c. 1_{3s}
    d. 4_{1s}
    e. a. and e.

20. The following Levy Jennings plot is for a single QC sample for 16 days. Which of the following is true?
    a. no violations
    b. violation of the 4_{1s} rule on day 8
    c. violation of the 4_{1s} rule on day 4
    d. violation of the 1_{3s} rule on days 1, 3, and 16
    e. violation of the 2_{2s} rule on days 1 and 2

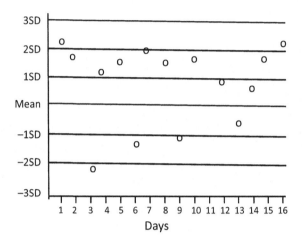

Answers

1. c. Biological variation studies are conducted on healthy subjects. Small number of subjects are needed. Variances in the absence of the disease are needed before a lab test can be interpreted for the presence of disease. A low index of individuality indicates that reference changes will be of little value. The reference change value refers to relative (or percent) change. Homeostatic set point refers to the number of values needed to establish a baseline concentration. *Fraser CS. Biological Variation, AACC Press, 2001.*

2. a. Sometimes it is difficult to achieve sufficient numbers of enrollments for a clinical trial at a single site. Meta-analyses can be conducted to evaluate the effectiveness of biomarkers for clinical trials, e.g., cardiac troponin for risk stratification. *Heidenreich PA, et al. The prognostic value of troponin in patients with non-ST elevation acute coronary syndromes: a meta-analysis. J Am Coll Cardiol 2001;38:478–85.*

3. d. Wider QC limits can be accepted when there is high biological variation given that there will be no impact on medical decisions. Wider limits can reduce the number of false rejections (QC outside of range due to statistical expectation rather than assay errors). *Fraser CS. Biological Variation, AACC Press, 2001.*

4. a. Creatinine has a low index of individuality, making it more useful for monitoring trends, while cystatin C has a high biological variation, making it more useful for diagnosis. *Keevil BG, et al. Biological variation of cystatin C: implications for the assessment of glomerular filtration rate. Clin Chem 1998;44:1535–9.*

5. e. The STARD checklist contains essential items used by authors, reviewers, and other readers to ensure that a report of a diagnostic accuracy

study contains the necessary information. The specifics of quality control materials used are not included. *Cohen JF, et al. STARD 2015 guidelines for reporting diagnostic accuracy studies: explanation and elaboration. BMJ Open 2016;6:e012799. doi:10.1136/bmjopen-2016-012799.*

6. b. The OR is the ratio of the number of subjects with adverse events (100) divided by the number without adverse events (400) among the cases (high cholesterol) over the number of adverse events (25) divided by the number without adverse events among the controls (low cholesterol): $100/400 \div 25/225 = 2.00$. Typically the 95% confidence interval is also calculated. https://www.statisticshowto.datasciencecentral.com/odds-ratio/.

7. a. In a Bland-Altman plot, the difference (y) is plotted against the mean (x) of two measurements. The $\pm 1.96SD$ is also plotted. *Giavarina D. Biochem Med (Zagreb). 2015;25(2):141–51. Figure from WikiMedia Commons.*

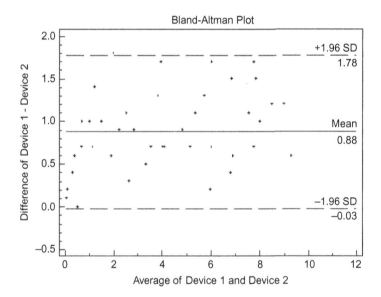

8. d. The Forest plot lists each study's odds ratio with the 95% confidence interval. The cumulative meta-analysis is shown at the bottom of the plot. *Cochran UK. How to read a Forest plot.* https://uk.cochrane.org/news/how-read-forest-plot *Figure from WikiMedia Commons.*

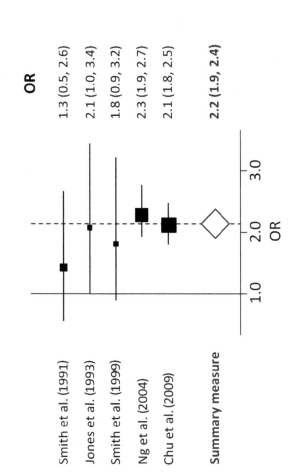

9. c. Machine learning uses tools such as image convolution, neural networks to learn from inputs. *Durant TJ. Machine learning and laboratory medicine: now and the road ahead. Clin Lab News 2019;45(2):20–5.*

10. a. In the absence of standardization, harmonization is the next desired goal. American Association for Clinical Chemistry. *The need to harmonize clinical laboratory test results July 2015. White paper.* https://www.harmonization.net/media/1086/aacc_harmonization_white_paper_2015.pdf.

11. b. If a material is not commutable, it cannot be the matrix used as a reference standard. American Association for Clinical Chemistry. *The need to harmonize clinical laboratory test results July 2015. White paper.* https://www.harmonization.net/media/1086/aacc_harmonization_white_paper_2015.pdf.

12. d. Equation a is the coefficient of variance. b and c are measurements of the limits of detection and quantitation, respectively. E is the calculation of the clinical sensitivity of a test. *Total analytical error.* https://www.aacc.org/publications/cln/articles/2013/september/total-analytic-error.

13. b. Odds ratios plot the probability of a test occurring to that of not occurring. For lab medicine, it is the odds that a lab test result predicts a certain outcome. *Perera R, et al. Systematic review and metaanalysis. In: Evidence-based laboratory medicine. Price CP et al. AACC Press, Washington DC 2007: pp. 266–7.*

14. c. The majority of questions asked by caregivers will relate to these issues. *Price CP. Evidence-based medicine and the diagnostic process. In: Evidence-based laboratory medicine. Price CP et al. AACC Press, Washington DC 2007: pp. 30–5.*

15. b. A line is drawn from 70 on the pretest probability scale, through 1 on the likelihood ratio to produce a 70% on the posttest probability. *Doust J, et al. Teaching evidence-based laboratory medicine. In: Evidence-based laboratory medicine. Price CP et al. AACC Press, Washington DC 2007: pp. 459–62.*

16. c. Serum-based tumor markers are most useful for monitoring disease course during therapy, but not for prediction of mortality. Increased concentrations are often seen in benign conditions, therefore cutoffs from a healthy reference population are usually not appropriate. *Bitterlich N, et al. Cut-off-independent tumour marker evaluation using ROC approximation. Anticancer Res 2007;27:4305–10.*

17. d. If the data is nonparametric and skewed to the left, the mean ± 2SD produces a negative lower limit. Use of the central 95th percentile is appropriate. *Horowitz GL. Reference intervals: practical aspects. eJIFCC 2008;19:95–105.*

18. b. The 4_{1s} indicates that within a run, one control exceeds the 2SD limit and the other is below the 2SD, and is a quality control violation. *Westgard rules and multi-rules.* https://www.westgard.com/mltirule.htm.

19. e. The 4_{1s} and $10 \times$ rules indicate overall trends over 4 and 10 days, respectively (4 consecutive above/below 1SD and 10 above/below the mean). https://www.westgard.com/mltirule.htm.

20. b. Four consecutive points above 1SD is the 4_{1s} rule, which is violated on day 8. https://www.westgard.com/mltirule.htm.

Chapter 35

Regulations and lab administration

Alan H.B. Wu
University of California, San Francisco, CA, United States

1. Which of the following is not required when implementing a test that is FDA cleared?
   a. verification of the reportable range
   b. accuracy (with proficiency testing)
   c. precision
   d. analytical sensitivity
   e. all of the above
2. Which is the difference between FDA cleared and FDA approved?
   a. Both terms mean the same thing.
   b. FDA cleared refers to tests approved via the 510(k) route process for new tests/devices.
   c. FDA cleared refers to tests that fall under the category of lab developed.
   d. Clearance refers to waived tests, approval refers to moderately- or high-complexity tests.
   e. Clearance refers to moderately complexity, approval refers to high complexity tests.
3. Which is not a failure of proficiency test result?
   a. passing only 75% of proficiency challenges
   b. testing by day-shift personnel only (when three shifts are used)
   c. sending samples to another lab for comparison
   d. performing testing in duplicate when not normally done
   e. not performing PT on every piece of equipment
4. Which of the following is incorrectly matched?
   a. waived: glucometer
   b. highly complex: LC-MS
   c. moderately complex: complete blood count
   d. moderately complex: protein electrophoresis
   e. highly complex: next-generation sequencing

Self-assessment Q&A in Clinical Laboratory Science, III. https://doi.org/10.1016/B978-0-12-822093-1.00035-1

5. Which of the following is can be performed by a nonlicensed individual?
   a. calculation to obtain a final result
   b. quantitative pipetting
   c. calibration of assays
   d. perform instrument maintenance and repair
   e. preparation of calibrators

6. The FDA has determined that blood glucose monitors (BGMs) must be separately approved for use among critically ill patients. Which of the following are not options for the laboratory?
   a. Discontinue use of BGMs among these patients and send samples to the central lab.
   b. Find and use FDA devices that have been cleared for critically ill patients.
   c. Validate the BGMs for use by care givers under the lab developed test statute.
   d. all of the above
   e. none of the above

7. Which of the following lab result is not considered a critical value by most hospitals?
   a. sodium 123 mmol/L
   b. calcium 13.5 mg/dL
   c. potassium 5.5 mmol/L
   d. pH 7.15
   e. glucose 510 mg/dL

8. Which of the following is the correct unit of measure for enzyme activity according to the International Systems of Units?
   a. nanokatals
   b. U/L
   c. mg/%
   d. mg/L
   e. U/g

9. Regarding proficiency testing, which of the following is permitted?
   a. Send the sample to a reference lab for confirmation of result.
   b. Use only the lab's most experienced lab technologist.
   c. Repeat testing of high values if initial results are abnormal.
   d. Omit proficiency testing for a moderately complex test.
   e. Perform testing at each site that owns a separate CLIA certificate.

10. Which of the following is not a requisite for an individual who has a Ph.D. to become a lab director for moderately complexity testing?
    a. The Ph.D. must be in chemical, physical, biological, or clinical laboratory science from an accredited institution.
    b. Certification by a recognized entity.
    c. One year of experience directing or supervision nonwaived testing.
    d. Where applicable, grandfathered in as a lab director prior to February 28, 1992.

e. One year of training and experience.

11. Which of the following individuals is not permitted to perform waived clinical lab tests?

  a. Board certified doctor of medicine or osteopathy licensed to practice in their state.

  b. Doctorate or master's degree in chemistry, physical, biological, or medical technology with 1 year of experience.

  c. B.S. degree in chemistry, physical, biological, or medical technology with 2 years of experience.

  d. High school graduate with 4 years of experience.

  e. no personnel requirements

12. Which of the following is not correctly listed with the typical designation of test complexity?

  a. urine pregnancy tests: waived

  b. flow cytometry: moderately complex

  c. mass spectrometry: highly complex

  d. heterogenous immunoassay analyzer: moderately complexed

  e. urine drugs of abuse screening: waived

13. Regarding reimbursement, what is the PAMA regulation?

  a. affects all lab tests equally

  b. affects radiographic procedures and the clinical laboratory equally

  c. reductions of up to reimbursement of inpatients only

  d. affects Medicare Part B Clinical Lab Fee Schedule for outpatients

  e. provides reimbursement to hospitals serving indigent populations

14. The FDA has established a scoring system for designating approved/cleared clinical laboratory tests as moderately vs. high complexity. Which of the following is not a criteria favoring high complexity?

  a. interpretation of raw data results

  b. sample extraction

  c. amplification of nucleic material

  d. external proficiency testing samples are labile

  e. knowledge required to perform the test may be obtained through on-the-job instruction

15. Which of the following is not an important element for laboratories' Individualized Quality Control Plan?

  a. Performing internal quality control testing at a frequency less than recommended by the manufacturer.

  b. Assessing the preanalytical, analytical, and postanalytical risks for errors.

  c. Incorporation of a quality assurance plan that could include investigation of complaints.

  d. Utilization of electronic quality control procedures.

  e. Evaluating the testing environment of the clinical laboratory.

16. Clinical laboratory directors must decide to offer new tests in house or send them to a reference lab. Which of the following is not a major consideration?
    a. availability of technology for the tests
    b. cost differential between in house and outsourced
    c. turnaround time needs
    d. computer support capabilities
    e. availability of point-of-care testing options

17. Two equally qualified candidates (one male, one female) are applying for an assistant lab director job. Which of the following attribute may be most important?
    a. select the woman due to affirmative action (women underrepresented in the clinical lab field)
    b. select the candidate who has published the most papers
    c. select the candidate who has the best communication skills
    d. select the candidate who has the most postdoctoral experience
    e. select the candidate who has the most grant support

18. Which test is most likely to be a sent to a reference laboratory from a hospital?
    a. procalcitonin
    b. molecular assay for *C. difficile*
    c. ammonia
    d. CO-oximetry
    e. testing for antinuclear antigen

19. An outpatient has a serum glucose of 550 mg/dL, which exceeds the critical value. The chemistry staff calls the attending in the evening and gets an answering service. The service is unwilling to take responsibility to the result. Which of the following options is not optimal?
    a. Contact the pathology resident on-call (or lab director) to assist in finding the doctor.
    b. Have the resident contact the patient by phone.
    c. Contact the fire department or paramedic and have someone go to the person's house.
    d. Call 9-1-1 for a medical emergency.
    e. Investigate the test order request to find an alternate physician.

20. Your laboratory is interested in generating income for the lab and has agreed to conduct a clinical trial for a company wishing to get their test approved by the FDA. Which of the following is false?
    a. All clinical trials involving patient samples must be reviewed and approved by a local or central institutional review board.
    b. Collection of extra blood requires patient consent.
    c. All laboratory testing must be conducted by certified clinical lab scientists and conducted in CLIA certified laboratories.

    d. Research subjects, including healthy donors, are subject to HIPAA privacy policies.

    e. The IRB provides advice as to whether or not abnormal results should be returned to participating subjects.

21. Which of the following is not a HIPAA violation?

    a. stolen or lost laptop containing personal health information

    b. patient photographs posted on social media posts

    c. patient records for someone who died 60 years ago

    d. faxing medical records to a location to a machine that has open access

    e. disposal of laboratory records into the regular trash bins

22. Which of the following scenarios may not be an accepted practice for a mislabeled or missing labeled specimen?

    a. samples that cannot be recollected may be accepted

    b. a disclaimer is written on the medical records

    c. the individual signs for the specimen and may be held legally responsible

    d. accept ABO blood typing results

    e. the laboratory director can supersede policies on a case-by-case basis

23. What is the difference between laboratory inspections by the College of American Pathologists (CAP) versus The Joint Commission (TJC), COLA, and individual states?

    a. CAP is peer review, TJC and COLA are full-time staff reviews

    b. CAP inspects labs, TJC inspects hospitals, COLA inspects point-of-care testing

    c. Only CAP has received deemed status by the Center for Medicare and Medicaid

    d. Washington, New York, and Florida operate their own accreditation programs and are exempt from any oversight or disclosure to Centers for Medicare and Medicaid.

    e. CAP inspects academic hospitals, TJC inspects private hospitals, COLA inspects physician office labs

24. Which of the following is the top deficiency surveyors find when filing inspection reports?

    a. problems with proficiency testing

    b. lack of documentation for lab developed tests

    c. lack of documentation of personnel qualifications and competency

    d. use of expired reagents or calibrators

    e. analytical measurement range verification

25. Who is the owner of remnant blood samples once testing has been completed?

    a. donor of the sample

    b. laboratory performing the test

    c. doctor who ordered the test

    d. insurance company who paid for the test

    e. medical examiner in situations of the patient's death

26. The medical examiner of your city is working on a postmortem case involving toxicology. He comes to the lab to request premortem samples as part of his investigation. Which of the following is true?

    a. The decedent must have given permission for this prior to death

    b. This information is rarely useful

    c. Permission from the immediate family is needed

    d. Samples must be sent deidentified

    e. Premortem results may be more useful than postmortem for cause of death

Answers

1. d. Analytical sensitivity is required if the test undergoes modification by the laboratory. *Jenning L, et al. Recommended principles and practices for validating clinical molecular pathology tests. Arch Pathol Lab Med. 2009;133:743–55.*

2. b. Clearance refers to tests approved via the 510(k) route, equivalence to an FDA test that has already been approved. *Jennings L, et al. Recommended principles and practices for validating clinical molecular pathology tests. Arch Pathol Lab Med. 2009;133:743–55.*

3. d. Passing is 80% of proficiency challenges (typically 4 out of 5). If there are multiple pieces of equipment of the same type, not all have to be tested within a PT challenge. *Clinical Laboratory Improvement Amendments (CLIA) Proficiency testing. ftp://ftp.cdc.gov/pub/CLIAC_meeting_ presentations/pdf/Addenda/cliac0908/Addendum%20C.pdf.*

4. d. Electrophoresis is under the category of high complexity. *Clinical Laboratory Improvement Amendment. https://careers-exactsciences.icims. com/jobs/3867/non-certified-clinical-laboratory-technician-i— saturday—tuesday%2C-9%3A30am—8%3A00pm-shift/job.*

5. d. Nonlicensed individuals cannot perform duties that can directly affect the result of a clinical laboratory. Instrument maintenance and repair can indirectly affect clinical lab results. *Exact Science. Non-Certified Clinical Laboratory Technician I. Duties and responsibilities. https://careers-exactsciences.icims.com/jobs/3867/non-certified-clinical-laboratory-technician-i—saturday—tuesday%2C-9%3A30am—8%3A00pm-shift/job.*

6. c. Laboratory developed tests are considered high complexity by the Centers for Medicare and Medicaid and must be operated by licensed clinical laboratory scientists. *Ford A. Devices, decisions: POC glucose in the critically ill. CAP Today. https://www.captodayonline.com/devices-decisions-poc-glucose-critically-ill./.*

7. c. Most labs have a higher critical value list for potassium, typically 6.0 mmol/L. *Lab Corp Panic (critical) values. https://files.labcorp.com/*

labcorp-d8/2019-12/22095%20Panic%20%28Critical%29%20Values%20link%20-%20Dec%202019%20update_FINAL.pdf.

8. a. Units per liter is the most commonly used units for enzyme activity and represents the number of micromoles of substrate converted per minute. However, the International System of Units express activity in nanomoles of substrate converted per second, and nanokatal is the correct unit. *https://en.wikipedia.org/wiki/Katal.*

9. e. The lab must follow the same procedures for patient samples as proficiency samples. Proficiency testing for waved tests is preferred but not required. Centers for Medicare and Medicaid. *Clinical Laboratory Improvement Amendments (CLIA) Proficiency Testing and PT referral. Dos and Don'ts. https://www.cms.gov/Regulations-and-Guidance/Legislation/CLIA/Downloads/CLIAbrochure8.pdf.*

10. c. The one year of training is not a requirement. *Olea S. CLIA required personnel qualifications. https://www.jointcommission.org/-/media/deprecated-unorganized/imported-assets/tjc/system-folders/topics-library/clia_required_personnel_qualificationspdf.pdf?db=web&hash=EB925F656724FE6D610624CB0BDB2748.*

11. e. For labs doing waived testing, there are no personnel requirements. *CLIA Requirements for Lab Personnel. https://www.aafp.org/practice-management/regulatory/clia/personnel.html.*

12. b. Flow cytometry analyzers and mass spectrometry are considered highly complex. The FDA compiles a list of manufacturers, tests, and categories of approved text complexities. *https://www.accessdata.fda.gov/scripts/cdrh/cfdocs/cfCLIA/Search.cfm.*

13. d. PAMA refers to "Protecting Access to Medicare Act." New payor rates took affect January 1, 2018. It has resulted in significant reduction in reimbursement to outpatient clinical laboratory tests. PAMA is applicable to physician's offices, independent laboratories, or in limited circumstances by hospital laboratories for its outpatients or nonpatients. *https://www.nila-usa.org/nila/PAMA.asp.*

14. e. Knowledge to perform a moderately complex test can be achieved from instruction or experience. *CLIA Categorizations. https://www.fda.gov/medical-devices/ivd-regulatory-assistance/clia-categorizations.*

15. a. The three elements of an IQCP are risk assessment, quality control, and quality assessment. *Clinical labs must adhere to the minimum frequency of QC testing. Individualized Quality Control Plan. https://www.cdc.gov/labquality/docs/IQCP-Layout.pdf.*

16. d. Reference laboratories have invested heavily in information technology support such that interfacing to laboratory information systems is no longer a major consideration. *Weinert K, et al. Triangulating dynamic of clinical laboratory testing. Clin Chem 2015;61:1320–7.*

17. c. The clinical laboratory field has a significant educational and consultative role, which requires good oral communication skills. *Gallo C. Five stars. The communication secrets to get from good to great. St. Martin's Griffin, New York, NY, 2018.*

18. e. Procalcitonin and *C. difficile* produce results that are immediately actionable, therefore outsourcing is not practical. Ammonia and CO-oximetry have sample stability issues. ANA testing is for autoimmune disease, where the turnaround time is not an issue. *Weinert K, et al. Triangulating dynamic of clinical laboratory testing. Clin Chem 2015;61:1320–7.*

19. d. Of these options, calling 9-1-1 option is probably the least acceptable. *Genzen JR, et al. Pathology consultation on reporting of critical values. Am J Pathol 2011;135:505–13.*

20. c. Unless results are to be reported to subjects and/or their physicians, CLIA certification is not required. University of California, clinical laboratory testing in human subjects research. *https://cphs.berkeley.edu/clia.pdf.*

21. c. HIPPA applies for deceased individuals up to 50 years after their death. *Health information of deceased individuals. 45 CFR 160.103. https://www. hhs.gov/hipaa/for-professionals/privacy/guidance/health-information-of-deceased-individuals/index.html.*

22. d. The clinical laboratory must have procedures for handling mislabeled or missing specimens and accept them only when they are deemed irreplaceable or may cause harm if recollected. It may be impossible for a lab to reject specimens in every case, but some labs have an absolute rejection policy for blood bank typing due to the potential medical consequences. Many labs do not accept specimens for ABO typing under any circumstance. A sample mislabeled specimen policy from Akron Children's Hospital is referenced. *https://www.akronchildrens.org/lab_test_specimen_procedures/Rejected_Specimen_Policy.html.*

23. a. These three out of the total of seven organizations have deemed status from CMS and inspect the majority of clinical labs in the United States today. They are the College of American Pathologists, The Joint Commission, and COLA. Some states are exempt to CLIA, but records are still subject to disclosure to CLIA. *Chittiprol S. Top Laboratory Deficiencies Across Accreditation Agencies. Problems with documenting personnel competency top the list. Clin Lab News 2018, July, https://www.aacc. org/publications/cln/articles/2018/july/top-laboratory-deficiencies-across-accreditation-agencies.*

24. c. This is particularly true for personnel conducting point-of-care testing. *Chittiprol S. Top Laboratory Deficiencies Across Accreditation Agencies. Problems with documenting personnel competency top the list. Clin Lab News 2018, July, https://www.aacc.org/publications/cln/articles/2018/july/top-laboratory-deficiencies-across-accreditation-agencies.*

25. b. Regulations have been stipulated that the laboratory owns pathology tissue blocks, and issue that has been debated. By a similar token, the clinical lab should have ownership of blood samples and can decide on disposal policies. *Dry S. Who owns diagnostic tissue blocks? Lab Med 2009;40:69–73.*

26. e. Premortem toxicology results may be more useful given the problems of postmortem interpretation of results, and cooperation with medical examiners or coroners is appreciated. *Dry S. Who owns diagnostic tissue blocks? Lab Med 2009;40:69–73.*

Chapter 36

Lab medicine informatics

Jason M. Baron and Anand S. Dighe
Massachusetts General Hospital, Boston, MA, United States

1. When moving laboratory data from a legacy information system to a new information system semantic interoperability would ensure that:
   a. The graphical display of a complete blood count will be the same in both systems.
   b. The metadata for the complete blood count (e.g., reporting lab name, report date) will be maintained without corruption.
   c. All of the regulatory required fields for a laboratory test report are present.
   d. Results' comments are included in their entirety in both systems.
   e. The results of a complete blood count, when viewed in the new system, will have the same meaning as the results did in the legacy system.

2. When installing a new Laboratory Information System (LIS) some users may be given additional training on the system and be termed "super users." What are super users typically asked to do during the implementation phase of the project?
   a. Adjust the configuration of the system based on user feedback.
   b. Create documentation for system maintenance activities.
   c. Train and support other users of the system.
   d. Prioritize system enhancement requests.
   e. Document defects with the new system.

3. A diagnosis-related group (DRG) is a patient classification system that standardizes reimbursement for inpatient care. DRG coding is based on factors including:
   a. actual cost of inpatient services provided
   b. diagnoses treated and procedures performed during the admission
   c. patient's age
   d. patient's medication lists
   e. complications the patient has during his/her inpatient stay

4. An organization upgrades their new blood bank system to the latest version, which includes new features for red blood cell antigen identification. The new features were all tested and worked properly. However, the ability to

Self-assessment Q&A in Clinical Laboratory Science, III. https://doi.org/10.1016/B978-0-12-822093-1.00036-3
427

post type and screen results, a feature that had worked in prior version of the system and was not tested with the update, was no longer functional after going live with the upgrade. What could have avoided this issue?

a. unit testing

b. more robust servers

c. Having more super users

d. improved training

e. regression testing

5. One of the main differences between a data warehouse and an information system database is:

a. A data warehouse is often optimized for analytic queries.

b. A database is read only while a data warehouse is not.

c. A data warehouse has a real time data feed while a database does not.

d. A database is mirrored and secure while a data warehouse is not.

e. A data warehouse has a robust audit trail while a database does not.

6. During a monthly update of an information system with some new advanced features, what is an essential process control step prior to the update to ensure the new features do not negatively impact other parts of the system:

a. training

b. defect tracking

c. process improvement

d. enhancement request process

e. change control

7. Health Level 7 version 2 messages are limited in their ability to connect healthcare systems due to:

a. their lack of structure.

b. limitations on message size.

c. variation from one vendor to the next regarding how messages are implemented.

d. lack of a human readable component.

e. their inability to be encrypted.

8. The following Health Level 7 message is a result message for a potassium value of 6.2. What do the vertical bars (|) and carats (ˆ) in the message represent?

MSH|ˆ~\&|LIS|SQTEST||1000|201907182122||ORU ʀ01|12332333
|P|2.3|
PID|1|12545|83838383||||TEST PATIENT||01/01/1999|M|||||||||||
OBR|1||83838383|C||201907181950|||||||||201907181952|||||||
|201907181952|||R||||
OBX|1|NM|6298-4 Potassium||6.2|mmol/L|3.4–5.3|H|||F

a. database fields

b. tabs and line breaks

c. check characters

d. database commands

e. data field delimiters

9. All of the following are stored in Laboratory Information System (LIS) maintenance tables (dictionaries) EXCEPT:

a. test reference ranges

b. allowable tube types for a given test

c. test units of measure

d. patient demographic information for specific patients

e. codes used for reporting

10. Patient demographic data found in the Laboratory Information System (LIS) typically originates from:

a. Technologist manual entry of demographics into the LIS.

b. message from the billing system.

c. An ADT message from the registration system.

d. An HL7 message from the instrument middleware.

e. An order from the Electronic Health Record.

11. The following are all reasons for implementing electronic interfacing between a Laboratory Information System (LIS) and a reference laboratory except:

a. more rapid turnaround time

b. fewer errors due to eliminating manual entry

c. improved test utilization

d. reduced labor due to decreased manual result entry

e. enhanced ability to include full, multicomponent results

12. A Laboratory Information System (LIS) printer located on a hospital floor prints bar-coded specimen labels for the collection of a basic metabolic panel (BMP) and CBC. The BMP label indicates the BMP should be collected in a green top tube. Typically encoded in the BMP specimen label bar code is:

a. The patient's medical record number.

b. The patient's encounter number.

c. The EHR order number of the specimen.

d. The container identifier.

e. The patient's first and last name.

13. Health Level 7 (HL7) messaging is commonly used for all of the following except:

a. Sending an Admit-Discharge-Transfer (ADT) message from a registration system to the LIS.

b. Electronic interfacing between a Laboratory Information System (LIS) and a reference laboratory for orders and results.

c. Sending results from the LIS to the EHR.

d. Sending results from an instrument to the LIS.

e. Sending orders from the EHR to the LIS.

14. LOINC (Logical Observation Identifiers Names and Codes) is an international standard for identifying medical laboratory observations. For LOINC codes that represent test results all the following characteristics of the test are defined by LOINC EXCEPT:
    a. manufacturer of the test
    b. interval of time over which specimen is collected (e.g., random urine vs 24 hour urine)
    c. units of measure
    d. methodology of the test
    e. test name

15. A clinical decision support system is developed to alert providers when a patient meets rule-based criteria for sepsis. When is the system less likely to provide appropriate advice?
    a. In a patient with classical signs and symptoms of sepsis.
    b. In a patient with an unusual presentation of sepsis.
    c. When the provider is asked to input data about the patient's symptoms before calculating the sepsis risk score.
    d. When triggering criteria are set so it fires in a highly specific manner.
    e. In a patient with a long length of stay and many laboratory results.

16. Two-dimensional barcodes (QR codes or DataMatrix) have all of the following advantages over linear bar codes (Code 128 or Code 39) EXCEPT:
    a. 2D bar codes take up less space for the amount of data encoded than linear bar codes.
    b. 2D bar codes can be more resistant to damage due to the presence of replicated data.
    c. 2D bar coded specimen labels can be more accurately scanned by most chemistry automation systems.
    d. 2D bar codes can encode more data types than linear bar codes.
    e. 2D bar codes can encode web addresses and long strings of text.

17. Which of the following is not a required data element in a laboratory test result report according to CLIA and good laboratory practice?
    a. reference ranges
    b. specimen source
    c. name and address of reporting lab
    d. units of measure
    e. date and time of the order

18. Which of the following is the best use case for the Fast Healthcare Interoperability Resource (FHIR)?
    a. Sending data from a blood gas analyzer to a Laboratory Information System.
    b. Sending laboratory orders from a Laboratory Information System to a reference laboratory.
    c. Laboratory order messages between the EHR and the Laboratory Information System.
    d. Accessing lab data from an EHR by a smartphone-based application.
    e. Sending point-of-care results from a glucose meter to middleware.

19. Which one of the following is a correct statement about Clinical Decision Support (CDS) alert fatigue in electronic health records (EHRs)?
    a. Alert fatigue is associated with alerting systems where there are a small number of critical alerts.
    b. With alert fatigue, clinicians may become desensitized to EHR alerts, and as a result may ignore or fail to respond appropriately to critical warnings.
    c. Alert fatigue has no impact on patient safety.
    d. Alert fatigue is highest for providers that see the fewest alerts.
    e. Alert fatigue can be minimized by making alerts less patient specific.

20. Six sigma-based process improvement programs:
    a. Are based on the removal of all non-value-added activities.
    b. Involve data-driven improvement cycles of a process with steps of defining, measuring, analyzing, improving, and controlling.
    c. Use Pareto charts to identify targets of improvement.
    d. Were developed at the Toyota motor company in the 1950s.
    e. Graphically represent a process as a series of steps.

21. In a software project which one of the following best describes the relationship between bugs (defects) and the cost to fix them:
    a. Bugs are easy to find and cheap to fix early in the development process.
    b. Later in the development process bugs are harder to find and more costly to fix.
    c. Later in the development process bugs are easier to detect but more costly to fix.
    d. Bugs are costly to fix at any point in the software development cycle.
    e. Early in the development cycle bugs are costly to fix.

22. Bidirectional electronic laboratory instrument interfacing with the laboratory information system (LIS) has all of the following advantages except:
    a. Reduces laboratory test turnaround time.
    b. Facilitates the reporting of results from the LIS to the electronic health record (EHR).
    c. Prevents manual result entry errors.
    d. Improves technologist productivity.
    e. Permits the instrument to know the tests that need to be performed on a given sample.

23. A highly elevated result for TSH (reported as ">50") is not flagged as high in the Electronic Health Record, leading to a patient safety issue. What is the most likely reason for the lack of flagging in the EHR?
    a. No reference range set for TSH in the laboratory information system (LIS).
    b. Mistranslation of the high result flag from the LIS to the EHR.

    c. The nonnumeric result value failed to flag in the LIS.
    d. Result message from the LIS failed transmission to the EHR.
    e. Elevated TSHs not set up to flag in the LIS.

24. Phishing email attacks, where email recipients are enticed into following the attacker's instructions, are best prevented by:
    a. external firewall configuration
    b. encryption of mobile devices
    c. user authentication procedures
    d. workforce education
    e. highly secure email servers

25. Best practice for Clinical Decision Alerts includes all of the following except:
    a. Increase alert specificity by reducing clinically low yield alerts.
    b. Tailor alerts to patient and provider characteristics.
    c. Classify alerts according to severity, with critical warnings clearly delineated from less important warnings.
    d. Only make low importance alerts interruptive.
    e. Use human factors engineering principles in alert design.

26. An important difference between supervised and unsupervised machine learning is that the training data used for supervised machine learning must include:
    a. randomly sampled data
    b. "ground truth" target results
    c. numerical predictors
    d. more predictors than cases
    e. a binary target variable

27. A researcher wishes to prepare a deidentified clinical dataset for a study of adult patients by removing patient identifiers from an extract of EHR data? Which of the following is a data element that the researcher can leave in the dataset and still have it likely meet the standards for deidentification?
    a. medical record number
    b. collection date of a laboratory test
    c. patient address as long as the street name is redacted
    d. patient employer and job title
    e. patient age in years

28. Which of the following genetic sequencing files contains annotations regarding differences between the patient and the reference sequence?
    a. image
    b. BAM (binary alignment map)
    c. SAM (sequence alignment map)
    d. VCF (variant call file)
    e. FASTQ

29. You perform an analysis of test ordering patterns for Primary Care Physicians (PCPs) in your health system. You observe that in the past year, some PCPs ordered CBC tests on most of their patients while others had ordered this test much less frequently. You intend to present these findings to PCP leadership. Which of the following additional analyses would be most useful prior to presenting the finding?
    a. Each PCP's CBC utilization adjusted for patient diagnoses and clinical status
    b. Each PCP's history of malpractice claims adjusted for practice volume
    c. Each PCP's gross revenue generated per patient
    d. Each PCP's rate of ED utilization adjusted for patient diagnoses and clinical status
    e. Each PCP's radiology test ordering adjusted for patient diagnoses and clinical status

30. You train a supervise machine learning model to predict which patients presenting to the emergency department will develop sepsis. You train the model on 200 training cases and then test it on 50 held out test cases. The model performs extremely well (AUC of 0.99) on the training cases, but only exhibits fair performance (AUC of 0.72) on the test cases. Had you instead trained your model on a larger training set consisting of 1000 cases, which of the following might you expect?
    a. better training data performance and better testing data performance
    b. better training data performance and worse testing data performance
    c. worse training data performance and better testing data performance
    d. worse training data performance and worse testing data performance
    e. better training data performance and similar testing data performance

Answers

1. e. Semantic interoperability is the ability of systems to exchange data with unambiguous, shared meaning. It is a requirement for healthcare organizations that wish to exchange laboratory results and other clinical data with each other. *Goossen WT. Detailed clinical models: representing knowledge, data and semantics in healthcare information technology. Health Inform Res 2014;20:163–72.*

2. c. Having a significant number of users of the new LIS system with early and advanced training is often important for successful implementation. Super users are primarily used for assisting with "at the elbow" training and support for the new system, once the system has been implemented. *Stud Health Technol Inform 2018;247:191–5.*

3. b. In general, a diagnosis-related group (DRG) payment provides a global reimbursement for an inpatient stay, with the exception of professional fees, which typically are paid on a fee for service schedule. Patients that share similar clinical conditions and procedures are grouped under the same DRG. The DRG system was designed to encourage efficient use of hospital resources by reimbursing a set amount for an episode of

inpatient care. *Revisions to hospital inpatient prospective payment system. Fed Regist 2006;71:47869-48351.*

4. e. Regression testing is a type of system testing used to confirm that a recent update has not adversely affected existing features. *Cowan DF et al. Validation of the laboratory information system. Arch Pathol Lab Med 1998;122:239–44.*

5. a. A data warehouse pulls together data from many different sources within an organization and is optimized for reporting and analysis. Data warehouses are often not updated in real time. Audit trails and security concerns may be just as important for data warehouses, depending on the data stored in the data warehouse. *Tuthil JM, et al. Pathology Informatics, Theory and Practice 2012, pp. 62–3.*

6. e. A formal change control process should be used to ensure that changes to a system are introduced in a controlled and coordinated manner. Documentation of all changes is critical to be able to understand what has been changed and what else that change could impact. It reduces the possibility that changes will be introduced to a system that could lead to errors in other parts of the system. *Cowan DF et al. Validation of the laboratory information system. Arch Pathol Lab Med 1998;122:239–44.*

7. c. Health Level 7 messaging, although considered a "standard," has sufficient flexibility built into the implementation such that different vendors and organizations have implemented it differently, limiting its ability to be used to easily share laboratory data between organizations without significant customization of the interfaces. *Tuthil JM, et al. Pathology Informatics, Theory and Practice 2012, pp. 135–7.*

8. e. The vertical bars (|) and carats (ˆ) are delimiters or separators that are used between data fields (vertical bar) or within a single data field (caret) to separate data. *Tuthil JM, et al. Pathology Informatics, Theory and Practice 2012, pp. 135–7.*

9. d. LIS maintenance tables store information needed to define tests, tube types, codes, and worksheets. These tables do not contain patient information. *Tuthil JM, et al. Pathology Informatics, Theory and Practice 2012, pp. 88–9.*

10. c. In most systems the LIS receives HL7-based Admit-Discharge-Transfer (ADT) messages from registration systems that serve to update the LIS with patient demographic information. *Tuthil JM, et al. Pathology Informatics, Theory and Practice 2012, pp. 78–79.*

11. c. Test utilization would not be expected to significantly change when interfacing a LIS to a reference laboratory. *Tuthil JM, et al. Pathology Informatics, Theory and Practice 2012, pp. 102–3.*

12. d. The bar code of a specimen label contains a unique identifier for that container. All the other choices would not necessarily be unique identifiers for that tube. *Tuthil JM, et al. Pathology Informatics, Theory and Practice 2012, pp. 92–3.*

13. d. Health Level 7 (HL7) messaging provides specifications for a variety of health and medical transactions including laboratory orders, laboratory results, and demographic updates between LIS systems or between LIS and EHR systems. Instrument to LIS interfaces do not generally utilize HL7 30. c. A significant issue in machine learning is overfitting. An overfit model identifies random patterns within a set of training data that do not generalize. An overfit model will perform better on training data than on an independent set of test data. The testing data performance usually represents the "expected" model performance should the model be used in practice and is usually the relevant performance metric. A model trained on a larger set of training data will be less likely to overfit; assuming adequate predictive information remains to be captured, using a larger set of training data will lead to a better fit and improved test data performance. However, a larger set of training data will tend to decrease the training data performance of an overfit model, since the model is less able to take advantage of random nuances to make predictions. *Rudolf JW, et al. Decision support tools within the electronic health record. Clin Lab Med 2019;39:197–213.*

14. a. LOINC codes do not include test manufacturer information as a test characteristic. All of the other answers are items that LOINC specifies for each test result. *Tuthil JM, et al. Pathology Informatics, Theory and Practice 2012, pp. 182–3.*

15. b. Clinical decision support can be based upon a variety of criteria about the patient's condition, diagnoses, and test results. Asking users to input data into the alert will improve the specificity of the alert recommendations but may be intrusive to users. Using highly specific criteria would likely lead to underfiring, not overfiring. CDS generally performs poorly in patients with unusual presentations of the condition of interest since the criteria for alerting are usually based on the typical presentation of the disease. *Rudolf JW et al. Decision support tools within the electronic health record. Clin Lab Med 2019;39:197–213.*

16. c. 2D bar codes can encode more data and more types of data than linear bar codes. Since most specimen labels for chemistry automation systems are linear bar codes, the systems are only designed to read linear bar codes and they cannot scan 2D bar codes. *Tuthil JM, et al. Pathology Informatics, Theory and Practice 2012, pp. 286–90.*

17. e. CLIA requirements for a laboratory report include patient demographics, the name and address of the reporting laboratory, the test report date, test performed, test units of measure, reference range, and specimen source. *Tuthil JM, et al. Pathology Informatics, Theory and Practice 2012, pp. 330–1.*

18. d. Fast Healthcare Interoperability Resources (FHIR, pronounced "fire") is a standard describing data formats, elements (resources), and programming interfaces that facilitate exchange of electronic health record data.

It is used to access EHR data for use by other applications. *Mandel JC. SMART on FHIR: a standards-based, interoperable apps platform for electronic health records. Am Med Inform Assoc 2016;23:899–908.*

19. b. In the context of EHR decision support, alert fatigue can occur when a provider is exposed to a large number of alerts, especially when many of the alerts are inconsequential, and then the provider begins to ignore or be desensitized to alerts. Desensitization can lead to longer response times or missing critically important alerts. *Rudolf JW, et al. Decision support tools within the electronic health record. Clin Lab Med. 2019;39:197–213.*

20. b. Six sigma is a data-driven methodology for eliminating defects, aiming for six standard deviations between the mean and the specification limit (3.4 defects per million opportunities). It involves a data-driven process with five steps for improvement cycles: define, measure, analyze, improve, and control. *Tuthil JM, et al. Pathology Informatics, Theory and Practice 2012, pp. 305–6.*

21. c. Late in the development cycle software defects tend to be easier to detect, since many distinct parts of the software now must all work together to complete the task and defects are more easily apparent to testers. Defects that are found late in the process are often much more expensive to fix since the software is largely complete and the defects may uncover design flaws that require considerable engineering to address. Early in the software development cycle defects can be more challenging to find since much of the software is not functional so cannot be fully tested. *Clinical Informatics Study Guide (Springer), 2016, Finnell and Dixon, eds., pp. 268–71.*

22. b. Bidirectional interfacing of laboratory instruments increases the efficiency and productivity of the laboratory. The reporting to the EHR would not be expected to be impacted by a LIS-instrument interface as the LIS-EHR interface is a separate interface, using HL7 to send result messages from the LIS to the EHR. *Tuthil JM, et al. Pathology Informatics, Theory and Practice 2012, pp. 91–6.*

23. c. In some situations results that are reported with a greater than symbol (e.g., >50) may be misinterpreted by the LIS and fail to flag. For a common test like TSH the laboratory would likely have a reference range set and would be flagging elevated results. Testing outlier results such as results with greater than or less than values is important in the testing plan for a LIS. *Tuthil JM, et al. Pathology Informatics, Theory and Practice 2012, pp. 94–5.*

24. d. While there is a role for some types of antiviral software and other central systems to help prevent phishing attacks, the best defense against such attacks is workplace education regarding how to avoid and how to report phishing attacks. *Mayhorn CB, et al. Training users to counteract phishing. Work 2012;41:3549–52.*

25. d. Interruptive alerts stop a provider's workflow, and these should be reserved for only high importance alerts. Alerts should be tailored as much as possible to patient and provider characteristics such that they only display when likely to be useful to providers. *Rudolf JW et al. Decision support tools within the electronic health record. Clin Lab Med. 2019;39:197–213.*

26. b. Supervised machine learning involves training a model to predict a target (dependent) variable based on a set of predictors. For example, a model may be trained to predict whether a patient will develop sepsis (target) using a set of laboratory test results. When the target variable is binary (e.g., sepsis or no sepsis) or categorical a supervised machine learning problem is referred to as classification. When the target is numerical (e.g., blood pressure) the problem is referred to as regression. Supervised machine learning algorithms may but do not necessarily need to be trained on randomly sampled data. Unsupervised machine learning, in contrast, does not involve prediction of ground truth target variables. Rather, unsupervised machine learning algorithms find patterns within the predictor variables themselves; some unsupervised algorithms cluster cases into groups based on similarity. *Deo RC. Machine learning in medicine. Circulation 2015;132(20):1920–30.*

27. e. EHR data may be used for secondary research purposes, subject to appropriate approvals, institutional policies, and legal and administrative requirements. Using deidentified data is often required by institutional review boards or may be necessary to meet procedural requirements (e.g., external data sharing). Deidentifying data helps to protect patient privacy. For data to be deidentified, it must have all elements removed that could be used in isolation or in combination to identify the actual patient. While some patient identifiers are obvious (e.g., name and medical record number), others may be more subtle. For example, dates with granularity less than a year are generally considered to be identifiers. Likewise addresses and even five-digit zip codes are generally considered identifiers. Employer and job title would often point to a specific person. Ages (in years) are generally not considered identifiers; however, ages 90 and older may be identifiers and may require special treatment. For research, the relative timing of events is often needed (e.g., collection date relative to symptom onset). Various date shift approaches are available to obscure the actual date while maintaining relative time. In the United States, HIPAA regulations provide deidentification standards. In general, data may be deidentified using a safe harbor method in which all elements from a list of identifiers is removed and there is no reason to believe the data can be reidentified. Data may also be classified as deidentified using a statistical approach with an expert determination. *Tuthil JM, et al. Pathology Informatics, Theory and Practice 2012, pp. 323–5.*

28. d. Depending on the type of sequencer, primary sequence data may output as a FASTQ file (or as a uBAM file), containing the sequence of each individual read and the quality of the base call. Primary sequence data (e.g., FASTQ files) are then typically aligned against a reference genome; the alignment can be stored as a SAM file (or the binary equivalent, the BAM file). Variants between the patient and the reference genome are then annotated and can be stored in a VCF file. *Pereira R. Bioinformatics and computational tools for next-generation sequencing analysis in clinical genetics. J Clin Med 2020;3;9(1). PMID 31947757.*

29. a. Variation analysis, comparing utilization between clinicians with similar practices can be a useful utilization management tool. However, sometimes important clinically relevant differences exist between the patient populations seen by clinicians within the same specialty with superficially similar practices. These differences may help to explain the apparent variation. Thus, it might be useful to adjust the CBC utilization for patient factors. While ED and radiology utilization might also be useful metrics, they do not directly address CBC utilization. *Lewandrowski K. Utilization Management in the Clinical Laboratory and Other Ancillary Services, 2017, Springer, p. 49.*

30. c. A significant issue in machine learning is overfitting. An overfit model identifies random patterns within a set of training data that do not generalize. An overfit model will perform better on training data than on an independent set of test data. The testing data performance usually represents the "expected" model performance should the model be used in practice and is usually the relevant performance metric. A model trained on a larger set of training data will be less likely to overfit; assuming adequate predictive information remains to be captured, using a larger set of training data will lead to a better fit and improved test data performance. However, a larger set of training data will tend to decrease the training data performance of an overfit model, since the model is less able to take advantage of random nuances to make predictions. *Rudolf JW, et al. Decision support tools within the electronic health record. Clin Lab Med 2019;39:197–213.*

Chapter 37

Point-of-care testing

Adil I. Khan[a] and Alan H.B. Wu[b]
[a]*Temple University, Philadelphia, PA, United States,* [b]*University of California, San Francisco, CA, United States*

1. Which of the following is false regarding point-of-care testing?
   a. small blood volume used
   b. more expensive than traditional laboratory tests
   c. small menu of tests available
   d. whole blood, urine, and other bodily fluids analyzed without processing
   e. quality is questionable as anyone can run the analysis
2. What is an advantage of point-of-care testing?
   a. more expensive than traditional laboratory testing
   b. quality is questionable as anyone can run the analysis
   c. difficulties with regulatory compliance, billing, and documentation
   d. whole blood, urine, and other bodily fluids analyzed without processing
   e. patient-focused staff with little formal education or experience in laboratory testing
3. Which of the following statements is false?
   a. CLIA applies to any point-of-care tests even if no patient care decision is made on the result
   b. point-of-care sites are treated like independent laboratories regardless of location
   c. a license is required for a physician to start testing patients using a point-of-care testing device.
   d. documentation of point-of-care testing results is required for accreditation
   e. all of the above
4. What are the functions of a point-of-care data manager?
   a. can store lists of operators that are competent to perform testing
   b. update any reagent and control lots, expiration dates to the device
   c. sends test results to the laboratory or hospital information system
   d. receives quality control results from the device to document successful completion
   e. all of the above

Self-assessment Q&A in Clinical Laboratory Science, III. https://doi.org/10.1016/B978-0-12-822093-1.00037-5

5. What is not true regarding the POCT connectivity standard developed by the Clinical Laboratory Standards Institute?
   a. The POCT1 standard is based on the Institute of Electronic Engineers and Health Level Seven standards for medical information transfer.
   b. Purpose was to have seamless information exchange between point-of-care testing devices, electronic medical records, and laboratory information systems.
   c. The standard defines the physical connections between point-of-care testing devices and data managers.
   d. The standard defines communication protocols between point-of-care testing devices and data managers.
   e. The standard defines the maximum size of point-of-care testing devices.

6. What is true about the purpose of a point-of-care committee?
   a. Reviews and approves requests for point-of-care testing.
   b. Has an interdisciplinary composition to address all hospital issues regarded with testing.
   c. Can help balance opinions, and depersonalize judgments.
   d. Helps maintain quality assurance at all aspects of the testing process.
   e. all of the above

7. What are examples of quality improvement performance monitors for point-of-care testing?
   a. documentation of quality control (QC)
   b. expired reagents and QC/reagent bottles and kits dated appropriately
   c. temperature checks for storage of point-of-care reagents, QC, and devices
   d. handling changes in lots for QC and reagents according to manufacturer recommendations
   e. all of the above

8. What is essential for a successful point-of-care testing service?
   a. Performance and documentation of quality control according to policy.
   b. Participation in an external quality assessment program.
   c. Effective interdisciplinary communication.
   d. Self-management of point-of-care testing service.
   e. Ensure testing is performed by laboratory scientists.

9. After implementing a new point-of-care service, the laboratory in conjunction with the hospital should perform a post hoc evaluation. Which of the following is the least important criteria?
   a. Compliance with training and daily quality controls.
   b. Satisfaction by the POCT operators.
   c. Economic analysis to determine if pretesting estimates have not been greatly exceeded.
   d. Determine if the quality of results were achieved.
   e. Demonstrate an improved medical outcome (morbidity and mortality).

10. The Clinical Laboratory Standards Institute recommends that glucose meters agree with a comparative central laboratory method within:
    a. ±6 mg/dL for results <100 mg/dL and ±10% for results >100 mg/dL
    b. ±15 mg/dL for results <75 mg/dL and ±20% for results >75 mg/dL
    c. ±20% for all results
    d. ±12 mg/dL for results <100 mg/dL and ±12.5% for results >100 mg/dL
    e. ±5 % for all results

11. Which statement is true, glucose meters:
    a. are used in the screening for diabetes
    b. are used in the management of diabetes
    c. can be used on plasma, serum, or whole blood
    d. are not affected by hematocrit
    e. are affected less by interferences than laboratory instruments

12. What are the factors that can affect a glucose result taken with a glucose meter?
    a. humidity
    b. temperature
    c. altitude
    d. hematocrit
    e. all of the above

13. Which of the following analytes cannot be measured transcutaneously?
    a. neonatal bilirubin
    b. hemoglobin oxygen saturation
    c. blood gases
    d. glucose
    e. hemoglobin A1c

14. Which of the following is not an etiology of kernicterus?
    a. prematurity
    b. Rh incompatibility
    c. glucose-6-phosphate dehydrogenase deficiency
    d. liver cirrhosis of the mother
    e. hemolytic disease

15. What is the appropriate therapy for neonatal hyperbilirubinemia?
    a. exchange transfusion
    b. plasmapheresis
    c. phototherapy with blue-green light
    d. antibilirubin antibodies
    e. renal dialysis

16. Certain tests such as the pulse oximeter, alcohol breathalyzer, and bilirubinometers are exempt from CLIA oversight. What is the best explanation for this?
    a. erroneous results have little impact on medical care
    b. no sample is taken
    c. tests are available over the counter

    d. they are considered as lab developed tests (LDTs)

    e. they were approved under the FDA's Emergency Use Authorization

17. The Clarke Grid is a statistical visualization of biases produced by glucose meters as compared to central lab glucose measurements. Which of the following is true?

    a. Zones A and B are within 20% of the reference

    b. Zone E indicates that treatment for hypoglycemia or hyperglycemia may be inappropriate

    c. Zone C indicates that hypoglycemia or hypoglycemia may go unrecognized

    d. Zone D indicates that a result in this region may lead to unnecessary treatment

    e. Zone B indicates that the patient has abnormally high or low hematocrit

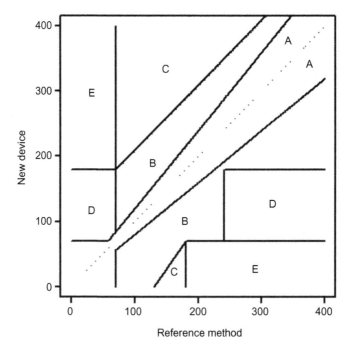

18. Which of the following is false regarding bilirubinometers?

    a. results are less with children of color

    b. correlation is good against indirect bilirubin

    c. testing is exempt from CLIA regulations

    d. does not correlate with delta bilirubin

    e. reduces blood volume loss of neonates

19. Which of the following is not a limitation of POC testing for PT-INR versus the central laboratory?
    a. abnormal hematocrits
    b. coadministration of other anticoagulants, e.g., low molecular weight heparin
    c. presence of antiphospholipid inhibitors
    d. venous blood collections are necessary
    e. less accurate at high PT-INR values (>3.0)

20. Which of the following is a limitation of qualitative urine pregnancy tests relative to quantitative serum hCG?
    a. Both suffer from heterophile antibody interference.
    b. Urine tests are less sensitive therefore are not useful for early pregnancy.
    c. Negative urine results can occur with high-dose hook effect.
    d. Urine pregnancy tests are not waived.
    e. Urine pregnancy tests cannot be used in clinical trials.

21. In 2009 the FDA recalled glucose meters that use the glucose dehydrogenase-pyrroloquinoline quinone method due to interferences with patients treated with icodextrin used in peritoneal dialysis. Which of the following is the mechanism of action for the interference?
    a. competitive inhibition of GDH with icodextrin
    b. noncompetitive inhibition of GDH with icodextrin
    c. icodexdrin metabolizes to maltose producing a false positive result
    d. icodextrin is highly pigmented causing a photometric interference
    e. interference also affects the hexokinase method

22. Which of the following is not an interferent with glucose meters?
    a. ascorbic acid
    b. acetaminophen
    c. uric acid
    d. mannitol
    e. ibuprofen

23. Point-of-care testing for cardiac troponin has received only minimal interest over the years for triage of patients with chest pain. What is the primary reason?
    a. Assays have lower analytical sensitivity than the corresponding central lab assays.
    b. POCT assays are not standardized to the central lab.
    c. The specificity of POCT assays is inferior to the central lab.
    d. Rapid turnaround time from the ED is no longer needed for NSTEMIs.
    e. Testing platforms are not affordable.

24. Which of the following is true for most lateral flow point-of-care assays for drugs of abuse testing?
    a. Most assays make use of two antibodies whereby a line in the test zone indicates a positive result.

    b. Most assays make use of two antibodies whereby a line in the test zone indicates a negative result.

    c. Most assays make use of one antibody whereby a line in the test zone indicates a positive result.

    d. Most assays make use of one antibody whereby a line in the test zone indicates a negative result.

    e. The control line is placed proximal to the sample application relative to the test line.

25. What of the following types of testing is correct?

    a. *In vitro*: outside the body, *in vivo*: in a test tube; *ex vivo*: tissue testing, wearable: worn testing device

    b. *In vitro*: worn testing device, *in vivo*: in a test tube, *ex vivo*: tissue testing outside, wearable: outside the body

    c. *In vitro*: in a test tube, *in vivo*: worn device, *ex vivo*: tissue testing outside, wearable: outside the body

    d. *In vitro*: in a test tube, *in vivo*: inside the body, *ex vivo*: tissue testing outside, wearable, worn testing device

    e. *In vitro*: outside the body, *in vivo*: inside a tst tube, *ex vivo*: worn testing device, wearable: tissue testing outside

Answers

1. c. A wide menu is available for point-of-care tests. There are several portable POCT devices on the market, and the available menu of analytes is extensive, including glucose, urinalysis, pregnancy, occult blood, electrolytes, blood gases, creatinine, urea, microalbumin, hemoglobin A1c, drugs of abuse, therapeutic drug monitoring (lithium), as well as testing for infectious diseases such as streptococcus, mononucleosis, influenza, and HIV. Even coagulation testing [prothrombin/International Normalized Ratio (INR)], enzymes (alanine transaminase/aspartate transaminase), and cardiac markers [B-type natriuretic peptide (BNP)] are available on waived devices. *Clarke W, et al. eds., Contemporary Practice in Clinical Chemistry, Clarke W, et al., eds., 2016, p. 277.*

2. d. Difficulties with regulatory compliance, billing, and documentation. Being able to analyze whole blood, urine, and other bodily fluids without processing helps in getting the result faster and is one of the advantages of point-of-care testing. *Clarke W, et al. eds., Contemporary Practice in Clinical Chemistry, Clarke W, et al., eds., 2016, p. 278.*

3. a. CLIA applies to any point-of-care tests even if no patient care decision is made on the result. Research testing where the point-of-care test results are not used for patient management decisions is exempt from CLIA regulations. *Clarke W, et al. eds., Contemporary Practice in Clinical Chemistry, Clarke W, et al., eds., p. 279.*

4. e. All of the above are important functions of a point-of-care data manager. *Clarke W, et al. eds., Contemporary Practice in Clinical Chemistry, Clarke W, et al., eds., p. 280.*

5. e. The POCT1 connectivity standard does not define the size of point-of-care devices. All of the other options are correct. It was developed so that any device could connect seamlessly and transfer data to any data management system and onto a laboratory information system or electronic medical record. *Clarke W, et al. eds., Contemporary Practice in Clinical Chemistry, Clarke W, et al., eds., p. 280.*

6. e. The point-of-care committee reviews and approves requests for point-of-care testing, has an interdisciplinary composition to address all hospital issues regarded with testing, can help balance opinions, depersonalize judgments and deflect political heat that may arise from decisions to remove testing, and helps maintain quality assurance at all aspects of the testing process. *Clarke W, et al. eds., Contemporary Practice in Clinical Chemistry, Clarke W, et al., eds., p. 281.*

7. e. Quality assurance of point-of-care testing services can be monitored by assessing performance indicators such as documentation of quality control (QC), expired reagents and QC/reagent bottles and kits dated appropriately, temperature checks for storage of point-of-care reagents, QC, and devices, handling changes in lots for QC and reagents according to manufacturer recommendations. *Clarke W, et al. eds., Contemporary Practice in Clinical Chemistry, Clarke W, et al., eds., p. 281.*

8. d. Effective interdisciplinary communication between stakeholders is important to resolve issues and implement changes. Performance and documentation of quality control according to policies and participation in an external quality assessment program are all vital in ensuring the quality of testing. However, all quality processes are futile if the clinical staff entrusted to performing point-of-care testing do not self-manage and ensure that corrective actions are being followed through. The laboratory-clinical partnership needs to be based on mutual respect and shared responsibility. *Clarke W, et al. eds., Contemporary Practice in Clinical Chemistry, Clarke W, et al., eds., p. 282.*

9. e. After implementation, many labs and hospitals fail to perform a post hoc analysis for effectiveness. Very few studies have demonstrated a reduction in morbidity and mortality, as this is highly multifactorial and it is difficult for POCT to impact this metric. *Lewandrowski EL, et al. Perspectives on cost and outcomes for point-of-care testing. Clin Lab Med 2009;29:479–89.*

10. d. The American Diabetes Association had recommended that results between glucose meters and central laboratory instruments should be within 5%, and the 2003 International Organization for Standardization 15197 guidelines had recommended $\pm 15\,mg/dL$ for results $<75\,mg/dL$ and $\pm 20\%$ for results $>75\,mg/dL$. Because glucose meters were being used in hospital for tight glycemic protocols, The Clinical and Laboratory Standards Institute published POCT12-A3 recommendations called for tighter performance standards from manufacturers. To put more pressure on manufacturers to improve their glucose meters, the Food and Drug

Administration has lately made their approval process for glucose meters more rigorous. *Clarke W, et al. eds., Contemporary Practice in Clinical Chemistry, Clarke W, et al., eds., p. 282.*

11. b. Glucose meters are used in the management of diabetes. Although they may provide a faster glucose measurement they have greater imprecision and are not suitable for diagnosis of diabetes. *Clarke W, et al. eds., Contemporary Practice in Clinical Chemistry, Clarke W, et al., eds., p. 283.*

12. e. Glucose meter strips need to be kept at certain humidity and temperature specifications. If the test method is based on colorimetry, then humidity can damage colorimetric test strips. Altitude can affect some test strips that are sensitive to oxygen. Enzymatic methods are sensitive to extremes of temperature. High and low hematocrit can also introduce a strong bias. *Clarke W, et al. eds., Contemporary Practice in Clinical Chemistry, Clarke W, et al., eds., p. 285.*

13. e. Bilirubinometers, transcutaneous blood gases, and pulse oximeters have been available for many years. Transcutaneous glucose monitors are in development where a needle is inserted into the transdermal layer for sampling. Hemoglobin A1c requires red cells not present directly under the skin. *Garg S, et al. Improvement in glycemic excursions with a transcutaneous, real-time continuous glucose sensor. Diabetes Care 2006;29:44–50.*

14. d. The inability to conjugate bilirubin leads to the accumulation of indirect bilirubin that is toxic to the brain. Hepatic cirrhosis can produce hyperbilirubinemia but does not cause kernicterus. There can be congenital liver cirrhosis of the infant that can produce kernicterus. *Mariano da Rocha C, et al. Neonatal liver failure and congenital cirrhosis due to gestational alloimmune Liver disease: a case report and literature review. Case Reports Pediat 2017. http://downloads.hindawi.com/journals/cripe/2017/7432859.pdf.*

15. c. Irradiation of bilirubin through the thin skin of a neonate is sufficient to photoisomerize bilirubin at the meso double-bound location to a less lipophilic form. *Sisson T. Molecular basis of hyperbilirubinemia and phototherapy. J Invest Dermatol 1981;77:158–61.*

16. b. CLIA applies to samples that are collected and tested. The Emergency Use Authorization was designed for approval of tests to protect citizens against chemical, biological, radiological, and nuclear threats. *Wu AHB. On vivo and wearable clinical laboratory testing devices for emergency and critical care laboratory testing. J Appl Lab Med 2019;4:254–63.*

17. b. Zone A of the grid indicates concordance within 20%. B indicates values outside 20% but would not lead to inappropriate treatments. C indicates points that might lead to inappropriate treatment. D indicates failure to detect hypoglycemia or hyperglycemia. E would confuse treatment of hypoglycemia or hyperglycemia. *Clarke error grid. Wikipedia https://en. wikipedia.org/wiki/Clarke_Error_Grid Image from Wikimedia Commons.*

18. e. Bilirubinometers measure total bilirubin by measuring the color through the skin and does not require any blood collection, and therefore does not qualify as a lab test. There is some correlation with neonatal bilirubin but not conjugated bilirubin. *Karen T, et al. Comparison of a new transcutaneous bilirubinometer (Bilimed®) with serum bilirubin measurements in preterm and full-term infants. BMC Pediatr 2009;9:70–84.*

19. d. POCT for PT-INR using a fingerstick is convenient and when tested at home can lead to adjustments of warfarin dosing. However, there are analytical limitations. *Johnson SA. Point-of-care or clinical lab INR for anticoagulation monitoring: Which to believe? Clin Lab News 2017 April: https://www.aacc.org/publications/cln/articles/2017/april/point-of-care-or-clinical-lab-inr-for-anticoagulation-monitoring-which-to-believe.*

20. b. Urine tests have a detection limit of about 25 mIU/mL significantly higher than the 2–5 mIU/mL for serum. Therefore early pregnancy, i.e., <4 weeks cannot be detected. *Greene DN, et al. Limitations in qualitative point of care hCG tests for detecting early pregnancy. Clin Chim Acta 2013;415:317–21.*

21. c. Icodextrin metabolizes to maltose and other glucose forms producing a false positive result. This polymer is used for peritoneal dialysis and some adhesion surgeries. *Flore KM, et al. Alanytical interferences in point-of-care testing lgucometers by icodexrin and its metabolites: an overview. Perit Dial Int 2009;29:377–83.*

22. e. Interferents can be oxidated through the glucose oxidase method causing a false positive result. Of those listed, there are no reports of ibuprofen interference. *Erbach M, et al. Interferences and limitations in blood glucose self-testing: An overview of the current knowledge. J Diabetes Sci Technol 2016;10:1161–8. doi:10.1177/1932296816641433.*

23. a. Highly sensitive troponin assays enable a faster rule out of AMI than conventional assays. While a POCT can produce results sooner, they cannot be used to detect troponin during the first samples. Therefore later samples are needed obviating the turnaround time advantage. While POCT assays are not standardized to the central lab, nevertheless they have been used. An ED can justify the expense if a patient can be discharged earlier. *Boeddinghaus J, et al. Clinical validation of a novel high-sensitivity cardiac troponin I assay for early diagnosis of acute myocardial infarction. Clin Chem. 2018;64(9):1347–60.*

24. d. Tests for hCG, where the analyte is large enough so that two antibodies can be used and a positive result is indicated by the presence of a line in the test zone (sandwich immunoassay). Drugs are too small to use two antibodies that measure different epitopes, therefore only one detection antibody is used in a competitive format and the absence of a line indicates a positive result. *Li et al. Development and clinical validation of a sensitive*

lateral flow assay for rapid urine fentanyl screening in the emergency department. Clin Chem 2020;66:324032.

25. d. The majority of clinical lab testing and point-of-care testing is *in vitro*. *In vivo* are implantable devices, such as glucose. *Ex vivo* is testing of organs and tissues outside their native environment. Wearable devices can be considered next-generation point-of-care testing where the testing device is applied to the skin. *Wu AHB. "On vivo" and wearable clinical laboratory testing devices for emergency and critical care laboratory testing. J Appl Lab Med 2019;4:254–63.*

Index

Printed in the United States
By Bookmasters